Exploiting the Rice Germplasm for Health-Promoting and Value-Added Foods

Exploiting the Rice Germplasm for Health-Promoting and Value-Added Foods

Editors

Piebiep Goufo
John C. Beaulieu
Ming-Hsuan Chen
Ida Wenefrida

MDPI • Basel • Beijing • Wuhan • Barcelona • Belgrade • Novi Sad • Cluj • Manchester

Editors

Piebiep Goufo
University of Minnesota Twin Cities
Saint Paul
USA

John C. Beaulieu
USDA Agricultural Research Service
New Orleans
USA

Ming-Hsuan Chen
USDA Agricultural Research Service
Stuttgart
USA

Ida Wenefrida
Louisiana State University
Rayne
USA

Editorial Office
MDPI AG
Grosspeteranlage 5
4052 Basel, Switzerland

This is a reprint of articles from the Special Issue published online in the open access journal *Foods* (ISSN 2304-8158) (available at: https://www.mdpi.com/journal/foods/special_issues/Rice_ValueAdded_Foods).

For citation purposes, cite each article independently as indicated on the article page online and as indicated below:

Lastname, A.A.; Lastname, B.B. Article Title. *Journal Name* **Year**, *Volume Number*, Page Range.

ISBN 978-3-7258-2363-5 (Hbk)
ISBN 978-3-7258-2364-2 (PDF)
doi.org/10.3390/books978-3-7258-2364-2

Contents

About the Editors

Piebiep Goufo

Piebiep Goufo is currently a Research Associate in the Department of Agronomy and Plant Genetics at the University of Minnesota Twin Cities, Saint Paul, USA. He earned a Master's degree in Plant Biochemistry and Pathology from the University of Yaoundé I in 2005, followed by a Ph.D. in Crop Physiology from South China Agricultural University in 2010. Following his Ph.D., he was a research fellow at the Centre for Research and Technology of Agro-Environmental and Biological Sciences in Portugal from 2011 to 2018, and later served as faculty in the Department of Agronomy at the University of Trás-os-Montes e Alto Douro, Portugal, from 2019 to 2022. His research primarily investigates how environmental factors, plant diseases, and management practices influence the phytochemical composition of crops. His work aims to enhance crop resilience to stresses, promote the production of nutrient- and bioactive compound-enriched fruits, and reduce reliance on synthetic chemicals in agriculture. Piebiep has an impressive publication record, including 34 peer-reviewed journal articles, eight conference proceedings, and four contributions to agricultural magazines, with 17 publications dedicated to rice research. His impactful work on rice and climate change has earned notable recognition, including the African Academy of Sciences Award in 2011 and the Elsevier Global Food Security Prize in 2013.

John C. Beaulieu

John C. Beaulieu retired from the USDA ARS as an Adjunct Professor at the Louisiana State University School of Nutrition and Food Sciences (Baton Rouge, LA, USA) in late 2022. Throughout his graduate and professional career, he specialized in postharvest quality, maturity-dependent and flavor volatile investigations, and innovative product R&D. He parlayed this into flavor chemistry and "on-the-job training" food science and functional foods projects. Dr. Beaulieu worked on hydroponic broccoli, ripening inhibition in tomatoes, fresh-cut kale, fresh-cut fruits, modified-atmosphere packaging, functional 100% not-from-concentrate fruit beverages, ideation leading to 100% natural sprouted brown rice beverages using green technologies, and several multidisciplinary collaborative projects. Career-wise, his best-known contributions are maturity-dependent plant physiology food processing studies on quality and volatiles (blueberry, cantaloupe, mango, tomato), overall postharvest quality work, and an array of aroma volatile/chemistry publications (blueberry, cantaloupe, honeydew, mango, pomegranate, pigmented rice, satsuma, and seedless watermelon) and anthocyanin phenolic profiles. His lab's work has resulted in over 50 peer-reviewed publications, 30 invited lectures, and 80 conference presentations. Dr. Beaulieu navigated three R&D tracks while at the USDA ARS. Regardless of the new subject matter, he developed deep understandings of topics through reviews of the literature, market analyses, and stakeholders, to build networks and collaborative teams, serving on several trade organization and professional committees and generating esteemed peer-review publications. Beaulieu became an adaptive, seasoned expert, utilizing novel ideation and cross-disciplinary projects. Ultimately, his scientific impact, acumen, creativity, and overall track record landed him on the prestigious Stanford University career-long 2020-2023 lists of the *"Top 2% of World Scientists"*.

Ming-Hsuan Chen

Ming-Hsuan Chen retired from the United States Department of Agriculture Agricultural Research Service (USDA-ARS) in 2022, after 21 years of federal service. She holds a Master's degree in Agricultural Chemistry from California State University, Fresno, and a Ph.D. in Entomology with a specialization in Insect Molecular Biology from Texas A&M University, College Station. Ming-Hsuan began her career with the USDA-ARS in 2001 as a Research Chemist at the Rice Research Unit in Beaumont, TX, and later moved to the Dale Bumpers National Rice Research Center in Stuttgart, AR, in 2012. Throughout her career, she has authored or co-authored over 47 peer-reviewed journal articles. Her research has focused on two key areas: the functional properties of rice grain and the health benefits of whole grain rice. In addition to her basic research, Ming-Hsuan has significantly contributed to the development of new rice varieties by evaluating hundreds of breeding lines, ensuring that new releases from U.S. breeding programs meet or exceed industry standards for grain quality, both domestically and internationally. Her work has earned her numerous accolades, including the 'Distinguished Rice Research and Education Team Award,' presented by her peers at the 38th Rice Technical Working Group Meeting in 2020.

Ida Wenefrida

Ida Wenefrida was an Associate Professor at the Louisiana State University AgCenter and a highly accomplished plant biologist at the H. Rouse Caffey Rice Research Station. In 2022, she was honored with the Kenneth W. Tipton Team Research Award for her groundbreaking work in developing Frontière, a high-protein, low-glycemic index rice variety. Frontière contains 53% more protein than conventional white rice and shows great promise for aiding diabetics and combating malnutrition in regions where rice is a staple. Her research on rice led to numerous publications and patents. In recognition of her achievements and philanthropic efforts, Wenefrida was named one of the Top 20 Global Women of Excellence in 2020 by the American Multi-Cultural Ethnic Coalition. She earned her Master's degree in Plant Pathology from Mississippi State University and her Ph.D. in Plant Health from Louisiana State University.

Preface

Rice is a staple food for nearly half the world's population, playing a crucial role in global food security. *Oryza sativa* was the first crop to be fully sequenced and is now a model for plant geneticists, with over 3,000 re-sequenced varieties and the largest publicly available single-species germplasm collection globally. This vast genetic diversity results in numerous varieties with different morphological, physical, and physicochemical traits, offering unique value-added food applications. Similarly, *Oryza glaberrima* possesses underexplored characteristics worthy of further development. Despite its ubiquity, the predominant consumption of white rice poses nutritional challenges, as much of the essential nutrients are lost during the milling process. In contrast, whole grain rice varieties—such as brown, red, purple, and black rice—retain most of their bran and germ layers, where vital nutrients and bioactive compounds are concentrated. These whole grain rice varieties are increasingly being recognized for their rich content of health-promoting phytochemicals and antioxidants, delivering significant nutritional benefits.

This Special Issue explores the untapped potential of whole grain rice, underutilized varieties, and rice byproducts. It aims to introduce new genetic insights and value-added strategies to enhance the health-promoting properties of this important crop. The Issue highlights cutting-edge research, from the genetic manipulation of rice to agricultural practices that enhance rice's nutritional value, to innovative processing methods that retain and exploit its health-beneficial compounds.

A major goal of the Issue is to gather novel insights and foster collaborations to pave the way for future developments. Contributions from a diverse group of researchers highlight different dimensions of the topic, including genetic approaches, agricultural practices, and value-added methods that enhance the health benefits of brown, red, purple, and black rice. Additional research has focused on improving rice's nutritional profile and its application in various health-related contexts.

The contributions are categorized into four main areas:

1. Genetic Factors Impacting Health-Promoting Compounds in Rice

The Impact of the Alteration of Starch Branching Enzyme IIb (BEIIb) and Soluble Starch Synthase IIa (SSIIa) Alleles on Rice Starch: Hu et al. examined the genetic effects of BEIIb and SSIIa on starch properties, providing insights into how these genetic factors could be used to breed rice with amylose content and resistant starch.

2. Preharvest Treatments and Their Effects on Health-Promoting Compounds in Rice

The Grain Quality, Bioactivity, and Safety of Organically Produced Rice: Bergman and Pandhi assessed the impact of organic versus conventional rice farming on grain quality and safety, finding minor differences in quality and nutritional traits but notable advantages in reducing pesticide residues.

3. Processing Effects on Rice's Physical and Nutritional Qualities

The Effects of Hard Water Boiling on Chalky Rice vs. Whole Rice: Nakamura and Ohtsubo investigated how hard water affects enzymatic activities, as well as the texture and calcium content of chalky rice, revealing that hard water can mitigate texture deterioration and enhance calcium absorption. Salinity Treatment as a Way to Boost the Accumulation of Bioactive Compounds in Germinated Rice: Hasan et al. explored how different salinity levels during germination can boost the levels of momilactones and phenolic compounds in brown rice, potentially enhancing its health benefits.

Improving Bioactive compounds in Amazake, a Naturally Sweet Beverage, using Germinated Rice: Castanho et al. demonstrated that germinated rice improves γ-oryzanol and γ-aminobutyric acid content in Amazake, highlighting the benefits of rice fermentation for health-promoting compounds.

Far-Infrared Radiation Drying and Rice Quality: Ratseewo et al. assessed how far-infrared radiation drying affects starch digestibility and phenolic compound levels in pigmented rice, noting its potential benefits for enhancing rice's nutritional profile.

Lipid Profiles in Germinated Brown Rice Beverages: Beaulieu et al. compared lipid profiles in beverages obtained from germinated brown, non-germinated brown, and white rice using green technologies. The authors noted the highest health-beneficial lipid concentrations in the germinated brown rice-derived beverage and potential for improving beverage quality.

4. Health Impacts of Rice Consumption

Cardiac Benefits of Parboiled Germinated Brown Rice: On-Nom et al. showed that parboiled germinated brown rice reduces blood pressure and improves cardiac health in hypertensive rats, suggesting its potential for managing hypertension

Prevention of Diabetes and Dementia with Brown Rice Blends: Nakamura et al. investigated the potential of high-pressure-treated brown rice blends for preventing type 2 diabetes and dementia, showing promising results in reducing insulin secretion and improving cognitive function.

The Treatment of Overlapping Gastroesophageal Reflux Disease and Irritable Bowel Syndrome with Rice Noodles vs. Wheat Noodles: Patcharatrakul et al. evaluated the effects of rice versus wheat on GERD symptoms, finding that rice induces fewer symptoms and may be better suited for patients with overlapping GERD and IBS.

5. Utilization of Rice Byproducts

A Review of Methods for the Stabilization of Rice Bran: Yılmaz Tuncel reviewed methods for stabilizing rice bran, addressing the challenges of rancidity and suggesting strategies to enhance its use as a nutritious food ingredient.

While these studies have expanded understanding of rice's nutritional and health potential, further research is needed to optimize these practices and explore additional health benefits. Key areas for future investigation include the value-added use of rice husks. Overall, this Special Issue provides a foundation for future innovation in rice production and utilization, encouraging cross-disciplinary collaboration in genomics, agronomy, food science, and biology to tackle the remaining challenges.

We thank all contributors and reviewers for their invaluable efforts in making this reprint a success. We look forward to seeing how this research continues to shape the field.

This reprint is dedicated to the memory of Dr. Ida Wenefrida, whose remarkable contributions to rice research and tireless dedication to advancing agricultural science have left a lasting impact. Her friendly personality, availability, and willingness to collaborate on this topic, along with her generosity in sharing knowledge and unwavering support for her colleagues, made this Special Issue possible. We hope that Ida's work and legacy will continue to inspire future generations of cereal researchers.

Piebiep Goufo, John C. Beaulieu, Ming-Hsuan Chen, and Ida Wenefrida
Editors

foods MDPI

Article

Combined Effects of *BEIIb* and *SSIIa* Alleles on Amylose Contents, Starch Fine Structures and Physicochemical Properties of Indica Rice

Yaqi Hu [1,2], Yanni Zhang [1], Shouwu Yu [3], Guofu Deng [4], Gaoxing Dai [4,*] and Jinsong Bao [1,2,*]

1 Institute of Nuclear Agricultural Sciences, College of Agriculture and Biotechnology, Zhejiang University, Zijingang Campus, Hangzhou 310058, China
2 Hainan Institute of Zhejiang University, Yazhou Bay Science and Technology City, Yazhou District, Sanya 572025, China
3 Institute of Crops and Nuclear Technology Utilization, Zhejiang Academy of Agricultural Sciences, Hangzhou 310021, China
4 Rice Research Institute, Guangxi Academy of Agricultural Sciences, Nanning 530007, China
* Correspondence: dgx@gxaas.net (G.D.); jsbao@zju.edu.cn (J.B.); Tel.: +86-571-86971932 (J.B.)

Abstract: Starch branching enzyme IIb (BEIIb) and soluble starch synthase IIa (SSIIa) play important roles in starch biosynthesis in cereals. Deficiency in the BEIIb gene produces the *amylose extender* (*ae*) mutant rice strain with increased amylose content (AC) and changes in the amylopectin structure. The SSIIa gene is responsible for the genetic control of gelatinization temperature (GT). The combined effects of *BEIIb* and *SSIIa* alleles on the AC, fine structures, and physicochemical properties of starches from 12 rice accessions including 10 recombinant inbred lines (RIL) and their two parents were examined in this study. Under the active *BEIIb* background, starches with the *SSIIa*-GC allele showed a higher GT than those with the *SSIIa*-TT allele, resulting from a lower proportion of A chain and a larger proportion of B1 chains in the amylopectin of *SSIIa*-GC. However, starch with the *BEIIb* mutant allele (*be2b*) in combination with any *SSIIa* genotype displayed more amylose long chains, higher amylose content, B2 and B3 chains, and molecular order, but smaller relative crystallinity and proportion of amylopectin A and B1 chains than those with *BEIIb*, leading to a higher GT and lower paste viscosities. These results suggest that *BEIIb* is more important in determining the structural and physicochemical properties than *SSIIa*. These results provide additional insights into the structure-function relationship in indica rice rather than that in japonica rice and are useful for breeding rice with high amylose content and high resistant starch.

Keywords: rice; starch synthase; branching enzyme; starch structure; functional property

Citation: Hu, Y.; Zhang, Y.; Yu, S.; Deng, G.; Dai, G.; Bao, J. Combined Effects of *BEIIb* and *SSIIa* Alleles on Amylose Contents, Starch Fine Structures and Physicochemical Properties of Indica Rice. *Foods* **2023**, *12*, 119. https://doi.org/10.3390/foods12010119

Academic Editors: John C Beaulieu, Ming Hsuan Chen, Ida Wenefrida and Piebiep Goufo

Received: 4 November 2022
Revised: 20 December 2022
Accepted: 21 December 2022
Published: 26 December 2022

1. Introduction

The emergence of hybrid rice technology has greatly increased the yield potential of rice and solved the problem of food shortages. However, people's demand for a better life and tastier rice is increasing with the gradual improvement of people's living standards. Starch is the most important element of rice, accounting for about 90% of the milled rice on a dry weight basis, and its physicochemical properties mainly determine rice cooking and eating quality (CEQ) [1,2]. Starch is composed of amylose and amylopectin molecules. Amylose is a relatively long linear α-glucan linked by α-(1→4) glycosidic bonds [3]. Amylopectin is a highly branched polymer with linear chains linked with α-(1→4) glycosidic bonds and branch points linked by α-(1→6) glycosidic bonds [3,4]. Starch biosynthesis is a complicated process. Amylose is synthesized under the sole action of the granule-bound starch synthase (GBSS) encoded by the *Waxy* (*Wx*) locus [5]. The synthesis of amylopectin requires a combination of multiple enzymes, including four soluble starch synthase (SS) isoforms (SSI, SSII, SSIII and SSIV), two starch branching enzyme (BE) isoforms (BEI and BEII), and starch debranching enzymes [5].

Natural variations in starch biosynthesizing genes are responsible for the different starch physicochemical properties and CEQ. Although many genes involved in starch biosynthesis have natural variations, the variation in *Wx* and *SSIIa* is well known and has great effects on the improvement in grain quality since they are responsible for the genetic basis of amylose content (AC) and gelatinization temperature (GT), respectively [6]. A G/T single nucleotide polymorphism (SNP) (AG<u>G</u>TAT/ AG<u>T</u>TAT) at 5′ splice site of the first intron of *Wx* pre-mRNA can differentiate indica type Wx^a (G SNP) and japonica type Wx^b (T SNP) alleles [7]. A simple sequence repeat (SSR) or microsatellite, cytosine-thymine (CT)n, in the *Wx* gene can be used to classify rice with different AC classes, which can explain >75% of the variation in AC among various rice accessions [8–10]. Two nonsynonymous SNPs, i.e., G/A at the 4198 bp, and GC/TT at the 4329/4330 bp of *SSIIa* have been discovered to affect the GT [11–13]. The G/GC combination (or haplotype) rice has an intermediate or high GT, whereas rice with the A/GC or G/TT combination has a low GT, but the A/TT haplotype does not exist in nature [12–16].

Mutations in the starch biosynthesizing genes may lead to enzyme deficiency and starches with altered structure and functionality. The *waxy* or glutinous rice with a characteristic of <2% AC is derived from a common mutation of the *Wx* gene. The knocking out of the *Wx* gene with physical and chemical mutagenesis, or a new tool such as the clustered regularly interspaced short palindromic repeats (CRISPR)/associated protein-9 (Cas9) can easily produce the *waxy* mutant [17]. The *amylose extender* (*ae*) mutant is derived from a defect in the *BEIIb* gene [18,19] that specifically changes the amylopectin structure by decreasing the number of short chains with a degree of polymerization (DP) of 17 or less, with the greatest decrease in the DP8-12 chains [18]. In addition, AC is significantly positively correlated with resistant starch content, which is defined as the sum of starch and products of starch degradation not absorbed in the small intestine of healthy individuals [20]. Therefore, *ae* starch is a kind of resistant-starch resource, which can offer a wide array of health benefits to humans [21].

The combination of alleles of different starch biosynthesizing genes has been reported to change the physicochemical properties and starch fine structures. For example, Kubo et al. [22] indicated that *ae* and *wx/ae* double mutant starches display no difference in the chain-length distribution (CLD) of amylopectin and morphology of the starch granule, but *wx/ae* starch showed a higher pasting temperature and higher peak viscosity. The role of the *Wx* and *SSIIa* combination on the amylose fine structure was reported by Wang et al. [23], who found that *Wx* SNPs can affect AC but they are unable to alter the CLD of both amylopectin and amylose. Itoh, et al. [21] introduced the *SSIIa* from an indica rice cultivar to an *ae* mutant, which led to a higher proportion of amylopectin chains with DP11–18 than those in *be2b*, and the introduction of *Wx* from an indica rice cultivar significantly increased AC in the endosperm starch. Ida et al. [24] found that the AC and starch crystallinity in the japonica *ss2a/be2b* mutant were significantly higher than those in the *be2b* single mutant. Although previous studies have reported the effects of allele combinations of different genes on physical and chemical properties and amylopectin chain length distribution (CLD) [21,22,24], most of these studies were focused on japonica rice. However, there is limited information about the effects of *SSIIa* and *BEIIb* on the AC, starch fine structure, crystal structure, and functional characteristics of indica rice starch.

In this study, we hypothesize that the starch structure-function relationship would be altered in different combinations of *BEIIb* and *SSIIa* genotypes. Indica rice recombinant inbred lines (RIL) with different *BEIIb* and *SSIIa* genotypes were produced by cross-breeding assisted with a selection of molecular markers. The objective of this study is to characterize the structural and functional properties of the RILs and elucidate the effects of *BEIIb* and *SSIIa* genotype combinations on the structure-function relations of starch.

2. Materials and Methods

2.1. Materials

An *indica* rice cultivar Longtefu B (LTFB) with *SSIIa* allele-TT/*BEIIb* genotype and an *amylose extender (ae)* mutant BP577 with *SSIIa* allele-GC/*be2b* genotype was crossed to obtain F_1. Each F_2 breeding line was advanced by the single seed descent method to generate a recombinant inbred line (RIL) population. Molecular markers were applied in the F_4 generation to select the recombination of *SSIIa* and *BEIIb* alleles. Ten breeding lines (BL01~BL10) were selected and advanced to F_7. In this study, a completely randomized design with two replications was conducted at the Zhejiang University farm, Hangzhou, China. Mature seeds were harvested in late September.

2.2. Genotyping of Breeding Lines

The leaves harvested from seedlings of each RIL were used for genomic DNA isolation with the CTAB method of Doyle [25]. The genotypes of each RIL were analyzed by molecular marker [26,27]. The sequences of primers used in this study are listed in Table S1. The PCR products were separated by running agarose or polyacrylamide gel electrophoresis (PAGE) gels [26,27]. To confirm whether all the rice lines carried the same *Wx* allele, the microsatellites (CT)n were amplified with a primer, as described in Bao et al. [9].

2.3. Starch Isolation and Debranching

Starch was extracted from rice flour following the methods of Syahariza, Li, and Hasjim [28]. The purified starch was debranched by the addition of 2.5 μL isoamylase (1000 U/mL) from Pseudomonas sp. (Megazyme International Ltd., Bray, Co. Wicklow, Ireland), and mixed with 100 μL acetate buffer solution (0.1 M, pH 3.5) and 5 μL sodium azide solution (0.04 g/mL), and then incubated at 37 °C for 3 h. The starch solution was neutralized with 100 μL NaOH (0.1 M) and heated at 80 °C for 1 h, then freeze-dried overnight. The debranched starch fraction was dissolved in DMSO/LiBr (5%) for further size-exclusive chromatography (SEC) analysis.

2.4. Size-Exclusion Chromatography (SEC)

The chain-length distributions (CLDs) of the debranched starch were separated with the SEC using an LC20AD system (Shimadzu Corporation, Kyoto, Japan) equipped with three columns (pre-column, Gram 100, and Gram 1000) (PSS, Mainz, Germany) in sequence according to the method described in Zhang et al. [26]. The weight CLD of the debranched chains with the degree of polymerization (DP) or X, denoted as $w_{de}(\log X)$, obtained from the DRI signal, was converted to the corresponding number distribution $N_{de}(X)$ by $w_{de}(\log X) = X^2\, N_{de}(X)$ (a relation that holds only for linear polymers) [29]. X_{AP1} and X_{AP2} are the DP value at the peak 1 & 2 of amylopectin, and X_{AM} represents the DP value at the amylose peak, h_{AP2}/h_{AP1} is the ratio of the peak heights of amylopectin, and h_{AM} is the peak height of amylose.

2.5. Amylose Content Measurement

The AC was calculated from the SEC weight distributions of debranched starch by dividing the area under the curve (AUC) for $100 < X < 10{,}000$ by the whole area, i.e., the whole weight distribution of the debranched starch molecules.

2.6. FACE Analysis

The debranched starches were labeled with the fluorescence 8-amino-1,3,6-pyrenetrisulfonic acid according to the method of Wu et al. [30]. The amylopectin chain-length distribution was separated in an MDQ Plus Fluorophore-assisted carbohydrate electrophoresis (FACE) System, coupled with an argon-ion laser as the excitation source and a solid-state laser-induced fluorescence detector. The side chains of amylopectin can be divided into four groups according to their DP: DP ≤ 12 (A chain), 13 ≤ DP ≤ 24 (B1 chain), 25 ≤ DP ≤ 36 (B2 chain), and DP $\geq$ 37 (B3 and ultra-long chain) [31]. The proportions of the different groups were calculated from the amylopectin CLDs denoted as fa, fb1, fb2, and fb3, respectively.

2.7. X-ray Diffraction

X-ray diffraction analysis of the starch granules was conducted on an X-ray powder diffractometer (D8, Bruker, Karlsruhe, Germany), at a voltage of 40 kV and a current of 40 mA. The starches were tightly packed into the glass sample holder, and data were collected over an angular range of 2θ from 3° to 40° with a step of 0.05°. The relative crystallinity (RC, %) was calculated using Origin software (OriginLab Co., Northampton, MA, USA) following the method of Zhang, et al. [26].

2.8. Attenuated Total Reflectance-Fourier Transform Infrared Spectroscopy

A Varian 7000 Fourier transform infrared spectrometer equipped with a DTGS detector and an ATR single-reflectance cell containing a germanium crystal (45° incidence angle) (PIKE Technologies, Madison, WI, USA) was used to measure the short-range ordered structure of the starch granules [32]. The sample was scanned 64 times from 4000 to 800 cm^{-1} with a resolution of 4 cm^{-1}. The relative absorbances at 1045, 1022, and 995 cm^{-1}, which represent the ordered regions, amorphous regions, and the bonding in the hydrated carbohydrate helices in starch, respectively [33], were extracted from the deconvoluted spectra and measured from the baseline to the peak height. The 1045/1022 cm^{-1} ratio represents the degree of order in the starch external region, while the 1022/995 cm^{-1} ratio represents an index for the ratio of amorphous to carbohydrate structure starch.

2.9. RVA Pasting Viscosity

Three grams of rice flour was weighed in an aluminum can and then mixed with deionized water (25 g). The pasting properties were analyzed using a rapid visco-analyzer (RVA) (Model 4500, Perten Instrument, Hägersten, Sweden) according to the method of Bao et al. [1]. The idle temperature was set to 50 °C, held for 1.0 min, and then linearly ramped up to 95 °C until 4.8 min had elapsed, held at 95 °C until 7.5 min had elapsed, before being linearly ramped down to 50 °C after 11 min had elapsed, and held at this temperature until 12.5 min had elapsed. The RVA trace was analyzed by TCW software to obtain the peak (PV), host paste (HPV), and cold paste (CPV) viscosities. The breakdown, (BD = PV − HPV), consistency (CS = CPV − HPV), and setback (SB = CPV − PV) were derived from the PV, HPV, and CPV. The unit of all viscosity parameters is Rapid Visco Units (RVU).

2.10. DSC Thermal (Gelatinization) Properties

The thermal (gelatinization) characteristics were measured by using a DSC 2920 thermal analyzer (TA Instruments, Newcastle, DE, USA) equipped with DSC standard and dual sample cells, according to the method of Bao et al. [34]. Rice flour (2.0 mg, d.b.) was weighed into an aluminum pan to which 6 µL of distilled water was added. The pan was hermetically sealed, equilibrated at room temperature for 1 h, and then heated at a rate of 10 °C/min from 30 °C to 110 °C. A sealed empty pan was used as a reference. Onset (T_o), peak (T_p), conclusion (T_c) temperature, and enthalpy (ΔH) of gelatinization were calculated automatically using the Universal Analysis 2000 program (Version 4.4A) software (TA Instruments, Newcastle, DE, USA).

2.11. Statistical Analysis

All analyses were carried out in duplicate and reported as mean ± SD. Analysis of variance (ANOVA) was conducted in SAS 9.0 (SAS Institute Inc., Cary, NC, USA) with Tukey's pairwise comparisons ($p < 0.05$). Clustering analysis of the genotypes based on the structural and physicochemical properties was conducted using IBM SPSS Statistics 25.

3. Results and Discussion

3.1. Genotyping of The Breeding Lines

The use of PAGE gel can easily identify the two alleles of *BEIIb*. The PCR products of *BEIIb* or *be2b* were digested with *MboI* restriction endonuclease (Figure 1), which can separate the *BEIIb* or *be2b* (mutant) allele. Two pairs of primers facing each other were used in a PCR reaction to amplify *SSIIa* GC/TT alleles (Figure 1), which are denoted as *SSIIa*-GC (GC) and *SSIIa*-TT (TT) alleles. To confirm whether all rice carried the same *Wx* allele, the (CT)n microsatellites were amplified, displaying the same allele (Figure 1). The parent BP577 is an *amylose extender (ae)* mutant, so it has the *be2b* allele and also harbors the *SSIIa*–GC allele (Figure 1; Table 1). The parent LTFB has the *BEIIb* and *SSIIa*-TT alleles. Theoretically, there are four combinations of *SSIIa* and *BEIIb* alleles. As a result, four combinations were found in the RILs, including *SSIIa*-GC/*BEIIb*, *SSIIa*-TT/*BEIIb*, *SSIIa*-GC/*be2b*, and *SSIIa*-TT/ *be2b* (Table 1).

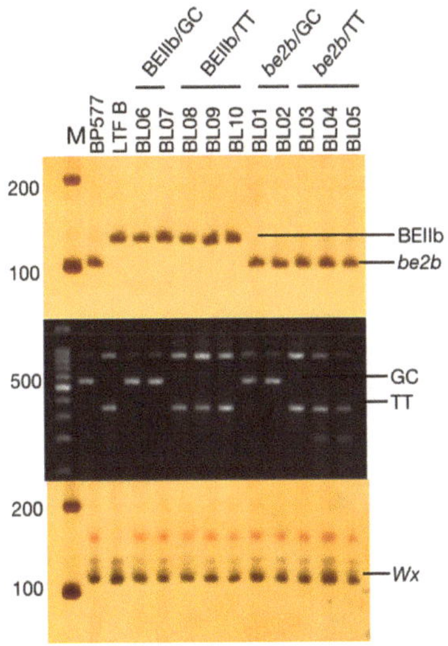

Figure 1. Genotyping of *Wx*, *BEIIb*, and *SSIIa* alleles in RILs.

Table 1. Structural parameters of chain length distribution (CLD) of debranched starches obtained from SEC.

Sample	Genotype	X_{AP1}	X_{AP2}	X_{AM}	h_{AM}	h_{AP2}/h_{AP1}	AC (%)
BP577	GC/*be2b*	17.26 ± 0.10 a	44.04 ± 0.08 ab	729.47 ± 4.76 a	0.42 ± 0.01 cd	0.86 ± 0.00 b	33.16 ± 0.011 bc
LTFB	TT/*BEIIb*	14.09 ± 0.19 c	39.68 ± 0.14 e	580.61 ± 12.17 e	0.25 ± 0.00 ef	0.58 ± 0.00 c	26.85 ± 0.007 e
BL01	GC/*be2b*	17.29 ± 0.07 a	43.89 ± 0.22 bc	710.71 ± 4.62 a	0.40 ± 0.01 d	0.86 ± 0.00 b	32.27 ± 0.002 c
BL02	GC/*be2b*	17.10 ± 0.06 a	43.52 ± 0.15 c	718.95 ± 24.96 a	0.45 ± 0.00 b	0.87 ± 0.00 b	33.84 ± 0.002 ab
BL03	TT/*be2b*	17.19 ± 0.10 a	44.42 ± 0.00 a	704.57 ± 1.52 ab	0.44 ± 0.01 bc	0.89 ± 0.00 a	32.94 ± 0.004 bc
BL04	TT/*be2b*	16.87 ± 0.03 a	43.60 ± 0.22 bc	743.92 ± 3.25 a	0.48 ± 0.02 a	0.88 ± 0.00 ab	34.84 ± 0.009 a
BL05	TT/*be2b*	17.06 ± 0.03 a	43.97 ± 0.15 abc	698.55 ± 7.54 abc	0.42 ± 0.00 bcd	0.86 ± 0.00 b	32.31 ± 0.003 c
BL06	GC/*BEIIb*	15.13 ± 0.06 b	40.50 ± 0.14 d	639.70 ± 2.72 d	0.24 ± 0.01 ef	0.54 ± 0.00 d	28.26 ± 0.005 d
BL07	GC/*BEIIb*	15.16 ± 0.09 b	40.23 ± 0.14 d	642.54 ± 10.95 d	0.22 ± 0.00 f	0.53 ± 0.00 d	26.82 ± 0.000 e
BL08	TT/*BEIIb*	14.65 ± 0.54 b	38.67 ± 0.20 f	651.70 ± 33.37 bcd	0.25 ± 0.00 e	0.53 ± 0.00 d	27.78 ± 0.001 de
BL09	TT/*BEIIb*	14.01 ± 0.00 c	38.60 ± 0.00 f	643.96 ± 41.11 cd	0.25 ± 0.00 e	0.53 ± 0.00 d	27.35 ± 0.003 de
BL10	TT/*BEIIb*	14.09 ± 0.03 c	39.00 ± 0.13 f	617.15 ± 11.76 de	0.24 ± 0.00 ef	0.53 ± 0.01 d	27.18 ± 0.009 de

Values with different letters in the same column are significantly different at $p < 0.05$. X_{AP1} and X_{AP2} are the degrees of polymerization (DP) of amylopectin peaks, X_{AM} represents the DP of amylose peak, h_{AP2}/h_{AP1} is the ratio of the peak heights of amylopectin, h_{AM} is the peak height of amylose, and AC is amylose content (%).

3.2. SEC Chain-Length Distributions of the Debranched Starch

The SEC weight distribution of debranched starch from the GC/*BEIIb*, TT/*BEIIb*, GC/*be2b*, and TT/*be2b* series is shown in Figure 2A. The maximum SEC weight distribution of each sample is normalized to an arbitrary value of 1. Three peaks exist in the SEC-weight CLDs (Figure 2A). Debranched amylopectin has two peaks: the first peak at DP from 14 to 17, or X_{AP1}, indicates short amylopectin chains, while the second peak at DP from 38 to 44, or X_{AP2}, indicates a long amylopectin chain (Table 1; Figure 2A). Debranched amylose has only one peak at DP 580 (LTFB) to 729 (BP577). Two parents showed distinctive CLD parameters, and most of the parameters of their offspring were between two parents with some transgressive segregations that had larger or smaller values than the parents (Table 1). The parent BP577 and the breeding lines with the *be2b* allele mostly had larger X_{AP1}, X_{AP2}, and X_{AM}, indicating that their amylopectin and amylose had larger molecular sizes than those carrying the *BEIIb* allele. The lines with *SSIIa*-GC/*BEIIb* had slightly larger X_{AP1} and X_{AP2} than those with *SSIIa*-TT/*BEIIb*. Rice with the *be2b* allele had much higher h_{AP2}/h_{AP1} than those with the *BEIIb* allele, which is in agreement with the result of Tappiban, et al. [19], who showed that the mutant deficiency in *BEIIb* had the h_{AP2}/h_{AP1} of 0.947. The value of h_{AP2}/h_{AP1} of the *be2b* genotypes was as high as that of potato starches [35], and a little larger than that of cassava starches and common rice starches [23,36].

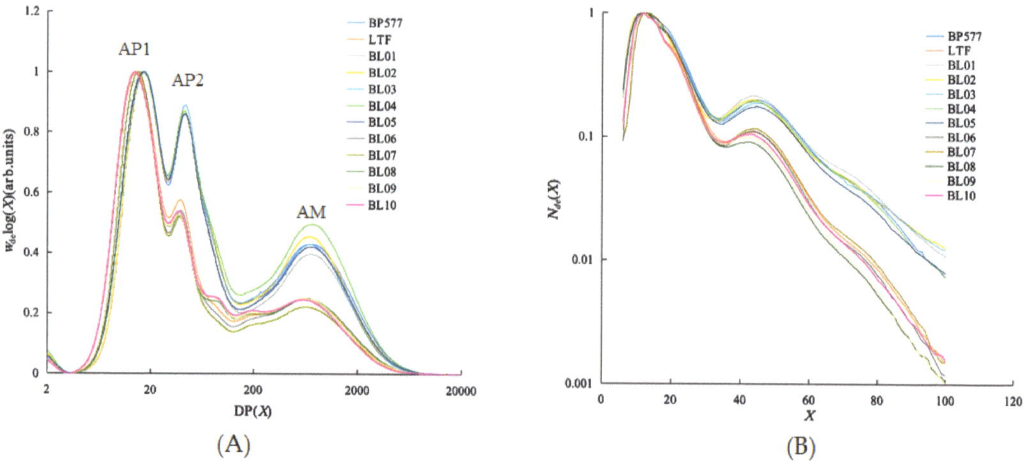

Figure 2. The chain length distribution (CLD) of debranched starch. (**A**): SEC weight CLDs of the whole range of debranched starch; (**B**): FACE number CLDs of debranched amylopectin branches. All distributions are normalized to the global maximum peak.

Both parents carry the same Wx^a allele with high AC, so all the breeding lines had AC larger than 27%. The AC of the rice with the *be2b* allele (32.27%~34.84%) was significantly higher than those with the *BEIIb* allele (26.82%~28.26%). However, there is no significant difference between the *SSIIa* GC and TT alleles (Table 1). In previous reports, the *be2b* allele represents the lack of BEIIb and can increase the AC of rice starches [19,37–39] and corn starches [40]. Since GBSS is responsible for amylose synthesis, these results may suggest that the GBSS might be inhibited by an active BEIIb.

The CLD number of the debranched starch from FACE is shown in Figure 2B, and the maximum value is normalized to 1. The maximum peak is at DP ~ 12, and a small bump is found at DP30~40. The amylopectin chains can be divided into four groups according to DP: DP $\leq$ 12 (A chain), 13 $\leq$ DP $\leq$ 24 (B1 chain), 25 $\leq$ DP $\leq$ 36 (B2 chain), and DP $\geq$ 37 (B3 and ultra-long chain) [31]. The proportion of amylopectin CLD showed a significant difference between starches from different *BEIIb* and *SSIIa* genotypes (Table 2), which was in agreement with the results from the SEC weight distribution (Table 1 and Figure 2A).

The proportion of A chain and B1 chain of starches with the *be2b* genotype ranged from 20.29% to 21.73% and 41.74% to 44.06%, which was much lower than those with the *BEIIb* genotype. Similarly, the proportions of the B2 and B3 chains of *be2b* were much higher than those of the *BEIIb* allele, leading to the average length of the *be2b* starch amylopectin being significantly higher than that of its *BEIIb* counterpart. The structural changes derived from BEIIb deficiency were in agreement with previous reports [18,21,22,26]. These results confirmed that the *ae* mutant not only prolongs the branch chain length of amylopectin but also significantly increases amylose content in starch [37]. Under the same *BEIIb* allele background, the A chain content of the SSIIa-GC starch was significantly lower, while the B1 chain content was significantly higher than that of SSIIa-TT (Table 2). In common rice accessions, it was proven some time ago that rice with the GC allele has lower fa chain content and fa/fb1 ratio than that with the TT allele [14]. However, under the *be2b* allele background, rice materials with SSIIa-GC starch had a little higher A chain content than those with the SSIIa-TT allele, while the reverse was found for the content of the B1 chain (Table 2).

Table 2. Structural parameters of amylopectin CLDs obtained from FACE.

Sample	Genotype	fa	fb1	fb2	fb3	$\overline{X}$
BP577	GC/*be2b*	21.36 ± 0.17 fg	42.53 ± 0.04 f	11.55 ± 0.09 c	24.56 ± 0.05 bc	26.25 ± 0.03 c
LTFB	TT/*BEIIb*	27.27 ± 0.25 c	46.59 ± 0.32 c	11.20 ± 0.07 d	14.95 ± 0.51 ef	21.73 ± 0.22 f
BL01	GC/*be2b*	21.05 ± 0.25 gh	41.74 ± 0.50 g	11.34 ± 0.07 d	25.76 ± 0.82 a	26.94 ± 0.35 a
BL02	GC/*be2b*	21.73 ± 0.21 ef	41.74 ± 0.17 g	11.01 ± 0.04 e	25.52 ± 0.34 ab	26.76 ± 0.15 ab
BL03	TT/*be2b*	20.50 ± 0.08 hi	43.29 ± 0.40 e	12.31 ± 0.11 a	23.90 ± 0.42 c	26.45 ± 0.16 abc
BL04	TT/*be2b*	20.29 ± 0.28 i	43.35 ± 0.07 e	12.01 ± 0.08 b	24.35 ± 0.27 c	26.43 ± 0.11 bc
BL05	TT/*be2b*	22.05 ± 0.61 de	44.06 ± 0.59 d	11.87 ± 0.09 b	21.02 ± 1.11 d	25.34 ± 0.54 d
BL06	GC/*BEIIb*	22.43 ± 0.14 d	52.15 ± 0.34 a	10.69 ± 0.01 fg	14.73 ± 0.47 ef	21.96 ± 0.19 ef
BL07	GC/*BEIIb*	22.09 ± 0.31 de	51.74 ± 0.20 a	10.82 ± 0.13 f	15.35 ± 0.03 e	22.29 ± 0.04 e
BL08	TT/*BEIIb*	29.17 ± 0.24 a	47.68 ± 0.46 b	10.49 ± 0.03 h	12.66 ± 0.73 h	20.57 ± 0.27 h
BL09	TT/*BEIIb*	28.71 ± 0.15 ab	47.46 ± 0.03 b	10.64 ± 0.05 gh	13.19 ± 0.14 gh	20.80 ± 0.08 gh
BL10	TT/*BEIIb*	28.16 ± 0.05 b	47.26 ± 0.08 cb	10.64 ± 0.05 gh	13.93 ± 0.08 gf	21.17 ± 0.03 g

Values with different letters in the same column are significantly different at $p < 0.05$. $\overline{X}$ indicates average chain length.

3.3. Crystalline Structure

The XRD patterns of the two parents and 10 breeding lines are shown in Figure 3A. The A-type crystallinity starch has the characteristic of containing more short branch chains of amylopectin, while the B-type crystallinity has the characteristic of containing more long branch chains, and the C-type starch is a mixture of A and B -type crystallinity. Starches with the *BEIIb* allele had a typical A-type crystalline with four peaks at diffraction angles (2θ) of 15°, 17°, 18°, and 23° in common with those of common rice and other cereal starches [19,37]. The BEIIb deficiency mutant starch generally has B-type crystallinity [19,22,24,37,40]. In this study, starches with the *be2b* allele had a C-type pattern with a strong peak at 17° and weak peaks at 5°, 15°, and 23°, which may be derived from incomplete suppression of the expression of the BEIIb [37]. The relative crystallinity (RC) showed significant differences between genotypes with different allele combinations (Table 3). Among starches with A-type crystalline, values of RC with the GC allele (BL06 and BL07) were a little higher than those of the TT allele (BL08-10). Starch with high RC was in alignment with higher B1 chain proportion, so in the common starches, those with the GC allele had higher fb1 content (Table 2), resulting in high RC. However, among the starches with C-type crystallinity, values of RC with the GC allele (BL01-BL02) were a little lower than those of the TT allele (BL03-05) (Table 3). It is possible that the short chains generated in the *BEIIb* mutant were inhibited from elongation due to the action of the SSIIa isoform. It is also suggested that SSIIa forms a heteromeric protein complex with SSI and BEIIb in cereal endosperm to synthesize amylopectin [6,41,42]. Therefore, the lack of BEIIb isoform may decrease the amount of functional protein complex, which results in less short-chain elongation, even though the SSIIa is abundant.

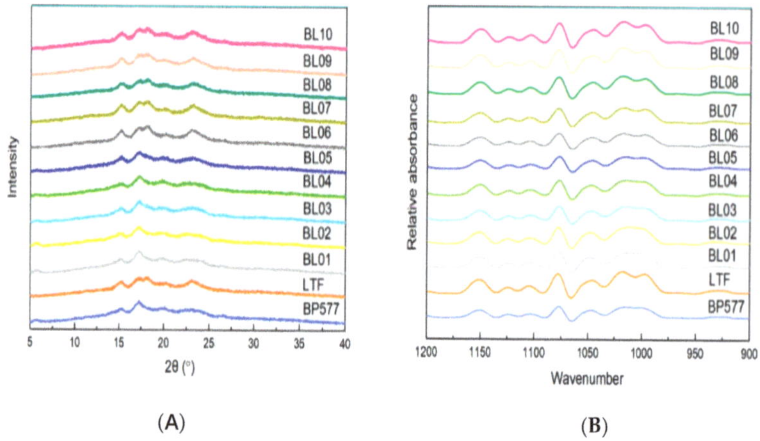

Figure 3. XRD spectra (**A**) and deconvoluted ATR-FTIR spectra (**B**) of rice starches.

Table 3. Relative crystallinity and IR ratios of rice starches.

Sample	Genotype	RC (%)	1045/1022	1022/995
BP577	GC/*be2b*	21.68 ± 0.33 cd	0.783 ± 0.010 b	0.998 ± 0.011 de
LTFB	TT/*BEIIb*	23.16 ± 0.01 b	0.654 ± 0.005 de	1.060 ± 0.011 a–e
BL01	GC/*be2b*	20.51 ± 0.13 e	0.812 ± 0.005 a	0.937 ± 0.011 e
BL02	GC/*be2b*	20.75 ± 0.27 de	0.782 ± 0.006 ab	1.020 ± 0.011 cde
BL03	TT/*be2b*	21.07 ± 0.16 cde	0.778 ± 0.002 ab	0.925 ± 0.011 e
BL04	TT/*be2b*	20.89 ± 0.34 de	0.753 ± 0.015 bc	0.998 ± 0.011 de
BL05	TT/*be2b*	21.88 ± 0.44 c	0.769 ± 0.001 b	0.967 ± 0.011 de
BL06	GC/*BEIIb*	24.27 ± 0.17 a	0.722 ± 0.026 c	1.102 ± 0.011 a–d
BL07	GC/*BEIIb*	24.22 ± 0.21 a	0.742 ± 0.008 bc	1.042 ± 0.011 b–e
BL08	TT/*BEIIb*	23.38 ± 0.14 ab	0.678 ± 0.007 d	1.164 ± 0.011 ab
BL09	TT/*BEIIb*	23.89 ± 0.25 ab	0.652 ± 0.012 de	1.186 ± 0.011 a
BL10	TT/*BEIIb*	23.09 ± 0.21 b	0.634 ± 0.005 e	1.140 ± 0.011 abc

Values with different letters in the same column are significantly different at $p < 0.05$; RC: relative crystallinity.

The 900–1200 cm^{-1} region of the FTIR spectra of the two parents and their breeding lines is shown in Figure 3B. The 1045/1022 cm^{-1} ratio displayed significant differences between the rice materials, ranging from 0.634 (BL10) to 0.812 (BL01) (Table 3). The higher ratio of 1045/1022 cm^{-1} may indicate that the starch has a higher degree of ordered structure. In either *BEIIb* or *SSIIa* allele backgrounds, it is clearly shown that starches with the *be2b* or GC allele had a higher degree of ordered structure than their counterparts with the *BEIIb* or TT allele. The ratio 1022/995 cm^{-1} indicates the proportion of amorphous structure. The starches with the *BEIIb* allele had a higher ratio of 1022/995 cm^{-1} than those with the *be2b* allele, which is not in agreement with the relative crystallinity data (Table 3).

3.4. Thermal Properties

The onset temperature (T_o), peak temperature (T_p), conclusion temperature (T_c), and the enthalpy of gelatinization (ΔH) of all the rice starches are presented in Table 4. For common starch with the *BEIIb* allele, it is well known that rice accessions with the GC allele had intermediate or high GT, while those with the TT SNP had a low GT [13,14,16]. BL06 and BL07 had a Tp of around 78 °C, and those of BL08-10 had a T_p of around 65 °C (Table 4). The difference in gelatinization temperature can be easily explained by the difference in amylopectin CLDs since the fa chain content and fa/fb1 ratio are negatively correlated with GT, whereas fb1 chain content is positively correlated with GT [14]. The ΔH of BL06 and BL07 was the highest among all samples (Table 4), partially because both these samples also had the highest RC (Table 3). It is plausible that the higher fb1 chain content in the BL06 and BL07 starches resulted in their highest ΔH and RC (Table 2). However, among the rice samples with the *be2b* allele, none of the thermal properties displayed a significant difference between the GC and TT alleles (Table 4). However, in japonica rice, Ida et al. [24] indicated that the GT of the *ss2a/be2b* double mutants was higher than that of the WT mutant line but lower than that of the *be2b* mutant lines. The *be2b* allele or the *ae* mutation caused the rice to synthesize a much longer amylopectin chain, leading to a higher gelatinization temperature [18,37]. Furthermore, *BEIIb* may suppress the expression of *SSIIa* to a certain extent, or else, SSIIa function is not so important when BEIIb is defect. Thus, no significant difference between the GC and TT alleles in the *be2b* background was found in the *be2b* background (Table 4).

Table 4. Thermal properties of rice starches.

Sample	Genotype	T_o	T_p	T_c	ΔH
BP577	GC/*be2b*	74.92 ± 0.06 b	83.06 ± 0.58 abc	92.96 ± 0.20 b	8.08 ± 0.24 bcd
LTFB	TT/*BEIIb*	61.24 ± 0.10 e	69.40 ± 0.08 f	77.46 ± 0.06 f	7.14 ± 0.12 cd
BL01	GC/*be2b*	74.73 ± 0.12 b	83.70 ± 0.13 a	93.96 ± 0.00 a	8.96 ± 0.35 ab
BL02	GC/*be2b*	74.01 ± 0.19 c	82.31 ± 0.44 c	92.58 ± 0.68 b	8.37 ± 0.12 abc
BL03	TT/*be2b*	73.74 ± 0.22 c	82.52 ± 0.03 bc	90.43 ± 0.39 c	8.41 ± 0.22 abc
BL04	TT/*be2b*	69.32 ± 0.11 d	80.03 ± 0.07 d	88.99 ± 0.25 d	5.22 ± 0.06 e
BL05	TT/*be2b*	75.52 ± 0.08 a	83.24 ± 0.10 ab	89.82 ± 0.35 cd	6.78 ± 0.52 d
BL06	GC/*BEIIb*	73.66 ± 0.26 c	78.70 ± 0.24 e	83.42 ± 0.28 e	9.38 ± 0.34 a
BL07	GC/*BEIIb*	73.97 ± 0.19 c	78.37 ± 0.16 e	83.67 ± 0.07 e	9.58 ± 1.22 a
BL08	TT/*BEIIb*	58.31 ± 0.06 g	65.04 ± 0.07 g	73.09 ± 0.13 g	6.88 ± 0.07 d
BL09	TT/*BEIIb*	58.24 ± 0.25 g	65.32 ± 0.29 g	73.56 ± 0.46 g	7.27 ± 0.11 cd
BL10	TT/*BEIIb*	58.81 ± 0.00 f	65.58 ± 0.22 g	73.84 ± 0.08 g	6.92 ± 0.03 d

Values with different letters in the same column are significantly different at $p < 0.05$. T_o: onset temperature; T_p: peak temperature; T_c: conclusion temperature; ΔH: enthalpy.

3.5. Pasting Viscosities

The RVA pasting traces of the rice materials are displayed in Figure 4 and Table 5. For the common rice with the *BEIIb* allele (BL06-10), all the rice samples generally had large PV, CPV, SB, CS, and small BD which are characteristic of high amylose rice. Furthermore, these rice samples with the GC allele (BL06 and 07) had larger PV and BD than those with the TT allele. Since SSIIa is generally responsible for the genetic basis of GT, the difference between the RVA profiles might not reflect the allele difference. Among the rice with the *be2b* allele, all the rice materials had small PV, BD, SB, and CS, showing resistance to heat shearing. This is because high amylose may restrict the swelling of starch during heating. Similarly, rice samples with different GC and TT alleles did not show different RVA profiles (Figure 4).

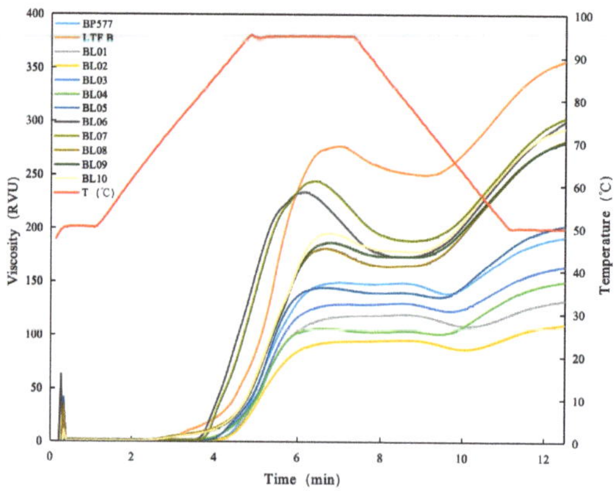

Figure 4. Rapid viscosity profiles of rice starches.

Table 5. Pasting properties of rice starches.

Sample	Genotype	PV	HPV	CPV	BD	SB	CS
BP577	GC/*be2b*	115.75 ± 0.00 fg	105.88 ± 0.38 f	142.55 ± 2.13 de	9.88 ± 0.38 cde	26.80 ± 2.13 ef	36.67 ± 1.75 bc
LTFB	TT/*BEIIb*	275.38 ± 6.38 a	246.88 ± 3.63 a	357.29 ± 10.04 a	28.50 ± 2.75 b	81.92 ± 3.67 bc	110.42 ± 6.42 a
BL01	GC/*be2b*	116.71 ± 2.54 fg	110.46 ± 4.79 ef	142.21 ± 13.38 de	6.25 ± 2.25 de	25.50 ± 10.84 ef	31.75 ± 8.59 c
BL02	GC/*be2b*	95.65 ± 4.15 h	89.59 ± 4.67 g	116.75 ± 9.67 e	6.07 ± 0.52 de	21.10 ± 5.52 f	27.17 ± 5.01 c
BL03	TT/*be2b*	126.42 ± 1.09 ef	119.92 ± 0.50 de	158.55 ± 1.38 d	6.50 ± 0.59 de	32.13 ± 2.46 ef	38.63 ± 1.88 bc
BL04	TT/*be2b*	104.46 ± 0.96 gh	100.92 ± 2.25 fg	149.96 ± 3.88 d	3.54 ± 1.29 e	45.50 ± 2.92 de	49.04 ± 1.63 bc
BL05	TT/*be2b*	139.96 ± 0.54 e	131.96 ± 0.13 d	197.09 ± 0.17 c	8.01 ± 0.42 cde	57.13 ± 0.38 d	65.13 ± 0.04 b
BL06	GC/*BEIIb*	231.88 ± 4.63 b	164.71 ± 4.13 c	300.13 ± 8.88 b	67.17 ± 8.75 a	68.25 ± 4.25 cd	135.42 ± 13.00 a
BL07	GC/*BEIIb*	241.50 ± 4.08 b	178.92 ± 4.67 b	306.96 ± 12.04 b	62.59 ± 8.75 a	65.46 ± 7.96 cd	128.05 ± 16.71 a
BL08	TT/*BEIIb*	183.38 ± 7.13 d	166.46 ± 6.96 c	282.71 ± 8.46 b	16.92 ± 0.16 bcd	99.34 ± 1.34 ab	116.25 ± 1.50 a
BL09	TT/*BEIIb*	185.42 ± 3.83 cd	167.79 ± 1.04 bc	294.92 ± 21.75 b	17.63 ± 4.88 bcd	109.51 ± 17.92 a	127.13 ± 22.79 a
BL10	TT/*BEIIb*	199.59 ± 9.58 c	179.67 ± 6.00 b	285.42 ± 0.00 b	19.92 ± 3.59 bc	85.84 ± 9.59 bc	105.75 ± 6.00 a

Values with different letters in the same column are significantly different at $p < 0.05$. PV: peak viscosity; HPV: host paste viscosity; CPV: cold paste viscosity; BD: breakdown; CS: consistency; SB: setback; RVU: Rapid Visco Units.

3.6. Relationship between Different SSIIa/BEIIb Genotypes

From the structural and physicochemical properties of the breeding lines derived from the parents BP577 and LTFB with different *SSIIa*-GC/TT and *BEIIb/be2b* alleles, it is clear that rice lines with the *be2b* allele displayed distinct structural and physicochemical properties from those with the *BEIIb* allele. However, in the *be2b* allele background, most parameters between the GC and TT alleles did not show differences, suggesting that the function of SSIIa is not important in the *be2b* allele background. From this aspect, it could be concluded that *BEIIb* is more important in determining the structural and physicochemical properties than *SSIIa*. By analysis of all the BE mutants, Tappiban, et al. [19] also indicated that BEIIb played a more important role in determining the structural and physicochemical properties than BEI and BEIIa. To further reveal the relative important functions of the *SSIIa* and *BEIIb* alleles, clustering analysis was carried out for different genotypes (Figure 5). As expected, two groups were formed as the breeding lines were derived from two parents, BP577 and LTFB. The two groups were formed according to different *BEIIb* alleles but not *SSIIa* alleles, further confirming that BEIIb is more important. For the *BEIIb*-related group (the upper one), rice materials with different GC/TT were revealed in two subgroups, and the parent LTFB was grouped with the BL08-10, indicating that the group represented the low GT with the TT allele. However, for the *be2b*-related group, although two or three subgroups could be revealed, they did not follow the GC/TT groups. Instead, parent BP577,

BL01, and BL02 formed a subgroup, BL03 and BL05 formed a second subgroup, and BL04 itself formed a third subgroup (Figure 5).

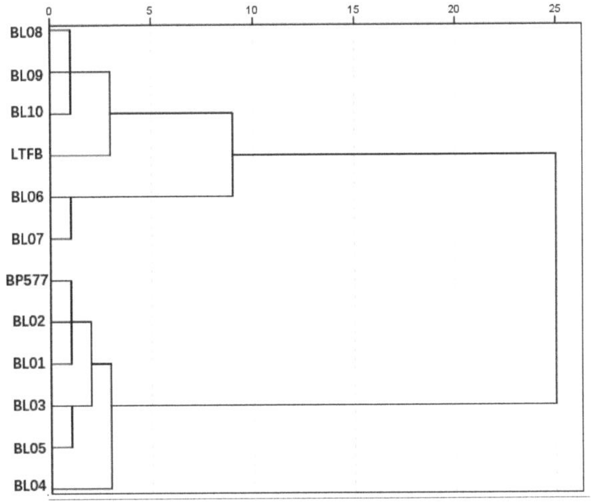

Figure 5. The clustering of genotypes with structural and physicochemical properties.

3.7. Relationships between Fine Structure and Physicochemical Properties

To explore the structure-function relationships, the correlation analysis between starch structural parameters and physicochemical properties is shown in Table S2. Most parameters had a significant correlation with each other, except for ΔH, which had no correlation with any other parameters, and breakdown (BD) viscosity, which showed correlations only with some structural parameters. Since there are two types of starch with distinct structural and physicochemical differences, the correlations may also differ from those in previous studies with common starches. The amylose content (AC) was positively correlated with h_{AM} (r = 0.99, $p < 0.01$) and also had a positive correlation with X_{AP1}, X_{AP2}, X_{AM}, $h_{AP2/AP1}$, fb2, and fb3, but a negative correlation with fa and fb1, suggesting the longer B chains and longer amylose chains led to a higher AC. AC is synthesized by the action of GBSS encoded by the *Wx* gene. All the rice materials contain the same *Wx* allele (Figure 1), so the difference in AC was attributed to the deficiency in BEIIb. The deficiency in BEIIb is responsible for the synthesis of larger amylopectin and amylose molecules with longer chains (Table 1) by which the synthesis of short chains of amylopectin was suppressed and the elongation of long chains was promoted [26,37].

BEIIb has an important effect on the crystalline structures of starches by modifying the synthesis of A and B1 chains [43]. The significant changes in the rice amylopectin CLDs in breeding lines with the *be2b* allele modified the RC of the starch granules. RC had a negative correlation with X_{AP1}, X_{AP2}, X_{AM}, h_{AM}, h_{AP2}/h_{AP1}, AC, fb3, and $\overline{X}$, but a positive correlation with fa and fb1, which was in agreement with the result of Zhang et al. [26]. The results indicated that lower AC and shorter amylose chain length, or more amylopectin A and B1 chains, would increase the RC.

The 1045/1022 cm^{-1} ratio had a negative correlation with fa (r = 0.92, $p < 0.01$), but had a positive correlation with X_{AP1}, X_{AP2}, X_{AM}, $h_{AP2/AP1}$, fb3, and $\overline{X}$ (Table S2), which indicated that longer amylopectin chains can form more double helices and increase the amount of short-range order [26]. RC had a negative correlation with the 1045/1022 cm^{-1} ratio, but a positive correlation with the 1022/995 cm^{-1} ratio (Table S2), which seemed to be contradictory. A possible explanation is that the FTIR data represent the ratio of the proportion of ordered structure to unordered structure, which is irrelevant to long-range order [33].

GT is genetically controlled by SSIIa, whose function is to elongate the short A and B1 chains of amylopectin with DP < 10 to form long B1 chains [15]. T_o, T_p, and T_c were found to be positively correlated with X_{AP1}, X_{AP2}, X_{AM}, $h_{AP2/AP1}$, fb2, fb3, $\overline{X}$, AC, and the 1045/1022 cm^{-1} ratio, but had a negative correlation with RC and the 1022/995 cm^{-1} ratio (Table S2). This result was similar to that of Zhang et al. [26], suggesting these correlations are derived from the *be2b* allele in the rice materials. It should be noted that for the common rice starches, GT generally had a negative correlation with the number of fa chains but a positive correlation with fb1 content [14]. In this study, the correlation between fa and fb1 was positive (r = 0.41, $p > 0.05$), indicating these starches are different from common starches.

The PV, HPV, CPV, SB, and CS had a negative correlation with X_{AP1}, X_{AP2}, X_{AM}, $h_{AP2/AP1}$, fb2, fb3, $\overline{X}$, and AC, but a positive correlation with fb1 (Table S2), which is in agreement with previous reports [26,44]. A large amylose chain content, or a large number of long B chains in amylopectin, is expected to extend through crystallites connecting multiple clusters, increasing the integrity of the starch granules, leading to the inhibition of starch swelling [43] and less resistance to shearing.

In conclusion, the structural and physicochemical properties of rice breeding lines changed with different *BEIIb* and *SSIIa* alleles. The *BEIIb*-deficient mutant starches had a higher AC with more amylose long chains, a larger amount of amylopectin long B chains, and a higher degree of molecular order, but a smaller amount of short chains and smaller RC, leading to a higher GT and lower ΔH and pasting viscosities. Therefore, *BEIIb* is more important in determining the structural and physicochemical properties than *SSIIa*. The combination of *be2b*/GC and *be2b*/TT showed no significant difference, suggesting that the function of SSIIa was not important in the *be2b* allele background. The resistant starch content was not measured in this study, but the fact that the *ae* mutant has high resistant-starch content is well known. In the near future, the breeding lines in the background *be2b* will be used for breeding new rice varieties with high resistant-starch content, which will benefit human health.

Supplementary Materials: The following supporting information can be downloaded at: https://www.mdpi.com/article/10.3390/foods12010119/s1, Table S1: The primer sequences; Table S2: Correlation coefficients between starch structural parameters and functional properties.

Author Contributions: Conceptualization, G.D. (Gaoxing Dai). and J.B.; methodology, Y.H., Y.Z., S.Y., J.B.; formal analysis, Y.H., Y.Z., S.Y., J.B.; investigation, Y.H., Y.Z., S.Y., J.B..; resources, G.D. (Guofu Deng), J.B.; data curation, Y.H.; writing—original draft preparation, Y.H., J.B.; writing—review and editing, J.B.; funding acquisition, G.D. (Guofu Deng), J.B. All authors have read and agreed to the published version of the manuscript.

Funding: This research was funded by the National Natural Science Foundation of China (U20A2032, 31871531), the Natural Science Foundation of Zhejiang Province (LZ21C130003), and the National Key Research and Development Program of China (2016YFD0400104).

Institutional Review Board Statement: Not applicable.

Informed Consent Statement: Not applicable.

Data Availability Statement: Data are contained within the article.

Acknowledgments: We thank Yaling Chen and Zhongwei Zhang for their assistance in genotyping and data analysis.

Conflicts of Interest: The authors declare no conflict of interest.

Abbreviations

AC	Amylose content
BD	Breakdown
BE	Branching enzyme
CLD	Chain length distribution
CPV	Cold paste viscosity
CS	Consistency viscosity
CEQ	Cooking and eating quality
DP	Degree of polymerization
FACE	Fluorophore-assisted carbohydrate electrophoresis
GT	Gelatinization temperature
HPV	Hot paste viscosity
PV	Peak viscosity
RC	relative crystallinity
RIL	Recombinant inbred line
RVA	Rapid Visco-Analyzer
RVU	Rapid Visco Unit
SB	Setback
SEC	Size-exclusion chromatography
SNP	Single nucleotide polymorphism
SS	Soluble starch synthase
T_c	conclusion temperature
T_o	Onset temperature
T_p	peak temperature
Wx	Waxy
ΔH	Enthalpy of gelatinization

References

1. Bao, J.; Shen, S.; Sun, M.; Corke, H. Analysis of genotypic diversity in the starch physicochemical properties of nonwaxy rice: Apparent amylose content, pasting viscosity and gel texture. *Starch-Starke* **2006**, *58*, 259–267. [CrossRef]
2. Singh, N.; Kaur, L.; Sandhu, K.S.; Kaur, J.; Nishinari, K. Relationships between physicochemical, morphological, thermal, rheological properties of rice starches. *Food Hydrocoll.* **2006**, *20*, 532–542. [CrossRef]
3. Tester, R.F.; Karkalas, J.; Qi, X. Starch—Composition, fine structure and architecture. *J. Cereal Sci.* **2004**, *39*, 151–165. [CrossRef]
4. Witt, T.; Doutch, J.; Gilbert, E.P.; Gilbert, R.G. Relations between molecular, crystalline, and lamellar structures of amylopectin. *Biomacromolecules* **2012**, *13*, 4273–4282. [CrossRef] [PubMed]
5. Jeon, J.-S.; Ryoo, N.; Hahn, T.-R.; Walia, H.; Nakamura, Y. Starch biosynthesis in cereal endosperm. *Plant Physiol. Biochem.* **2010**, *48*, 383–392. [CrossRef] [PubMed]
6. Bao, J.S. Towards understanding of the genetic and molecular basis of eating and cooking quality of rice. *Cereal Foods World* **2012**, *57*, 148–156. [CrossRef]
7. Hirano, H.Y.; Eiguchi, M.; Sano, Y. A single base change altered the regulation of the Waxy gene at the posttranscriptional level during the domestication of rice. *Mol. Biol. Evol.* **1998**, *15*, 978–987. [CrossRef]
8. Ayres, N.M.; McClung, A.M.; Larkin, P.D.; Bligh, H.F.J.; Jones, C.A.; Park, W.D. Microsatellites and a single-nucleotide polymorphism differentiate apparent amylose classes in an extended pedigree of US rice germ plasm. *Theor. Appl. Genet.* **1997**, *94*, 773–781. [CrossRef]
9. Bao, J.S.; Corke, H.; Sun, M. Microsatellites, single nucleotide polymorphisms and a sequence tagged site in starch-synthesizing genes in relation to starch physicochemical properties in nonwaxy rice (*Oryza sativa* L.). *Theor. Appl. Genet.* **2006**, *113*, 1185–1196. [CrossRef]
10. Chen, M.-H.; Bergman, C.J.; Pinson, S.R.; Fjellstrom, R.G. Waxy gene haplotypes: Associations with pasting properties in an international rice germplasm collection. *J. Cereal Sci.* **2008**, *48*, 781–788. [CrossRef]
11. Umemoto, T.; Aoki, N. Single-nucleotide polymorphisms in rice starch synthase IIa that alter starch gelatinisation and starch association of the enzyme. *Funct. Plant Biol.* **2005**, *32*, 763–768. [CrossRef]
12. Waters, D.L.E.; Henry, R.J.; Reinke, R.F.; Fitzgerald, M.A. Gelatinization temperature of rice explained by polymorphisms in starch synthase. *Plant Biotechnol. J.* **2006**, *4*, 115–122. [CrossRef] [PubMed]
13. Bao, J.S.; Corke, H.; Sun, M. Nucleotide diversity in starch synthase IIa and validation of single nucleotide polymorphisms in relation to starch gelatinization temperature and other physicochemical properties in rice (*Oryza sativa* L.). *Theor. Appl. Genet.* **2006**, *113*, 1171–1183. [CrossRef] [PubMed]
14. Bao, J.; Xiao, P.; Hiratsuka, M.; Sun, M.; Umemoto, T. Granule-bound SSIIa protein content and its relationship with amylopectin structure and gelatinization temperature of rice starch. *Starch-Starke* **2009**, *61*, 431–437. [CrossRef]

15. Nakamura, Y.; Francisco, P.B.; Hosaka, Y.; Sato, A.; Sawada, T.; Kubo, A.; Fujita, N. Essential amino acids of starch synthase IIa differentiate amylopectin structure and starch quality between japonica and indica rice varieties. *Plant Mol. Biol.* **2005**, *58*, 213–227. [CrossRef] [PubMed]
16. Song, J.; Hu, Y.; Deng, G.; Dai, G.; Bao, J. The origin of the A/G single nucleotide polymorphism of *starch synthase IIa* in rice and its relation to gelatinization temperature. *Cereal Chem.* **2021**, *99*, 275–285. [CrossRef]
17. Fei, Y.; Yang, J.; Wang, F.; Fan, F.; Wenqi, L.I.; Wang, J.; Yang, X.U.; Zhu, J.; Zhong, W. Production of two elite glutinous rice varieties by editing Wx gene. *Rice Sci.* **2019**, *26*, 118–127.
18. Nishi, A.; Nakamura, Y.; Satoh, T.H. Biochemical and genetic analysis of the effects of amylose-extender mutation in rice endosperm. *Plant Physiol.* **2001**, *127*, 459–472. [CrossRef]
19. Tappiban, P.; Hu, Y.; Deng, J.; Zhao, J.; Ying, Y.; Zhang, Z.; Xu, F.; Bao, J. Relative importance of branching enzyme isoforms in determining starch fine structure and physicochemical properties of indica rice. *Plant Mol. Biol.* **2021**, *108*, 399–412. [CrossRef]
20. Bao, J.; Zhou, X.; Xu, F.; He, Q.; Park, Y.-J. Genome-wide association study of the resistant starch content in rice grains. *Starch-Starke* **2017**, *69*, 1600343. [CrossRef]
21. Itoh, Y.; Crofts, N.; Abe, M.; Hosaka, Y.; Fujita, N. Characterization of the endosperm starch and the pleiotropic effects of biosynthetic enzymes on their properties in novel mutant rice lines with high resistant starch and amylose content. *Plant Sci.* **2017**, *258*, 52–60. [CrossRef] [PubMed]
22. Kubo, A.; Akdogan, G.; Nakaya, M.; Shojo, A.; Kitamura, S. Structure, physical, and digestive properties of starch from wx ae double-mutant rice. *J. Agric. Food Chem.* **2010**, *58*, 4463–4469. [CrossRef]
23. Wang, K.; Hasjim, J.; Wu, A.C.; Li, E.; Henry, R.J.; Gilbert, R.G. Roles of GBSSI and SSIIa in determining amylose fine structure. *Carbohydr. Polym.* **2015**, *127*, 264–274. [CrossRef] [PubMed]
24. Ida, T.; Crofts, N.; Miura, S.; Matsushima, R.; Fujita, N. Structure and properties of starch in rice double mutants lacking starch synthase (SS) IIa and starch branching enzyme (BE) IIb. *J. Appl. Glycosci.* **2021**, *68*, 31–39. [CrossRef]
25. Doyle, J. DNA protocols for plants CTAB total DNA isolation. In *Molecular Techniques in Taxonomy*; Hewitt, G.M., Johnston, A., Eds.; Springer: Berlin, Germany, 1991; pp. 283–293.
26. Zhang, Z.; Hu, Y.; Zhao, J.; Zhang, Y.; Ying, Y.; Xu, F.; Bao, J. The role of different Wx and BEIIb allele combinations on fine structures and functional properties of indica rice starches. *Carbohydr. Polym.* **2021**, *278*, 118972. [CrossRef] [PubMed]
27. Jin, L.; Lu, Y.; Shao, Y.; Zhang, G.; Xiao, P.; Shen, S.; Corke, H.; Bao, J. Molecular marker assisted selection for improvement of the eating, cooking and sensory quality of rice (*Oryza sativa* L.). *J. Cereal Sci.* **2009**, *51*, 159–164. [CrossRef]
28. Syahariza, Z.A.; Li, E.; Hasjim, J. Extraction and dissolution of starch from cereal grains for accurate structural analysis. *Carbohydr. Polym.* **2010**, *82*, 14–20. [CrossRef]
29. Castro, J.V.; Ward, R.M.; Gilbert, A.R.G.; Fitzgerald, M.A. Measurement of the molecular weight distribution of debranched starch. *Biomacromolecules* **2005**, *6*, 2260–2270. [CrossRef]
30. Wu, A.C.; Li, E.; Gilbert, R.G. Exploring extraction/dissolution procedures for analysis of starch chain-length distributions. *Carbohydr. Polym.* **2014**, *114*, 36–42. [CrossRef]
31. Hanashiro, I.; Abe, J.-I.; Hizukuri, S. A periodic distribution of the chain length of amylopectin as revealed by high-performance anion-exchange chromatography. *Carbohydr. Res.* **1996**, *283*, 151–159. [CrossRef]
32. Zhang, Z.; Fan, X.; Ma, H.; Li, C.; Li, E.; Gilbert, R.G. Characterization of the baking-induced changes in starch molecular and crystalline structures in sugar-snap cookies. *Carbohydr. Polym.* **2020**, *256*, 117518. [CrossRef] [PubMed]
33. Sevenou, O.; Hill, S.; Farhat, I.; Mitchell, J. Organisation of the external region of the starch granule as determined by infrared spectroscopy. *Int. J. Biol. Macromol.* **2002**, *31*, 79–85. [CrossRef] [PubMed]
34. Bao, J.; Sun, M.; Corke, H. Analysis of genotypic diversity in starch thermal and retrogradation properties in nonwaxy rice. *Carbohydr. Polym.* **2007**, *67*, 174–181. [CrossRef]
35. Ahmed, S.; Ru, W.; Han, H.; Cheng, L.; Bian, X.; Li, G.; Jin, L.; Wu, P.; Bao, J. Fine molecular structure and its effects on physicochemical properties of starches in potatoes grown in two locations. *Food Hydrocoll.* **2019**, *97*, 105172. [CrossRef]
36. Tappiban, P.; Ying, Y.; Pang, Y.; Sraphet, S.; Srisawad, N.; Smith, D.R.; Wu, P.; Triwitayakorn, K.; Bao, J. Gelatinization, pasting and retrogradation properties and molecular fine structure of starches from seven cassava cultivars. *Int. J. Biol. Macromol.* **2020**, *150*, 831–838. [CrossRef]
37. Lin, L.; Huang, J.; Zhang, L.; Liu, Q.; Wei, C. Effects of inhibition of starch branching enzymes on starch ordered structure and component accumulation in developing kernels of rice. *J. Cereal Sci.* **2019**, *91*, 102884. [CrossRef]
38. Butardo, J.V.M.; Daygon, V.D.; Colgrave, M.L.; Campbell, P.M.; Resurreccion, A.; Cuevas, R.P.; Jobling, S.A.; Tetlow, I.; Rahman, S.; Morell, M.; et al. Biomolecular analyses of starch and starch granule proteins in the high-amylose rice mutant Goami 2. *J. Agric. Food Chem.* **2012**, *60*, 11576–11585. [CrossRef]
39. Butardo, V.M.; Fitzgerald, M.A.; Bird, A.R.; Gidley, M.J.; Flanagan, B.M.; Larroque, O.; Resurreccion, A.P.; Laidlaw, H.K.C.; Jobling, S.A.; Morell, M.K.; et al. Impact of down-regulation of starch branching enzyme IIb in rice by artificial microRNA- and hairpin RNA-mediated RNA silencing. *J. Exp. Bot.* **2011**, *62*, 4927–4941. [CrossRef]
40. Fuwa, H.; Inouchi, N.; Glover, D.V.; Fujita, S.; Sugihara, M.; Yoshioka, S.; Yamada, K.; Sugimoto, Y. Structural and physicochemical properties of endosperm starches possessing different alleles at the amylose-extender and Waxy locus in maize (*Zea mays* L.). *Starch-Starke* **1999**, *51*, 147–151. [CrossRef]

41. Crofts, N.; Abe, N.; Oitome, N.F.; Matsushima, R.; Hayashi, M.; Tetlow, I.J.; Emes, M.J.; Nakamura, Y.; Fujita, N. Amylopectin biosynthetic enzymes from developing rice seed form enzymatically active protein complexes. *J. Exp. Bot.* **2015**, *66*, 4469–4482. [CrossRef]
42. Chen, Y.; Pang, Y.; Bao, J. Expression profiles and protein complexes of starch biosynthetic enzymes from white-core and waxy mutants induced from high amylose Indica rice. *Rice Sci.* **2020**, *27*, 152–161.
43. Jane, J.; Chen, Y.Y.; Lee, L.F.; McPherson, A.E.; Wong, K.S.; Radosavljevic, M.; Kasemsuwan, T. Effects of amylopectin branch chain length and amylose content on the gelatinization and pasting properties of starch. *Cereal Chem.* **1999**, *76*, 629–637. [CrossRef]
44. Tao, K.; Li, C.; Yu, W.; Gilbert, R.G.; Li, E. How amylose molecular fine structure of rice starch affects functional properties. *Carbohydr. Polym.* **2018**, *204*, 24–31. [CrossRef] [PubMed]

foods

MDPI

Review

Organic Rice Production Practices: Effects on Grain End-Use Quality, Healthfulness, and Safety

Christine Bergman * and Mhansi Pandhi

Food & Beverage and Event Management, University of Nevada Las Vegas, Las Vegas, NV 89154, USA
* Correspondence: christine.bergman@unlv.edu

Abstract: Demand for rice labeled as organic is growing globally. Consumers state that foods labeled as organic are nutritionally superior and safer than their conventionally produced equivalent. The research question for this systematic review is as follows: is there a difference between the effects of sustainable agriculture and conventional farming methods on rice grain end-use quality, healthfulness, and safety? The studies (n = 23) examined for this review suggest that organic production practices don't influence most end-use quality (e.g., chalk, milling yield, pasting properties) and healthfulness (e.g., gamma-oryzanol fraction and tocols) traits studied, or if it does, it will be to a small degree. If differences in end-use quality traits are found, they will be associated with grain protein content, which varies along with the dose of nitrogen applied during rice growth. We conclude that the studies evaluated in this review found that organically produced rice grain was less likely to contain residues of the pesticides (e.g., organochlorine) examined in the study than the rice is grown using conventional methods. There was some evidence that organically grown rice is more likely to be contaminated with mycotoxin-producing fungi and some mycotoxins. Common shortcomings of the studies were that they were poorly designed, with limited to no details of the cultural management practices used to grow the rice studied, the length of time fields was under organic management not stated, cultivars were not named, and the data wasn't analyzed statistically.

Keywords: rice; organic; sustainable; quality; safety; nutrition; pesticides; mycotoxins; gamma-oryzanol; tocols

Citation: Bergman, C.; Pandhi, M. Organic Rice Production Practices: Effects on Grain End-Use Quality, Healthfulness, and Safety. *Foods* **2023**, *12*, 73. https://doi.org/10.3390/foods12010073

Academic Editor: Carla Brites

Received: 8 December 2022
Revised: 18 December 2022
Accepted: 19 December 2022
Published: 23 December 2022

1. Introduction

Rice is the staple food for more than half of the world's population and is grown in over 100 countries [1]. The International Rice Information System indicates there are at least 5000 released rice varieties and many more if traditional varieties are considered [2]. Consumers generally choose a rice type with the cooking and sensory properties they are accustomed to or that suits the meal they will be preparing. For example, jasmine-types with their buttery popcorn-like aroma and soft texture are preferred by many people in Thailand and are served along with Thai-inspired meals prepared across the world. The cooking and sensory properties of rice and how it performs in processed products, such as canned soups, are known as rice end-use quality [3].

Most people across the globe eat rice in its milled (or polished) form. Rice that has been milled provides consumers with kilocalories, protein, vitamins, and minerals. Those that choose to eat unmilled (i.e., threshed or brown rice), commonly referred to as thrashed or brown rice, also obtain significant amounts of dietary fiber, lipids, and various phytonutrients [4]. An increasing number of consumers are choosing to eat unmilled rice and rice that has been grown using the principles of organic farming practices [5].

It has been proposed that there are two distinct schools of thought on how farming is practiced across the globe: the industrial and the agrarian philosophies [6]. Farmers and agronomy researchers tend to use the following categories to describe rice production methods: conventional production methods, sustainable agriculture techniques (e.g., organic, biodynamic, and regenerative), and the system of rice intensification.

The classifications of industrial or conventional farming "typically use synthetic pesticides, herbicides, and fertilizers, may use organic soil amendments; fields are frequently planted in short rotations" and generally uses monocropping systems [7]. Sustainable agriculture, as legally defined in U.S. Code Title 7, Section 3103 is an integrated system of plant and animal production techniques that have a site-specific application that will over the long term: Satisfy human food and fiber needs, enhance environmental quality and the natural resource base upon which the agricultural economy depends; make the most efficient use of nonrenewable resources and on-farm resources and integrate, where appropriate, natural biological cycles and controls, sustain the economic viability of farm operations and enhance the quality of life for farmers and society as a whole.

The UN Food and Agriculture Organization has defined organic agriculture as "a unique production management system which promotes and enhances agro-ecosystem health, including biodiversity, biological cycles, and soil biological activity, and this is accomplished by using on-farm agronomic, biological and mechanical methods in exclusion of all synthetic off-farm inputs" [8]. Thus, organic agriculture is a type of sustainable agriculture. Another form of sustainable agriculture is known as biodynamic farming. These types of farms generally grow several different crops, avoid the use of conventional inputs, produce and distribute the food in a decentralized manner, and take into consideration celestial and terrestrial influences on biological organisms [9]. A definition for regenerative farming has been proposed to be as follows: "an approach to farming that uses soil conservation as the entry point to regenerate and contribute to multiple provisioning, regulating and supporting services, with the objective that this will enhance not only the environmental, but also the social and economic dimensions of sustainable food production" [10].

Farmers that produce rice using sustainable production methods are often smallholders that perform low-input farming because it is their traditional way of farming, and they have limited resources to invest in conventional inputs [11]. Others, utilize sustainable methods to prevent the negative effects of conventional production methods that have been in use since the Green Revolution. Lastly, others have converted to using sustainable farming methods due to increased consumer demand for these foods and their willingness to pay premium prices for them [12]. These farmers in general adhere to their nation's regulations on production practices required to allow foods to be labeled organic.

Between 2019 and 2025, the global organic rice market is expected to increase at a compound annual growth rate of 8% [5]. This increased demand is occurring globally, with the greatest increase in individual demand being in the European Union and North America. Consumers report that they purchase foods produced organically for the following reasons: healthier, helping the environment, and convenience [13]. For organic rice, the emotional route (e.g., I will feel happy if I buy organic rice) had a greater impact on its purchase intention than did the rational route (e.g., buying organic rice can form a good impression for me) [14]. Review articles report inconsistent evidence that foods produced using organic methods are significantly different in nutrient content compared to conventionally produced foods. However, organic foods are generally considered safer for consumption due to containing lower levels of pesticides and antibiotic residues [15].

The research question for this systematic review is as follows: is there a difference between the effects of sustainable agriculture and conventional farming methods on rice grain end-use quality, healthfulness, and safety?

2. Methods

This review was created according to the Preferred Reporting Items for Systematic Reviews and Meta-Analyses (PRISMA) Checklist and Guidance of the European Food Safety Authority [16]. This type of systematic review uses a clearly formulated question along with explicit steps to identify, select, and critically appraise previously published research and summarize the data from the studies found during the review.

2.1. Inclusion Criteria

The inclusion criteria for this study were experimental studies that evaluated the effects of sustainable farming production methods on rice grain end-use quality, healthfulness, and safety. Studies included those published during the previous 25 years (i.e., 1996–2021) and complete articles written in English. This time period was selected because it was in 2002 that the regulations under the U.S. Organic Foods Production Act were implemented and other countries such as Brazil adopted similar regulations sometime after this [15].

2.2. Information Sources

Five electronic databases that house different journals were searched; specifically, Academic Search Premier© (EBSCO Industries, Birmingham, AL, USA), Directory of Open Access Journals (Licensed under CC BY-SA 2.0), PubMed (US National Library of Medicine and National Institute of Health, https://pubmed.ncbi.nlm.nih.gov/, ScienceDirect ® (Elsevier B.V., https://www.elsevier.com/solutions/sciencedirect, and Scopus® (Elsevier B.V., https://www.elsevier.com/en-gb/solutions/scopus. All databases were accessed between 1 June 2022 and 5 June 2022.

2.3. Study Selection Process

Two authors followed the following protocol to identify articles for use in this study. The following keywords and Boolean operators were used to search the databases: #1. rice AND (#2. organic OR sustainable OR biodynamic or regenerative) AND (#3. nutrition OR health OR Vitamins OR minerals OR phytochemical OR quality OR cooking OR processing OR amylose OR safety OR Pesticide OR Herbicide OR Fungicide). Each search included one term from categories #1, #2, and #3.

The titles and abstracts of all articles identified from the selection process were screened for each one's relevancy to the research question by two authors separately. They were independently screened, coded, and evaluated for suitability to be included in the study. Specifically, a study was selected if it evaluated the effects of growing rice using any sustainable agriculture methods on any aspect of rice grain end-use quality, healthfulness, safety, and methodological quality. Full articles for each study selected in the previous step were then evaluated using the exclusion criteria. The authors discussed any differences in the articles identified for inclusion in the study and came to an agreement on which to include and which to not include based on the exclusion criteria.

2.4. Exclusion Criteria

The following exclusion criteria were used: (1) full paper not written in English, (2) not specifically related to the research question, (3) no conventional rice used for comparison to sustainably produced rice, (4) non-research articles, (5) duplicates, and (6) methodology very unclear.

2.5. Data Analysis

A meta-analysis was not conducted for this review as the studies didn't use similar experimental designs or analytical methods. The articles reviewed either didn't describe the cultural management practices or the ones used were very different from the others reported. Therefore, only a qualitative analysis of the data collected from the studies was conducted, as recommended by the Cochrane handbook for systematic reviews of intervention studies [17].

Research articles included in the study were evaluated to determine if suitable statistical methods were used to analyze the data collected during the study.

2.6. Risk of Bias

Bias in study design was decreased by using five databases to identify the studies for the systematic review. However, grey literature such as conference proceedings was not included. Selection bias was limited by the creation and use of inclusion and exclusion criteria. Two authors, separately, performed the literature identification and selection process to limit personal bias. The risk of assessment bias was reduced by critically evaluating all of the studies included in the systematic review and discussing them in the discussion section below.

3. Results

3.1. Identification of Included Studies

The study selection process yielded 12,757 records (Figure 1). After the deletion of duplicates, 1028 records remained. The titles and abstracts of these records were evaluated to determine if they contained descriptions of experimental studies that evaluated the effects of any sustainable farming production methods on rice grain end-use quality, healthfulness, or safety. In addition, the dates were evaluated to determine if the records were published in the previous 25 years and were written in English. Author one and two, using these inclusion criteria identified 60 and 49 articles, respectively. The additional articles chosen by author two were found by author one to be studies related to the nutrient content of organic soils used to grow rice, not in rice grain. Therefore, 49 full articles were selected for further evaluation since soil nutrient content wasn't one of the dependent variables being evaluated in this study. These 49 articles were then evaluated with the exclusion criteria. Of the 49 articles, 26 were deleted from the group due to one of the following exclusion reasons: full article not written in English, methods not specifically related to the research, no conventional agriculture farming dependent variable included, article duplication, or unclear methodology (Table 1). The primary reason for exclusion from the study was due to articles not being specifically related to the research topic (e.g., frog and rice co-culture using organic production methods). Thus, there were 23 articles found and deemed to be suitable for examination in this review study.

Table 1. Full-text articles deleted from the total records based on exclusion criteria.

Exclusion Criteria	# of Articles Identified
Full paper not written in English	3
Not specifically related to the topic	17
No conventional rice used for comparison to organic	2
Not a research article	2
Duplicate articles	1
Methodology unclear	1
TOTAL	26

3.2. Study Characteristics

The studies included in this review were performed in the following countries: Afghanistan, Brazil, India, Italy, Malaysia, Philippines, South Korea, Spain & Portugal, Thailand, and United States (Table 2). The top countries involved in sustainable rice grain research were found to be Brazil and Thailand. Of the sustainable rice, grain-focused studies eight evaluated effects on end-use quality, nine on healthfulness, and 14 examined safety issues. All of the studies focused specifically on rice described as being organic by the authors. No other terminology used to describe sustainable farming practices was found in the articles.

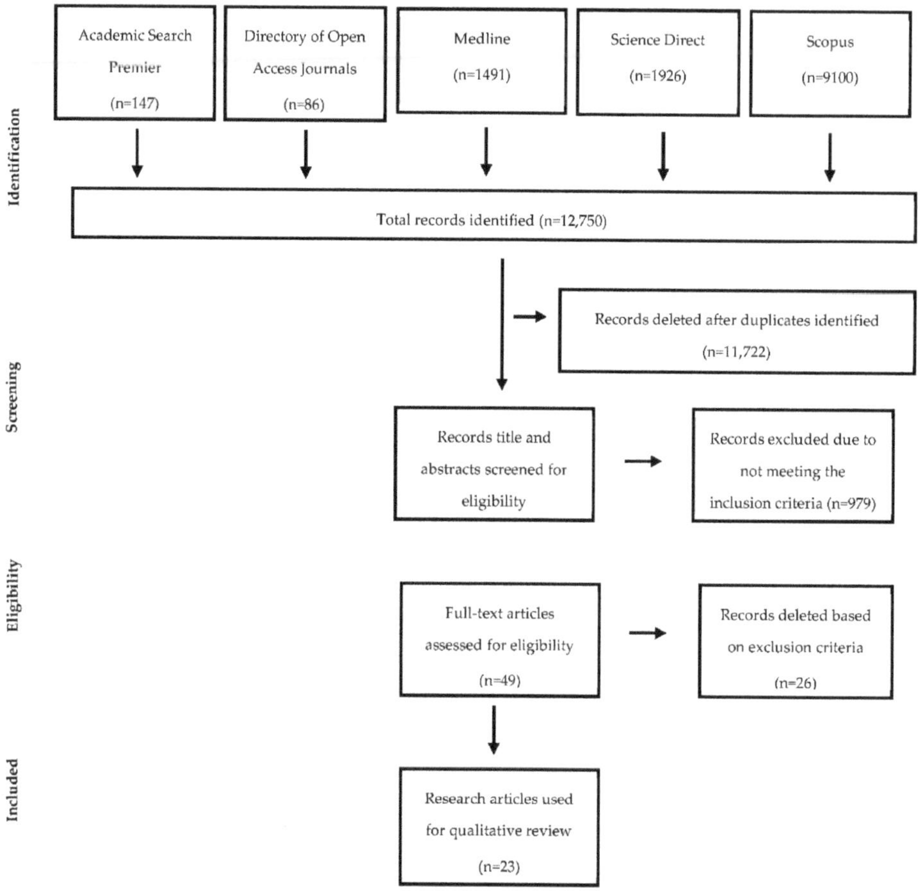

Figure 1. Flow chart created from a systematic review of organic versus conventional rice grain using the preferred reporting items for systematic reviews (PRISMA) methodology [16].

Table 2. Organic versus conventional rice grain-related studies were identified using the preferred reporting items for systematic reviews (PRISMA) methodology.

COUNTRY [1]	# of STUDIES	CATEGORIES OF TRAITS EXAMINED [2]		
		End-Use Quality	Healthfulness	Safety
Afghanistan	1	1		
Brazil	5	1	3	5
India	2	1		1
Italy	1			1
Malaysia	1	1		
Philippines	1	1	1	1
South Korea	2		1	1
Spain & Portugal	3			3
Thailand	4	2	3	1
United States	3	1	1	1
TOTAL	23	8	9	14

[1] Countries where the rice samples used in the studies were grown or purchased. [2] Some studies evaluated traits in more than one category.

Only Chen and McClung [18], Champagne et al. [19], Cho et al. [20], Tuano et al. [21], and VanQuyen et al. [22], reported in some detail the agronomic practices used to produce the rice for their study. Of these Chen and McClung [18], Champagne et al. [19], Tuano et al. [21], and VanQuyen et al. [22] also included the name of the rice cultivar(s) they studied and performed replications of their field treatments. Alves et al. [23] and Keawpeng et al. [24] reported the cultivar they studied, but limited information was included about the field management methods used. The other studies examined rice or rice products obtained from grocery stores and rice cooperatives.

Of the studies that evaluated end-use quality, four evaluated amylose content (Table 3) (Champagne et al. [19], Kaker et al. [25], Keawpeng et al. [24], Tuano et al. [21]), three examined protein content (Alves et al. [23], Kakar et al. [25], Keawpeng et al. [24]), and three examined lipid content (Alves et al. [23], Kakar et al. [25], Keawpeng et al. [24]. One study examined the sensory quality and volatile compounds (Champagne et al. [19]. Two studies determined the yield of head rice or whole kernels after milling (Alves et al. [23], Kakar et al. [25]) and one examined the percentage of stained grain (Kakar et al. [25]). Cooking time was measured by one study (Alves et al. [23]), as were elongation ratio, hardness, color, water uptake capacity, and starch crystallinity (Keawpeng et al. [24]). Kernel swelling power, H_2O solubility, and starch X-ray diffraction patterns were each examined by one study (Keawpeng et al. [24]). Pasting properties were evaluated by two studies (Champagne et al. [19], Keawpeng et al. [24]). Rice noodle color, tensile strength, elasticity, cooking loss, rehydration ratio, aerobic plate content, and total yeast and mold contents were discussed in one paper (Thomas et al. [26]). One study evaluated kernel length, kernel breadth, and the ratio between the two before and after cooking (VanQuyen et al. [22]).

Table 3. Organic versus conventional rice grain-related studies identified using the PRISMA methodology: materials, methods, and findings summary.

Materials and Study Design	Traits	Organic Rice Production Methods	Conventional Rice Production Methods	Main Findings	Citation
"IRGA 410" [1] grown in Rio Grande Do Sul, Brazil.	Brown rice proximate analysis, phenolics, amylose content, cooking time, head rice yield, and stained grains during storage. Mycotoxin analysis. Fungal incidence in rough rice.	Seeding rate 90 kg/h. Water management same as conventional No other inputs were reported.	Seeding rate 90 kg/h. Water management same as organic. Urea was applied at 140 kg/ha in dry soil, before the appearance of the 3rd leaf, and at 60 kg/ha at the beginning of panicle development. Two fungicide applications of PrimoR©300 mL/ha. One application of TalismanR© 250 mL/ha.	Conventional unmilled rice had greater protein, lipid, and ash content, and higher head rice yield. It also had greater Aspergillus sp. after storage. Organic thrashed rice had greater total carbohydrates, soluble protein, amylose content, free phenolics, and phenolic acids. It also had a greater percentage of stained grains and Bipolaris sp. after storage.	[23]
Organic (n = 17) and conventional (n = 33) rice samples were purchased from different Brazilian producers (50 different brands) in different cities.	Cu, Zn, Mg, B, P, Mo, As, Pb, Cd, Mn, Se, Co, Cr, Ba, Rb, Fe, Ca, La, and Ce contents.	Certified organic by the Brazilian IBD-Agricultural and Food Inspections and Certifications which is accredited by the International Federation of Organic Agriculture Movements.	No description of conventional methods was reported.	Ca levels are significantly higher in all org. samples but one. Cd is higher in conventional samples. No difference in As levels. Statistical differences in other minerals weren't analyzed.	[27]

Table 3. *Cont.*

Materials and Study Design	Traits	Organic Rice Production Methods	Conventional Rice Production Methods	Main Findings	Citation
Rice samples purchased from markets in Brazil. 12 milled rice, 10 parboiled milled rice samples, 2 unmilled, and 5 unmilled parboiled samples. One organic milled and one parboiled organic sample.	As	No organic production methods were described.	No conventional production methods were described.	The mean of total As for the milled samples was greater than in the organic milled sample (222.9 and 161.6 ng g^{-1}, respectively). The mean of total As for parboiled white samples was greater than the parboiled organic sample (214.9 and 174.1 ng g^{-1}, respectively).	[28]
Randomized block design, 4 replications during 3 years. "Cypress", "Bengal", "Jasmine 85", "Jacinto", and "Neches" grown in adjacent fields that had been fallow for 2 years.	Paste viscosity properties; amylose, protein, Ca, Fe, P, K, Mg, Mn, and Zn contents. Volatile compounds and sensory evaluation.	Chicken litter was applied (76 kg/ha N, 25 kg/ha P, and 25 kg/ha K), a microbial product of trace minerals (67 kg/ha), and a microbial soil activator (33.8 L/ha) was applied prior to planting. Seed treated with humic acid (5 mLJkg of seed), a microbial inoculant (mL/100 g of seed), and manganese sulfate (20 g/kg of seed). Before the flood, a side dress of chicken litter was applied (126 kg/ha N, 42 kg/ha P, and 42 kg/ha K). At panicle differentiation, a fish emulsion was applied as a foliar spray (16.5 L/ha) for insect control.	Urea nitrogen (56 kg/ha, 90 kg/ha, and 78 kg/ha) was applied at planting, flood, and panicle differentiation, for the 100% nitrogen treatment. The 50% nitrogen treatment was applied using half of the rate of urea (112 kg of N/ha). Standard chemical management practices were used to control weeds and insects in conventionally managed plots.	Milled kernel protein content in conventional 100% urea nitrogen samples > other treatments. Little to no difference in amylose and mineral contents between treatments. Differences in pasting properties were found associated with kernel protein content. No differences in flavor attributes were found via the sensory panel or in volatile compounds. No differences in microbially produced volatile compounds were found.	[19]
Randomized block design with two field replications and two years. "Cocodrie', 'Presidio' "Sierra", "Giant Embryo" (GSOR 25), "IL 121-1-1" and "Sigoendab"	Total phenolics, flavonoids, tocols, and γ-oryzanol	Certified organic fields followed two years of fallow and a clover/ryegrass winter cover crop. Nature Safe 13-0-0 fertilizer (1681 kg/ha) was applied just prior to planting in both years. Year one seed was drilled with two seeding rates, 112 and 135 kg/ha. In year two, water-seeding was done with 202 kg/ha.	Following two years of fallow, seeds were drilled at 112 kg/ha. A total of 224 kg/ha of nitrogen as urea (46-0-0) was applied with a three-way split: at planting (56 kg/ha), at permanent flood (90 kg/ha), and at panicle differentiation (78 kg/ha).	The growing environment affected the concentrations of most traits, especially the tocols and γ-oryzanol. The effect of conventional versus organic management systems had the lowest effect on the phytochemical levels of the year, replication, and management system.	[18]
"Dongjin" was obtained from one organic field and one conventional field. A sampling of organic rice was done in the central region of the block.	Total gamma-oryzanol compounds.	The field had been organically managed for 5 years. Independent groundwater was used. The green manure crop, Vicia villosa Roty), was used in the organic plot.	Field managed conventionally for 10 years. Fertilization included (N/P/K = 9:4.5:5.7, *w/w*, 202 kg/ha). Weed prevention performed using herbicide [1.0% fentrazamide and 0.07% pyrazosulfuron-ethyl, suspension concentrate, 321 green area index (gai)/ha] Pesticide (2% chlorantraniliprole, 16 gai/ha)	Organic brown rice had higher total gamma-oryzanol than conventional (65.6 ± 2.7 mg/100 g and 60.2 ± 1.8 mg/100 g respectively)	[20]

Table 3. *Cont.*

Materials and Study Design	Traits	Organic Rice Production Methods	Conventional Rice Production Methods	Main Findings	Citation
Rice-based foods from Italian stores: flours, biscuits, and rice flakes (13 conventional and 11 organic products).	Deoxynivalenol, fumonism B1, and fumonism B2	No organic production methods were described.	No conventional production methods were described.	Deoxynivalenol was identified in more organic rice foods than conventional (91% vs. 85%). Furmonism B1 occurred in 36% of organic rice foods versus 23% of conventional. Furmonism B2 occurred in 38% of conventional vs. 45% of organic rice food products.	[29]
Rice samples and rice-based foods were collected from cultivars and markets in Spain. 64 were conventional and 20 were organic.	Ochratoxin A	No organic production methods were described.	No conventional production methods were described.	Ochratoxin A was found in a greater % of organic rice and products compared to those grown using conventional methods.	[30]
10 baby cereals are made using conventional rice and 3 produced using organic rice obtained from manufacturers.	Hg and total As.	No organic production methods were described.	No conventional production methods were described.	Hg was higher in organic rice cereal compared to conventional rice cereal (4.54 and 4.39 µg/kg, respectively). As was higher in organic rice cereal compared to conventional rice cereal (154.9 and 96.3 µg/kg, respectively)	[31]
9 organic and 12 conventionally produced rice samples and rice foods bought from markets in Spain and Portugal.	Ochratoxin A	No organic production methods were described.	No conventional production methods were described.	Ochratoxin A found in 4/9 organic samples and 0/12 conventionally produced rice foods.	[32]
10 infant rice cereals made with commercially produced milled rice and 10 infant rice cereals made with organically produced unmilled rice.	Total As and inorganic As (i-As).	No organic production methods were described.	No conventional production methods were described.	All samples had identifiable As and iAs. No significant difference between organic and conventionally produced rice cereals.	[33]
Randomized block design, one year "Attai-1", five cultural management treatments, four replications.	Unmilled kernels: perfect grains, broken grains and amylose, protein, and lipid contents.	The recommended dose for traditional farming (120 kg/ha urea and 100 kg/ha diammonium hydrogen phosphate)	Animal manure (5 tons/ha) (AM), animal manure + 50% recommended dose of nitrogen and phosphorus (AMRD), sawdust + green leaves (5 tons per ha) (SD), sawdust + leaves and 50% recommended dose of nitrogen and phosphorus (SDRD).	Greater whole kernels in AMRD compared to other treatments. No significant difference in broken grains. Amylose content in AMRD and SDRD > AM and SD > RD. Protein content in AMRD and AM and SDRD > RD and SD. Lipid content in AMRD was > than in the other treatments.	[25]

Table 3. *Cont.*

Materials and Study Design	Traits	Organic Rice Production Methods	Conventional Rice Production Methods	Main Findings	Citation
Unmilled rice is grown organically and conventionally in Thailand.	Kernel: elongation ratio, hardness, and color, water uptake capacity, and starch crystallinity	No organic production methods were described.	No conventional production methods were described.	Higher elongation ratio for conventional than organic (1.10 and 1.06, respectively, after six months of storage). Conventional rice was harder and darker compared to organic rice. Higher water uptake capacity in cooked organic rice than conventional after six months (4.52% and 4.47%, respectively). Crystallinity increased for both organic and conventionally produced rice during ageing	[34]
One organic and one conventional rice system is used to produce unmilled "Sungyod" rice.	Kernel: size, weight, and color. Free fatty acids, proximate analysis. amylose and anthocyanin content. Kernel swelling power and H_20 solubility of starch. Pasting gelatinization enthalpy and temperature. X-ray diffraction patterns.	No organic production methods were described.	No conventional production methods were described.	Conventional rice had greater grain length and breadth than organic (0.62 and 0.61 cm, respectively). Conventional rice weight was higher than organic rice (1.44 and 1.42 g/100 grains, respectively). No difference in kernel color, free fatty acid content, and crystallinity pattern. Conventional rice protein content > organic (7.02 and 5.64%, respectively). Conventional rice lipid content > organic rice (2.59 and 2.48%, respectively). Conventional rice amylose content (16.27 and 15.32%, respectively). Conventional rice anthocyanin content > organic rice (15.6 and 14.66 mg cyanidin-3-glucoside/100 g, respectively). Organic rice showed higher swelling power and H_20 solubility than conventional rice. Conventional rice had a higher setback value than conventional rice (129.93 and 123.65 RVU, respectively). Convention rice had a higher transition temperature and gelatinization enthalpy than organic rice.	[24]

Table 3. *Cont.*

Materials and Study Design	Traits	Organic Rice Production Methods	Conventional Rice Production Methods	Main Findings	Citation
One unmilled organic Jasmine rice sample and one unmilled milled conventional Jasmine rice sample.	Serum cholesterol, triglycerides, HDL-C, and LDL-C levels in rats after a feeding trial.	No organic production methods were described.	No conventional production methods were described.	No significant difference in rat lipids between those fed conventional versus organic rice.	[35]
One organic unmilled rice sample and one conventional unmilled rice sample were supplied by a farming cooperative in Thailand.	Rat protein efficiency (PER) level. Pesticide residues (carbofuran, methyl parathion, p-nitrophenol, and beta-cyfluthrin) in rice and rat serum after a 28-day feeding trial.	No organic production methods were described.	No conventional production methods were described.	Carbofuran, methyl parathion, and B-cyfluthrin were not present in any rat serum samples or in rice samples. P-nitrophenol was found in both samples but not in rat blood serum. Data wasn't analyzed to determine if levels were different between the conventional and the organic sample. No significant effect of organic rice on PER compared to conventional rice was found.	[36]
39 milled conventional rice samples and 37 milled organic rice. 26 conventional unmilled rice and 22 organic unmilled samples. All samples were obtained from stores in Korea.	Five mycotoxins: 8-ketotrichothecenes (deoxynivalenol (DON), nivalenol (NIV), 3-acetyldeoxynivalenol (3ADON), 15-acetyldeoxynivalenol (15ADON) and fusarenone-X (FUS-X)	No organic production methods were described.	No conventional production methods were described.	Contamination of NIV was greater in organic samples compared to their conventional counterparts. DON was detected in 19% and 41% of organic milled and unmilled rice, respectively, and 10% and 27% in conventional milled and brown samples. 3ADON, 15 ADON, and FUS-X were low in all samples, and no difference was found between conventional and organic samples.	[37]

Table 3. *Cont.*

Materials and Study Design	Traits	Organic Rice Production Methods	Conventional Rice Production Methods	Main Findings	Citation
Samples purchased rice at stores in Brazil. Not enriched. Rice wasn't enriched. 5 conventional milled rice and 2 organic milled rice samples. 2 conventional unmilled and 3 organic unmilled samples.	As, Cd, Pb, Ti, Sb, Co, Cu, Mn, Se, Zn, Cr, Ni, and Mo.	No organic production methods were described.	No conventional production methods were described.	Hg, Sb, and Tl were not detected in any samples. Cr was highest in the milled conventional rice (641 μg kg^{-1}). Conventional milled rice mean were as follows: (As, Cd, Ni, Pb, Zn, Mn, Cu, Se, Co, Mn, and Mo μg/k (164, 18.9, 130, 57.4, 17.9, 14.4, 1.79, 66.9, 25.6, 14.4, and 511, respectively). Conventional unmilled rice means were as follows: (As, Cd, Ni, Pb, Zn, Mn, Cu, Se, Co, Mn, and Mo (293,16.8,140, 109, 23.5, 31.4, 2.34, 84.7, 36.2,31 and 4,344 μg/k, respectively). Organic unmilled rice means were as follows: (As, Cd, Ni, Pb, Zn, Mn, Cu, Se, Co, Mn, and Mo (215, 13.4 179, 119, 23.6, 29.8, 2.17, 107, 44.1, and 29.8,367 μg/kg, respectively). Organic milled rice means were as follows: As, Cd, Ni, Pb, Zn, Mn, Cu, Se, Co, Mn, and Mo (149, 19.6, 98.9, 39.7, 15.7, 8.2, 1.43, 57.1, 11.5, 8.2, and 361 μg/kg, respectively). Organic and conventional milled samples did not differ in the amount of any elements measured. Organic and conventional unmilled samples did not differ in the amount of any elements measured.	[38]
Rice samples were collected from 10 organic and 10 conventional farms from all 16 agro-climatic zones in India.	Four groups of pesticides: organochlorine, carbamates, organophosphorous, and pyrethrites.	Organic farms are certified by each local state government.v No description of production methods was provided.	Conventional farms were adjacent to organic farms. No description of production methods was provided.	Carbamates and pyrithroid were found in conventional rice samples. No traces of pesticides were found the in the organic rice samples.	[39]
Rice samples from grocery stores in Brazil. Organic milled rice (n = 18), conventional milled rice (n = 11), organic husked rice (n = 12), conventional husked rice (n = 15), and specialty types (n = 13).	Organic and inorganic As.	No organic production methods were described.	No conventional production methods were described.	No difference in total As between conventional husked and organic husked samples. No difference in total As between conventional milled and organic milled samples. Inorganic As is 45% greater in organic milled rice compared to conventional milled rice and 41% greater in organic husked versus conventional husked rice.	[40]

Table 3. *Cont.*

Materials and Study Design	Traits	Organic Rice Production Methods	Conventional Rice Production Methods	Main Findings	Citation
Rice was purchased from a supermarket in Malaysia. "Bario" is grown organically. Basmati rice is grown conventionally in Pakistan. Noodles are stored for 3 days.	Rice noodle color, tensile strength, elasticity, cooking loss, rehydration ratio, aerobic plate content (APC), and total yeast and mold contents (TYMC). Water activity and sensory evaluation.	No organic production methods were described.	No conventional production methods were described.	Both rice noodles became darker in color during storage. Bario noodles had higher tensile strength due to higher amylose content compared to Basmati noodles (46.33 and 36.33 kPa, respectively, on day 0). Bario noodle was higher in elasticity than basmati (13.19 and 7.89 kPa, respectively). Basmati rice noodles had a higher cooking loss compared to Bario (7.14% vs. 3.89% respectively). Bario rice noodles had a higher rehydration ratio than basmati rice noodles (3.89 and 3.71). Higher APC in Basmati rice than in Bario, but both were acceptable after three days of storage. TYMC was higher in Basmati than in Bario, but both were above recommended amount after day two. Water activity was higher in Basmati rice than in Bario rice (0.82–0.87 and 0.80–0.83, respectively, after three days). Bario rice was most accepted and concluded to have better quality than Basmati rice.	[26]
The study was conducted for two seasons at PhilRice Philipines. A splitplot experimental design was used: two main plots and three subplots with four replications. "NSIC Rc146" was planted in the dry season (1st crop of organic farming) and NSIC Rc160 in the wet season (2nd crop of organic farming).	Tocols, gamma-oryzanol, and total phenolics. Head rice yield, kernel length, breadth, and ratio. Amylose and protein content. Kernel color.	The two main plots consisted of "with pesticide" and "without pesticide" treatments. The subplots included control, organic fertilizer, and inorganic fertilizer with a quadruplicate plot size of 10 × 4 m per plot. The organic fertilizer used was compost at 3 tons/ha (13-2-17-16 S) applied 3 d before transplanting. The NSIC Rc160 crop was applied with Bayluscide EC 250 (250 g/L Niclosamide, 200 g/L methyl isobutylketone, 100 g/L isobutanol, 1 L/ha) molluscicide 1 DAT, and Brodan 3.51 EC insecticide at 39 and 83 DAT.	The two main plots consisted of "with pesticide" and "without pesticide" treatments. The subplots included control, organic fertilizer, and inorganic fertilizer with a quadruplicate plot size of 10 × 4 m per plot. Inorganic fertilizer was applied 21-0-0-24 S at 13 DAT, urea (45-0-0) at 28 DAT, 34-0-0 at 41 DAT, and 20-0-0 at 51 DAT for a total of 120-0-0-24 S. The NSIC Rc146 crop was treated with Furadan (3 g/kg Carbofuran, 16.7–33.3 kg/ha) 28 d after transplanting (DAT) and with Brodan 3.51 EC (210 g/L Chlorpyrifos + 105 g/L BPMC (Fenobucarb), 2.5–3.5 tablespoons (45 ± 7.5 mL) 16 L −1, 120 mL/ha) insecticide 69 DAT.	Pesticide application had no effect on tocols and gamma-oryzanol levels. Organic milled rice had lower total and γ-oryzanol than conventional milled rice with applied pesticides. NSIC Rc160, organic brown rice with pesticide had higher contents of total tocols than inorganic unmilled rice with pesticide. Organic milled rice had lower total and gamma-oryzanol compared to conventional rice. Organic fertilizer resulted in lower total phenolics. No difference in milling quality, grain color, apparent amylose content, and alkali spreading value between organic and conventional rice was found. Organic rice was lower in protein content compared to conventionally grown rice.	[21]

Table 3. *Cont.*

Materials and Study Design	Traits	Organic Rice Production Methods	Conventional Rice Production Methods	Main Findings	Citation
Randomized block design with three replications of "Pusa Basmati-1" Rice.	Head rice recovery (HRR). Kernel length (KL), kernel breadth (KB), and the ratio between the two (KL:KB) before and after cooking.	The soil of the experimental field was a sandy clay loam, having 52.8% sand, 21.5% silt, and 25.7% clay. It contained 0.56% organic C, 163.2 kg/ha 71 NaOH-$KMnO_4$ hydrolysable N, 15.5 kg/ha 71 0.5 N $NaHCO_3$, extractable P, and 232.4 kg/ha 71 N NH_4AOC extractable K and had a pH value of 8.2. Seven combinations of organic sources (Farm yard manure (FYM), Sesbania green manuring (SGM), FYM + blue-green algae (BGA), SGM + BGA, FYM + SGM, FYM + SGM + BGA and FYM + SGM + BGA + PSB). FYM was applied at 10 tons/h at the time of final puddling. Sesbania was grown for 60 days and incorporated 5 days before transplanting. BGA was inoculated 10 days after transplanting of rice, whereas PSB was inoculated by dipping the roots of rice seedlings in the slurry of Pseudomonas striata culture.	The soil of the experimental field was a sandy clay loam, having 52.8% sand, 21.5% silt, and 25.7% clay. It contained 0.56% organic C, 163.2 kg/ha 71 NaOH-$KMnO_4$ hydrolysable N, 15.5 kg/ha 71 0.5 N $NaHCO_3$, extractable P, and 232.4 kg/ha 71 N NH_4AOC extractable K and had a pH value of 8.2. Four rates of inorganic fertilizers (control, 60 kg N+13 kg P + 17 kg K/ha, 120 kg N + 26 kg P + 34 kg K/ha, and 180 kg N + 39 kg P + 51 kg K/ha).	The different conventional management treatments did not affect HRR, KL, KB, and KL/KB ratio before or after cooking. Organic manure increased HRR in comparison to conventional treatments. KL hasn't affected the organic treatments. KB in the organic treatments was greater than in the conventional treatments. The data wasn't analyzed statistically.	[22]

[1] Names in quotation marks denote that these are the names of rice cultivars.

Studies focused on healthfulness traits included three that evaluated macro- and micro-mineral (i.e., element) content (Barbosa et al. [27], Champagne et al. [19], Poletti et al., 2014). Total phenolics were examined in three of the studies (Chen and McClung [18], Alves et al. [21], Tuano et al. [21]). One study examined flavonoid content (Chen and McClung [18]), and serum cholesterol, triglycerides, HDL-C, and LDL-C contents in rats (Mesomya [35]). Tocols and gamma-oryzanol were evaluated in two studies (Cho et al. [21], Chen and McClung [18], Tuano et al. [21]). The protein efficiency ratio, measured in rats, was also examined in one study (Mesomya et al. [36]).

Safety-related studies include six that evaluated total As levels (Barbosa et al., [27], Batista et al., [28], Hernandez-Martinez et al. [33], Juskelis et al. [38], Poletti et al. [40], Segura et al. [40]). Two studies evaluated inorganic As (Juskelis et al. [38]), Segura et al. [40]) and Cr levels (Barbosa et al. [27], Poletti et al. [33]). Hg levels were examined in one study (Hernandez-Martinez et al. [31]). Pb (Poletti et al. [38]) and Cd (Barbosa et al. [27]) were examined in one study each. Infestation by fungal pathogens was evaluated only by Alves et al. [23]. Various mycotoxins were examined in five studies (Alves et al. [23], Cirillo et al. [29], Gonzales et al. [30], Juan et al. [32], Ok et al. [37]). Two studies evaluated pesticide residues (Mesomya et al. [36], Rekha [39]).

3.3. End-Use Quality Traits

Alves et al. studied one long-grain cultivar with high amylose content [23]. The sample produced using conventional farming methods had a greater ($p < 0.05$) head rice yield (i.e., milling yield) than the grain harvested from a field managed using an organic cropping system. Another study examined perfect grains, which, according to their definition, was equivalent to head rice yield [25]. The cultivar examined had a greater ($p < 0.05$) head rice yield when grown using animal manure (organic treatment) or 50% animal manure + 50% sawdust fertilizer compared to the other treatments (i.e., conventional, sawdust, and 50% sawdust + conventional fertilizer treatments).

Champagne et al. studied cultivars that varied in amylose content from a low of 0% to a high of 21% [19]. The study found no difference in milled rice amylose content between organic and conventionally grown rice. The conventional management practices in terms of pesticides and herbicides weren't described, however, the details of the organic practices were fully provided. Similarly, Tuano et al. found no effect of organic management on milled grain amylose content compared to conventional methods [21]. In this study amylose content of one of the cultivars examined across all treatments was 24.2% and the other was 13.8%. The study by Kakar et al. found amylose content (23%) to be higher in milled rice grown using animal manure plus 50% of the conventional treatment compared to all other treatments [25]. The lowest amyose content was reported in the rice grown using conventional management methods (20.9%). All of these studies evaluated samples collected from field experiments designed by the authors.

The protein content of milled rice grown using conventional methods was significantly greater ($p < 0.05$) than the rice grown using either 50% of the N in the conventional treatment or using organic methods [19]. Similarly, Keawpeng et al. found that conventionally produced rice had greater protein content compared to that produced using an organic method, 7.02 and 5.64%, respectively [24]. Alves et al. studied unmilled rice grown under organic methods and found that it was lower in protein (6.7%, $p < 0.05$) in comparison to its conventionally grown counterpart (7.8%) [23]. The protein content of unmilled rice was also examined by Kakar et al. [25]. They found that protein content was greater ($p < 0.05$) in grains produced using animal manure and 50% of the local recommended amount of N and P (8.75) and the organic treatment with 100% animal manure (8.0%) compared to the treatment using 100% of the recommended dose of N and P for conventional farming (7.6%). The study by Tuano et al. examined two rice crops, one in the wet season and one in the dry, the organically produced milled rice had less protein ($p < 0.05$, 6.3%) than those grown using conventional methods (8.8%) [21]. These studies all evaluated samples collected from field experiments designed by the authors.

Champagne et al. evaluated milled rice grown using three treatments: 100% conventional methods, 50% conventional N, and organic methods [19]. No significant difference ($p > 0.05$) was found for pasting viscosity properties between these treatments, except for one cultivar out of the five studied. That cultivar, Cypress, had a higher peak ($p < 0.05$) viscosity in rice grown organically (249 and 159 RVU, respectively) and using 50% of the conventional farming N level (244 and 156 RVU, respectively) in comparison to that grown using conventional methods (207 and 126, RVU, respectively). Keawpeng and Meenune reported that pasting peak temperature wasn't different ($p > 0.05$) between organic and conventionally grown milled rice [24]. However, peak viscosity and setback were lower in the organic (117 and 119 RVU, respectively) than in conventional rice (124 and 130 RVU, respectively). Both of these studies evaluated rice that was by the authors using field production trials.

The sensory properties of milled samples of five cultivars (i.e., Cypress, Bengal, Jasmin 85, Jacinto, and Neches) were examined using descriptive sensory analysis, which included the assessment of 12 flavors and 14 textural attributes [19]. These cultivars were different cooking quality types and were grown using the following field management treatments: 100% nitrogen/conventional, 50% nitrogen/conventional, and organic. No differences were found in aroma due to production methods. Of the textural properties, slickness, hardness, and chewiness were slightly different ($p < 0.05$) for some of the cultivars grown using 100% nitrogen/conventional or 50% nitrogen/conventional versus organic production methods. Another study found no significant difference ($p < 0.05$) in the instrumental hardness of unmilled rice from the cultivar Phatthalung Sungyod grown using organic versus conventional production methods [34]. The rice examined in these two studies was grown by the authors using field production trials.

The cooking time of unmilled samples of cultivar IRGA 410 reportedly was 29.0 and 26.0 min for those grown using conventional and organic growing conditions, respectively [23]. These values weren't analyzed statistically. No other studies evaluated the effects of field production methods on rice cooking time.

The kernel length of the cultivar Pusa Basmati 1 wasn't different under seven different organic treatments in a randomized block trial with three replications. However, the kernel length-to-kernel breadth ratio in the organic treatments was greater than in the conventional treatments. The data wasn't analyzed statistically, and the paper didn't mention whether unmilled or milled samples were evaluated [22].

3.4. Healthfulness Traits

This study reviewed articles to determine if the authors found differences in Ca, Fe, and Zn levels in rice grown organically versus conventionally. One common finding was that the studies that identified the cultivars being studied identified a wide variation in mineral content between cultivars.

Champagne et al. found no significant ($p < 0.05$) difference in the level of Ca between milled samples of cultivars grown using organic and two conventional treatments (described above) [19]. Similarly, no significant ($p < 0.05$) effect of year was found on the Ca content of these samples. Effects of cultivar on Ca levels weren't reported. In another study, Ca levels were significantly ($p < 0.05$) higher in all organic samples ($n = 17$; 103 mg kg^{-1}) compared to those milled and produced conventionally ($n = 33$; 39.5 mg kg^{-1}) [27]. These samples were all purchased from Brazilian grocery stores. No mention was made of whether the samples were unmilled or milled. Poletti et al. also studied samples purchased in Brazil [38]. They reported no significant difference in Ca levels in organic ($n = 5$) versus conventional ($n = 9$) samples.

Champagne et al. studied the Fe content of milled samples of several cultivars (described above) grown using organic and two conventional treatments (as described above) [19]. One cultivar, Jacinto, had significantly ($p < 0.05$) more Fe when grown under organic conditions (23 ppm) compared to the conventional one (15 ppm). No difference was found between the treatments for the other four cultivars. Similarly, Barbosa et al. reported no difference in Fe content between conventional and organically grown rice samples (as described above) [27]. However, in this case, the data wasn't analyzed statistically. It should also be noted that these authors indicated the samples they evaluated had not been industrially enriched with Fe, which is common in some countries [41].

The studies that evaluated Zn all reported that the content of this mineral varied within organically grown rice and also within conventionally produced rice. Out of four cultivars examined by Champagne et al., only one was reported to have a significant difference in the amount of Zn between them when grown using organic and conventional cultural management practices [19]. The cultivar Jacinto had 22 ppm when grown using organic management practices while when produced using conventional methods it had 27 ppm. Another study of organic and conventionally produce rice found the former to have (16.9 mg kg^{-1}) and the latter (21.1 mg kg^{-1}), but the data wasn't analyzed statistically [27]. Poletti et al. concluded that there were no differences in the Zn content of organic and conventionally grown rice, both unmilled and milled samples [38]. The data in this study weren't analyzed statistically.

Alves et al. reported that organically grown unmilled rice had 33% more ($p < 0.05$) free phenolics (48 mg GAE equiv/100 g) than were found in the same cultivar grown under conventional field practices (32 mg GAE equiv/100 g) [23]. Unmilled rice (six cultivars) with various bran colors was obtained from organic and conventionally managed field trials grown over two years [18]. This study reported no difference ($p > 0.05$) in the level of total phenolics between the different field management methods, except for a cultivar (i.e., IL 121-1-1) with red bran during one year. The organic sample contained 5.86 mg GAE equiv/g and the conventionally grown one had 6.76 mg GAE equiv/g. These authors also studied flavonoid levels in the same samples and found similar results. Specifically, the only significant difference ($p < 0.05$) in flavanoid content was also found for IL 121-1-1. When grown organically (0.6 mg +-catechin equiv/g) it had lower flavanoid content compared to when grown conventionally (0.79 mg +-catechin equiv/g). In a study by Tuano et al. (2011), an organic fertilizer treatment was reported to be associated with lower total phenolics ($p < 0.05$) (30.7 mg g^{-1}) for one cultivar in comparison to when it was grown using conventional field methods (37.7 mg g^{-1}) [21].

Mesomya et al. evaluated the protein efficiency ratio of unmilled rice grown under organic and conventional field management using rats (10 rats per treatment plus a control). models [35]. No significant difference ($p < 0.05$) in the growth of the animals or protein efficiency ratio was found. Rats were also used to examine serum cholesterol, triglycerides, HDL-C, and LDL-C contents [35]. These authors using the same experimental design as in the previous study evaluated the effects of organic and conventional unmilled rice on the following rice serum lipid levels: serum cholesterol, triglycerides, HDL-C, and LDL-C. No significant difference ($p > 0.05$) in these lipid levels was found.

Gamma-oryzanol levels were found to be significantly higher ($p < 0.05$) in unmilled rice samples of cultivar Dongjin that had been grown using organic conditions (65.6 mg/100 g) compared to the cultivar grown using conventional management methods (60.2 mg/100 g) [20]. Both fields were in the same general location, therefore the rice matured under similar weather conditions. The organic field had been under South Korean organic management standards for five years. Gamma-oryzanol content was also evaluated in unmilled samples of three U.S. cultivars (i.e., Cocodrie, Presidio, and Sierra), a U.S. breeding line (IL 121-1-1), a giant embryo mutant, and one Indonesian cultivar (i.e., Sigoendaba) [18]. They were grown over two years, in both organically and conventionally managed fields. Cultivar and year effects had a significantly greater effect ($p < 0.05$) on gamma-oryzanol levels than the field management methods. Of these three effects, the cultivar had the greatest impact on the levels of this trait. Gamma-oryzanol levels tended to be lower in the genotypes grown using organic compared to conventional field management. But, this wasn't true in all samples. For example, in one-year Sigoendaba had greater gamma-oryzanol levels in the organic versus conventionally grown samples. 'NSIC Rc146' and 'NSIC Rc 160' were grown over two years in the Philippines [21]. Organic fertilizer compared to conventional (e.g., inorganic) fertilizer and pesticide application versus none applied didn't have a consistent effect on the gamma-oryzanol content of the samples. Conventional NSIC Rc146 grown with pesticide application had significantly ($p < 0.05$) more gamma-oryzanol compared to the organic samples. NSIC Rc160 grown using organic and conventional fertilizer with and without pesticide had similar levels ($p < 0.05$) of gamma-oryzanol. Replicate plots contributed significantly ($p < 0.05$) to the variation in sample gamma-oryzanol levels.

Cultivar and year effects had a significantly greater effect ($p < 0.05$) on tocol levels than did the field management methods for the samples studied by Chen and McClung and discussed above [18]. Cultivar had the greatest effect on the levels of tocols compared to the year the samples were grown in and the type of cultural management practices. No consistent effect was found on the levels of tocols in the unmilled samples described above in the study by Tuano et al. [21]. No effect of fertilizer type or pesticide application or not was found for the levels of tocols in the NSIC Rc146 samples. However, NSIC Rc160 grown using organic (157 mg 100 g^{-1} wet basis) versus conventional (76 mg 100 g^{-1} wet basis) fertilizer with pesticides had significantly ($p < 0.05$) higher levels of tocols. In addition, replicate plots contributed significantly ($p < 0.05$) to the variation in sample tocol levels.

3.5. Safety Traits

Barbosa et al. reported that they found higher levels of total As in conventional rice grain (n = 33, median 0.208 mg kg^{-1}) in comparison to certified organic rice grain (n = 17, 0.158 mg kg ^{1}) that was purchased from grocery stores [27]. The organic rice was certified by Brazilian IBD-Agricultural and Food Inspections. The data in this study weren't analyzed statistically. Another study evaluating total As content found greater levels in organic milled rice (222.9 and 161.6 ng g-1, respectively) than in conventional milled rice [28]. In the same study, the mean of total As for the parboiled white samples was reported to be greater than the parboiled organic sample (214.9 and 174.1 ng g^{-1}, respectively). Total As was reported to be higher in organic rice infant cereals compared to conventional infant rice cereal (154.9 and 96.3 µg/kg, respectively) [31]. The data in this study weren't analyzed statistically. Another study that evaluated milled rice infant cereals reported indicated there was no difference between organic and conventionally produced cereals. The data in this study weren't analyzed statistically [33]. Poletti et al. [38] and Segura et al. [40] evaluated organic and conventional milled rice samples and found no difference in total As levels. In addition, these authors reported no difference in total As levels between organic and conventional unmilled rice samples. All of the studies that evaluated total As used samples collected from stores, and none of them analyzed the reported data using statistical methods.

The levels of inorganic As in infant rice cereals that were obtained from grocery stores reportedly had no difference between those made using milled rice grown under organic and conventional production techniques [33]. Inorganic As was reported by Segura et al. to be 45% greater in organic milled rice compared to conventional milled rice and 41% greater in organic husked versus conventional husked rice [40]. The data in both of these studies weren't analyzed statistically.

Cd was reported to be higher in conventionally (0.012 mg kg^{-1}) grown rice samples purchased in grocery stores compared to ones produced using organic (0.005 mg kg^{-1}) production methods [27]. However, this data wasn't analyzed statistically.

Milled rice Cr levels were reportedly greater in conventional (3.0 mg kg^{-1}) rice in comparison to that grown under organic conventional (2.3 mg kg^{-1}) production methods [27]. On the contrary, Poletti et al. reported no difference in the level of Cr in organic versus conventional unmilled and milled samples purchased at grocery stores [38]. The data in both of these studies weren't analyzed using statistical methodology.

Hg was higher in organic infant rice cereal compared to cereals made from conventionally grown rice (4.54 and 4.39 µg/kg, respectively) [31]. The Pb levels found in conventional and organic rice were similar [27]. On the contrary Poletti et al. reported that conventionally grown milled rice contained more PB (130 µg kg^{-1}) than organically grown milled rice (98 µg kg^{-1}). Barbosa et al. reported that the Cd content of conventionally produced conventional rice (0.012 mg kg^{-1}) was higher than that grown under organic conditions (0.005 mg kg^{-1}) [27]. All samples used in the studies described above were obtained from grocery stores and the data reported weren't analyzed using statistical methods.

Alves et al. studied fungal pathogen levels in rough rice during storage [23]. They found that *Penicillium* sp., in organic rice was 4, 9, and 4 times higher in rough rice stored for 0, 6th, and 12th months, respectively, compared to rice produced under conventional systems. *Aspergillius* sp. was slightly higher in organic rough rice compared to organic prior to storage, while after six months of storage, conventional rough rice contained 70% more. Before storage, *Bipolaris* sp. was found in organic rice but not in the rough rice grown conventionally. The rice used for these studies was collected from field trials.

Deoxynivalenol, and fumonisin B1 and B2 were evaluated in various rice-based foods such as biscuits and breakfast cereals [29]. Deoxynivalenol and fumonisin B1 were reported to be higher in products made using conventionally grown rice (207 µg kg^{-1} and 205 µg kg^{-1}, respectively) compared to that produces using organic rice (65 µg kg^{-1} and 150 µg kg^{-1}, respectively). On the contrary, rice-based food products made using organically grown rice contained more fumonisin B1 (145 µg kg^{-1}) than those produced using conventionally grown rice (30 μg kg^{-1}). None of the data in this study were examined statistically.

Gonzales et al. studied the levels of ochratoxin A in milled rice and rice-based food products [30]. More organic samples contained ochratoxin A (30.0%, range 1.0 to 7.1 Ag/kg) compared to those produced using conventional growing methods (7.8%, range 4.3 to 27.3 Ag/kg). Juan et al. also evaluated the levels of ochratoxin A in conventionally and organically grown rice [32]. They found that some organically produced rice (4/9 samples, mean 2.57 ng/g) contained ochratoxin A, but none of the conventionally produced samples did (0/4 samples). No difference between organic unmilled rice and milled rice was reported. The data in this study weren't examined statistically.

The occurrence of mycotoxins, specifically five 8-ketotrichothecene compounds, was studied in organically and conventionally produced rice [37]. They found no significant difference in the levels of deoxynivalenol, 3-acetyldeoxynivalenol, 15-acetyldeoxynivalenol, and fusarenone-X in conventionally and organically produced unmilled and milled rice. However, the levels of nivalenol were significantly higher ($p < 0.05$) in both organically produced unmilled and milled rice compared to the corresponding conventionally grown rice. The sample studies in this study were all obtained from grocery stores in South Korea.

Pesticide residues (i.e., p-nitrophenol, carbofuran, methyl parathion, and β-cyfluthrin) were measured in cooked rice samples of unmilled conventionally and organically produced rice that was obtained in Thailand [35]. The cooked rice samples contained p-nitrophenol (8.23 and 10.13 mg/kg, respectively). None of the other residues were found in the cooked rice sample. Also, none of the pesticide residues were identified in the serum from rats fed the cooked rice.

Rice obtained from organic and conventional farms in 16 regions of India was examined for pesticide residues [39]. The resides were from four groups of pesticides (i.e., organochlorine, carbamates, organophosphorous, and pyrethrites). The sites were chosen to represent all of India's rice-growing regions in the northern and central parts of the country. Residues of organochlorines were present in all the conventionally grown rice samples. Organochlorine pesticide residues were found in two out of ten organic farms. These farms had both been converted from conventional to organic practices a few years ago. The presence of carbamates and pyrithroid were found in conventional rice samples, while no trace of either of these was found in the organic rice samples. Nothing related to the organophosphorous levels in rice was discussed in the paper.

4. Discussion

Rice grain characteristics result from differences in the genetics of the rice variety and environmental effects [3]. These effects include such things as the influence of climate, soil quality, seeding rates, field in-puts, grain processing, and grain storage. Thus, a well-designed study evaluating the effects of rice production practices would keep all of the cultivars the same. The inputs' type, amount, and application time (e.g., soil amendments, soil type, fertilizers, pesticides, herbicides, and irrigation) would need to be reported in detail. Also, there must be cultivars and environmental replications.

This systematic review found only a relatively small number of studies examining sustainable agriculture practices' effects on rice grain end-use quality, healthfulness, and safety. They all examined organic practices specifically. The rice cultivars used varied from study to study. In some studies, the cultivars were named; in others, they weren't, as the rice samples were obtained from retail markets. Most of the studies lacked details on how the rice samples were grown, stored, milled, and packaged. Therefore, the results in this review should be taken with caution, as most of the studies have error rates from poor experimental design, limited or no reporting of how the rice kernels were processed and stored before analysis, or inadequate statistical analysis [42].

4.1. End-Use Quality

The translucence, shape, and uniformity are important aspects of rice end-use quality for consumers, millers, wholesalers, and retailers [43]. Surprisingly few studies in this review reported measuring aspects of grain appearance. Those that did examine rice grain appearance either found very small differences or none at all. Differences in grain color and chalk (i.e., opaque spots) between rice samples can often be seen using the naked eye. Therefore, had large differences in appearance occurred in the studies where these traits weren't measured using instrumentation they would have still likely been reported. No such differences were mentioned in the studies reviewed for this paper. Chalk is known to have a genetic link and also an environmental cause, temperatures during grain filling increase its levels [44,45]. Future studies need to examine cultivars that are susceptible to developing chalk, expose them to low and high temperatures during grain filling, and examine them for differences when grown under organic versus conventional conditions.

Consumers prefer rice that isn't broken. Therefore, head rice yield is an extremely important characteristic for rice farmers because millers are willing to pay more for rice kernels that are whole and not broken. Some cultivars consistently have greater head rice yield than others across different years and environmental conditions [3]. The two studies that examined head rice yield in this review found conflicting results for one cultivar each, grown in an organic versus the conventionally managed field. Thus, a conclusion on the effects of organic growing conditions on head rice yield can't be made.

People in different global regions prefer cooked rice with a particular suite of textural properties. Rice with the preferred texture demands a premium price. Thus, it is important to understand the effects of growing conditions on rice cooked rice texture. There are fourteen aspects of cooked rice texture that are evaluated using trained sensory panels as well as various instrumental methods that are predictive of some of these characteristics [3]. One study in this review reported that three aspects of texture (i.e., slickness, hardness, and chewiness) were found by a trained sensory panel to be slightly influenced by organic versus conventional cultural management [19]. These differences were associated with grain protein content, which varies along with the dose of nitrogen applied during rice growth [19,46]. Thus, if differences are seen due to cultural management practices, they will likely be small and not be caused by the practices per se, but rather by changes in protein content caused by the difference in nitrogen application rates.

The primary predictor of cooked rice texture is the amount of amylose the grains contain and to a lesser degree the protein and lipid fractions (Fitzgerald 2004) [46]. Rice grains are classified according to amylose content: waxy (0%); very low (3–9%); low (10–19%); intermediate (20–25%); or high (>25%) [3]. The greater the amount of amylose the firmer and the less sticky the cooked rice is. Cultivars in the same amylose category are expected to have similar textural properties. The studies examined in this review indicate that organic production practices either don't influence the quantity of amlyose in milled rice grain or have very little impact. When differences did occur, they were not large enough to move a cultivar from one amylose classification to another.

The food processing industry tests the viscous properties of rice using cooking stirring viscometers, typically a Rapid Visco Analyser (RVA) [47]. Some of these properties are related to aspects of cooked rice texture [48]. For example, the setback parameter is frequently used to predict cooked rice firmness/stickiness and pasting temperature is used when the rice will be included as a source of carbohydrates in brewing. The studies reviewed in this paper indicate that rice pasting properties will not likely be influenced by organic versus conventional farming practices for most cultivars. If differences are seen, they will probably be small and associated with grain protein content, which varies along with the dose of nitrogen applied during rice growth, as discussed above [46,48].

The aroma of 'fragrant rice' is an aspect of end-use quality of particular importance, as certain ethnic groups prefer it over nonfragrant rice; this impacts the market price of fragrant rice. For example, many different varieties of the jasmine style of rice are consumed in South East Asia, and many basmati styles of rice are consumed in South and Central

Asia. Fragrant rice is commonly reported to smell similar to popcorn or bread-like due to a compound it contains called 2-acetyl-1-pyrroline and likely due to other compounds yet to be identified [49]. Various effects, such as genetics, environment, and cultural management practices, are reported to impact the level of 2-acetyl-1-pyrroline in fragrant rice [50–52]. Thus, it was surprising that Jasmine 85, a U.S. fragrant cultivar, had a similar aroma when grown using different cultural management practices, specifically organic and conventional [19]. Perhaps the difference lies in that all of the studies mentioned above were performed by measuring 2-acetyl-1-pyrroline using laboratory instrumentation, except for the one that evaluated Jasmine 85. The latter used a trained sensory panel to evaluate the "popcorn" aroma and several other aspects of rice aroma. The differences identified via instrumentation may be too small for humans to sense, and humans may also be smelling aromatic compounds other than 2-acetyl-1-pyrroline in the fragrant rice.

4.2. Healthfulness

Micronutrients, vitamins, and minerals, are essential for human health, and their deficiency in the diet remains a widespread problem, especially in low- and middle-income countries (Bailey et al., 2015) [53]. The articles examined in this review evaluated three minerals of particular importance for human health globally; Ca, Zn, and Fe. The latter two are minerals of particular importance to this review since reliance on milled rice with minimal dietary diversity contributes to Zn and Fe deficiency in developing countries [54]. Considerable variation in Zn and Fe levels exists within rice germplasm that is genotype dependent [55]. Iron levels vary from 6.9 to 22.3 mg/kg and zinc concentration ranges from 14.5 to 35.3 mg/kg in unpolished, brown rice. No trend in the effect of organic versus conventional management practices on Zn and Fe levels was found in the studies examined in this review. However, the data indicates that if organic production methods impact Zn and Fe levels of rice, the effects are small and less than the variation that exists between cultivars.

Interest in the effects of cultural management practices on gamma-oryzanol and tocols is of interest because these compounds have been proposed to have human health-beneficial properties [4,56]. Previous work suggested that in general the environment rice is grown in has a greater effect on gamma-oryzanol and tocol levels than genotype ([57]. Similarly, the studies examined in this review found that gamma-oryzanol and tocol levels were influenced by the environment and cultivar. No trend was found for the effects of organic versus conventional management methods. Although the data is limited, when there are effects of cultural management on gamma-oryzanol and tocols they will likely be small and less than the variation caused by differences in environment and cultivar.

4.3. Safety

Soil contamination with heavy metals has increased in certain regions of the world because of the following anthropogenic activities: urbanization, industrialization, mining, transportation, and agriculture [58,59]. Reports from several countries indicate that heavy metal concentrations (i.e., As, Cd, Cr, Hg, Ni, and Pb) in rice often exceed guidance values [60–62]. This elevated exposure to these metals creates an elevated risk to humans that rely on rice for a significant portion of their food. Consequently, there is an increasing need to mitigate the phytoaccumulation of heavy metals in rice.

The repeated use of inorganic fertilizers and metallo-pesticides is associated with increased levels of heavy metals in rice-growing soil [63]. Thus, it has been proposed that using organic cultural management practices may result in lower levels of heavy metals in soils. However, soils that are already contaminated with heavy metals may still pose a risk of contaminating rice with heavy metals even when under organic management. Increasing soil organic matter is one of the primary cultural management goals of organic farming, as it provides benefits such as an increase in the biodiversity of soil microflora, which in turn helps the soil retain nutrients. However, a potential drawback of an enhanced amount of organic matter in soil was reported by Zeng et al.) [64]. They reported that the bioavailability of Pb was positively correlated with the level of organic matter in the soil used to grow rice because of the effects of organic matter on element mobilization and bioavailability in soil.

In this review, none of the studies evaluated the same cultivars grown in the same environment. Rice purchased from grocery stores can be a blend of several cultivars that have been grown in fields with different levels of heavy metals, thus there is no way to fairly evaluate the effects of cultural management practices. In addition, these studies didn't use statistical methods to evaluate the levels of heavy metals in the rice samples. Therefore, no conclusions about the effects of organic versus conventional cultural management methods on rice heavy metal content could be drawn.

Since 1960, the average yield of rice globally has more than doubled, as pesticides have increased by 15 to 20-fold [65]. Significant evidence exists that the use and especially the overuse of pesticides is associated with adverse effects on human health and non-target organisms such as birds, bees, and fish [66–68]. We conclude that the studies evaluated in this review found that organically produced rice grain was less likely to contain residues of the pesticides examined in the study than the rice grown using conventional methods. However, a problem with the design of one of the studies was that the reader wasn't informed of how long the field studied had been under organic management. Over time some pesticides degrade and become nontoxic or less so [69]. The degradation rates of pesticides vary along with the soil microbial composition and other soil factors, such as pH and temperature [70]. Therefore, the length of time a field has been under organic management needs to be recorded in studies designed to evaluate the effect of this cultural management technique in comparison to conventional methods.

Mycotoxins are naturally occurring toxic contaminants found in cereal grains and other foods [71]. These secondary metabolites are made by fungi which, when consumed, have acute and long-term health risks for humans [72]. These fungi reportedly grow in rice when certain conditions occur during "particular crop seasons, cultivation regions, and agricultural practices (pre-harvest: paddy variety, crop residue management, and fertilizer application; post-harvest: means of transportation and delayed drying time)" [73,74]. A multi-year survey reported that the mycotoxin-producing fungi, *Fusarium proliferatum,* and *Aspergillus flavus*, were found more commonly in fields managed using a combination of fertilizers (organic and inorganic) or with crop debris compared to those fields that had only inorganic fertilizer applied [73]. This review found that most studies related to fungi and mycotoxin levels in rice-based food products and milled rice were difficult to interpret because the data wasn't analyzed statistically. However, Juan et al. found ochratoxin A in some of the organically grown rice and rice products, but not in any of the products or rice from conventionally managed fields [32]. Also, the levels of nivalenol were significantly higher ($p < 0.05$) in organically grown rice (unmilled and milled) compared to the corresponding conventionally grown rice [37]. Although the evidence is limited, the literature suggests that rice and rice-based products may contain some fungi or some mycotoxins in particular global regions while their conventionally produced rice counterparts don't.

5. Conclusions

This review provides tentative conclusions that food processing companies and consumers will not likely notice any aroma or processing quality differences between the same rice cultivars grown using organic farming practices compared to conventional methods. However, slight differences in cooked rice texture may be sensed due to differences in kernel protein content which is known to impact rice texture. Differences in rice grain protein content occur from exposure to different amounts of nitrogen, not likely due to organic sources of nitrogen versus conventional sources. There was insufficient evidence to evaluate the effect of organic production methods on chalk or milling yield. We conclude that the studies evaluated in this review found that organically produced rice grain was less likely to contain residues of the pesticides examined in the study than the rice grown using conventional methods. There was some evidence that organically grown rice is more likely to be contaminated with mycotoxin-producing fungi and some mycotoxins. Common shortcomings of some of the studies evaluated in this review were that they were poorly designed, with limited to no details of the cultural management practices used to grow the rice studied, cultivars were not named, and the data wasn't analyzed statistically. Future related research should use fields that have been under organic management for more than two years, take place in more than one environment, and have a variety of soil types.

Author Contributions: Conceptualization, C.B.; methodology, C.B.; formal analysis, C.B. and M.P.; investigation, C.B. and M.P.; writing—original draft preparation, C.B.; funding acquisition, C.B. All authors have read and agreed to the published version of the manuscript.

Funding: This research received no external funding.

Institutional Review Board Statement: Not applicable for studies not involving humans or animals.

Informed Consent Statement: Not applicable.

Data Availability Statement: Not applicable.

Acknowledgments: The authors thank Lateka Grays (Research and Education Librarian, UNLV) for her contribution to the selection of the databases for use in this study.

Conflicts of Interest: The authors declare no conflict of interest.

References

1. Muthayya, S.; Sugimoto, J.D.; Montgomery, S.; Maberly, G.F. An overview of global rice production, supply, trade, and consumption. *Ann. N. Y. Acad. Sci.* **2014**, *1324*, 7–14. Available online: https://onlinelibrary.wiley.com/doi/abs/10.1111/nyas.12540 (accessed on 1 June 2022). [CrossRef]
2. Hamilton, R. How many rice varieties are there? *Rice Today* **2006**, 3, 50. Available online: https://ricetoday.irri.org/how-many-rice-varieties-are-there/ (accessed on 1 June 2022).
3. Bergman, C.J. Rice end-use quality analysis. In *Rice*, 4th ed.; Bao, J., Ed.; AACC International Press: Washington, DC, USA, 2019; pp. 273–337. [CrossRef]
4. Dipti, S.S.; Bergman, C.; Inasari, S.D.; Herath, T.; Hall, R.D.; Lee, H.; Habibi, F.; Zaczuk Bassinello, P.; Graterol, E.; Ferraz, J.; et al. The potential of rice to offer solutions for malnutrition and chronic diseases. *Rice* **2012**, *5*, 16. [CrossRef]
5. Childs, K.; Ouyang, S. Rice Genome Initiative. *Princ. Pract. Plant Genom.* **2019**, 3, 205. [CrossRef]
6. Thompson, P.B. The Agrarian Vision: Sustainability and Environmental Ethics. *J. Environ. Stud. Sci.* **2010**, *1*, 271. [CrossRef]
7. NRC. *Toward Sustainable Agricultural Systems in the 21st Century*; Council CoT-FCSABoAaNRDoEaLSNR; National Academies Press: Washington, DC, USA, 2010; Available online: https://nap.nationalacademies.org/catalog/12832/toward-sustainable-agricultural-systems-in-the-21st-century (accessed on 1 June 2022).
8. Rebsdorf, S.O. ICROFS News 3/2009-Newsletter from ICROFS 2009 3. Available online: http://www.icrofs.org/pdf/icrofsnews/2009_3.pdf (accessed on 1 June 2022).
9. Lotter, D.W. Organic Agriculture. *J. Sustain. Agric.* **2003**, *21*, 59–128. [CrossRef]
10. Schreefel, L.; Schulte, R.P.O.; De Boer, I.J.M.; Schrijver, A.P.; Van Zanten, H.H.E. Regenerative agriculture—The soil is the base. *Global Food Security* **2020**, *26*, 100404. [CrossRef]
11. Patlavath, R.; Albert, S. Organic basmati rice cultivation in India: Challenges from field to flight–A review article. *Int. J. Agric. Technol.* **2021**, *17*, 607–626. Available online: http://www.ijat-aatsea.com (accessed on 1 June 2022).
12. Winter, C.K.; Davis, S.F. Organic Foods. *J. Food Sci.* **2006**, *71*, R117–R124. [CrossRef]
13. Nearly Two-Thirds of Americans Have Tried Organic Foods and Beverages. Available online: https://www.webwire.com/ViewPressRel.asp?aId=5889 (accessed on 2 December 2022).
14. Zheng, Q.; Zeng, H.; Xiu, X.; Chen, Q. Pull the Emotional Trigger or the Rational String? A Multi-Group Analysis of Organic Food Consumption. *Foods* **2022**, *11*, 1375. Available online: https://www.ncbi.nlm.nih.gov/pubmed/35626945 (accessed on 1 June 2022). [CrossRef]
15. Smith-Spangler, C.; Brandeau, M.L.; Hunter, G.E.; Bavinger, J.C.; Pearson, M.; Eschbach, P.J.; Sundaram, V.; Liu, H.; Schirmer, P.; Stave, C.; et al. Are Organic Foods Safer or Healthier Than Conventional Alternatives? *Ann. Intern. Med.* **2012**, *157*, 348–366. [CrossRef] [PubMed]
16. Moher, D.; Shamseer, L.; Clarke, M.; Ghersi, D.; Liberati, A.; Petticrew, M.; Shekelle, P.; Stewart, L.A. Preferred reporting items for systematic review and meta-analysis protocols (PRISMA-P) 2015 statement. *Syst. Rev.* **2015**, *4*, 1. Available online: https://www.ncbi.nlm.nih.gov/pubmed/25554246 (accessed on 1 June 2022). [PubMed]
17. Higgins, J.P.; Thomas, J.; Chandler, J.; Cumpston, M.; Li, T.; Page, M.J.; Welch, V.A. (Eds.) *Cochrane Handbook for Systematic Reviews of Interventions*; John Wiley & Sons: Hoboken, NJ, USA, 2019.

18. Chen, M.; McClung, A.M. Effects of Cultivars, Organic Cropping Management, and Environment on Antioxidants in Whole Grain Rice. *Cereal Chem.* **2015**, *92*, 364–369. [CrossRef]
19. Champagne, E.T.; Bett-Garber, K.L.; Grimm, C.C.; McClung, A.M. Effects of Organic Fertility Management on Physicochemical Properties and Sensory Quality of Diverse Rice Cultivars. *Cereal Chem.* **2007**, *84*, 320–327. [CrossRef]
20. Cho, J.; Lee, H.J.; Kim, G.A.; Kim, G.D.; Lee, Y.S.; Shin, S.C.; Park, K.; Moon, J. Quantitative analyses of individual γ-Oryzanol. (Steryl Ferulates) in conventional and organic brown rice (*Oryza sativa* L.). *J. Cereal Sci.* **2012**, *55*, 337–343. Available online: http://www.sciencedirect.com/science/article/pii/S0733521012000173 (accessed on 1 June 2022). [CrossRef]
21. Tuaño, A.P.P. Content of tocols, gamma-oryzanol. and total phenolics and grain quality of brown rice and milled rice applied with pesticides and organic and inorganic nitrogen fertilizer. *Philipp. Agric. Sci.* **2011**, *94*, 211–216.
22. van Quyen, N.; Sharma, S. Relative effect of organic and conventional farming on growth, yield and grain quality of scented rice and soil fertility. *Arch. Agron. Soil Sci.* **2003**, *49*, 623–629. [CrossRef]
23. Alves, G.H.; Paraginski, R.T.; Lamas, N.D.S.; Hoffmann, J.F.; Vanier, N.L.; de Oliveira, M. Effects of Organic and Conventional Cropping Systems on Technological Properties and Phenolic Compounds of Freshly Harvested and Stored Rice. *J. Food Sci.* **2017**, *82*, 2276–2285. [CrossRef]
24. Keawpeng, I.; Meenune, M. Physicochemical properties of organic and inorganic Phatthalung Sungyod rice. *Int. Food Res. J.* **2012**, *19*, 857. Available online: https://www.proquest.com/docview/1315524718?pq-origsite=gscholar&fromopenview=true (accessed on 1 June 2022).
25. Kakar, K.; Xuan, T.D.; Noori, Z.; Aryan, S.; Gulab, G. Effects of Organic and Inorganic Fertilizer Application on Growth, Yield, and Grain Quality of Rice. *Agriculture* **2020**, *10*, 544. [CrossRef]
26. Thomas, R.; Yeoh, T.K.; Wan-Nadiah, W.A.; Bhat, R. Quality Evaluation of Flat Rice Noodles (Kway Teow) Prepared from Bario and Basmati Rice. *Sains Malays.* **2014**, *43*, 339–347. Available online: https://core.ac.uk/download/pdf/19790502.pdf (accessed on 1 June 2022).
27. Barbosa, R.M.; de Paula, E.S.; Paulelli, A.C.; Moore, A.F.; Souza, J.M.O.; Batista, B.L.; Campiglia, A.D.; Barbosa, F. Recognition of organic rice samples based on trace elements and support vector machines. *J. Food Compos. Anal.* **2016**, *45*, 95–100. [CrossRef]
28. Batista, B.L.; Souza, J.M.O.; De Souza, S.S.; Barbosa Jr., F. Speciation of arsenic in rice and estimation of daily intake of different arsenic species by Brazilians through rice consumption. *J. Hazard. Mater.* **2011**, *191*, 342–348. [CrossRef] [PubMed]
29. Cirillo, T.; Ritieni, A.; Visone, M.; Cocchieri, R.A. Evaluation of conventional and organic Italian foodstuffs for deoxynivalenol. and fumonisins B(1) and B(2). *J. Agric. Food Chem.* **2003**, *51*, 8128–8131. [CrossRef] [PubMed]
30. González, L.; Juan, C.; Soriano, J.M.; Moltó, J.C.; Mañes, J. Occurrence and daily intake of ochratoxin A of organic and non-organic rice and rice products. *Int. J. Food Microbiol.* **2006**, *107*, 223–227. [CrossRef]
31. Hernández-Martínez, R.; Navarro-Blasco, I. Survey of total mercury and arsenic content in infant cereals marketed in Spain and estimated dietary intake. *Food Control* **2013**, *30*, 423–443. [CrossRef]
32. Juan, C.; Moltó, J.C.; Lino, C.M.; Mañes, J. Determination of ochratoxin A in organic and non-organic cereals and cereal products from Spain and Portugal. *Food Chem.* **2008**, *107*, 525–530. [CrossRef]
33. Juskelis, R.; Li, W.; Nelson, J.; Cappozzo, J.C. Arsenic speciation in rice cereals for infants. *J. Agric. Food Chem.* **2013**, *61*, 10670–10676. Available online: http://www.scopus.com/inward/record.url?eid=2-s2.0-84887713709&partnerID=40&md5=af026d5e589cb3094056e6ba00c32e15 (accessed on 1 June 2022). [CrossRef]
34. Keawpeng, I.; Meenune, M. Changes in cooking behavior of organic and inorganic Phatthalung Sungyod rice during ageing. *Thai J. Agric. Sci.* **2011**, *44*, 348–353.
35. Mesomya, W.; Sutthivaiyakit, P.; Cuptapun, Y.; Hengsawadi, D. Effects of organic rice compared with conventional rice on serum lipids in rats. *Kasetsart J.-Nat. Sci.* **2009**, *43*, 703–708. Available online: https://li01.tci-thaijo.org/index.php/anres/article/view/244747/167225 (accessed on 1 June 2022).
36. Mesomya, W.; Sutthivaiyakit, P.; Cuptapun, Y.; Hengsawadi, D. Effects of organic and conventional rice on protein efficiency ratio and pesticide residue in rats. *Maejo Int. J. Sci. Technol.* **2012**, *6*, 470–482. Available online: https://www.proquest.com/docview/1268828977?pq-origsite=gscholar&fromopenview=true (accessed on 1 June 2022).
37. Ok, H.E.; Choi, S.; Chang, H.J.; Chung, M.; Chun, H.S. Occurrence of five 8-ketotrichothecene mycotoxins in organically and conventionally produced cereals collected in Korea. *Food Control* **2011**, *22*, 1647–1652. [CrossRef]
38. Poletti, J.; Pozebon, D.; Barcellos de Fraga, M.V.; Dressler, V.L.; de Moraes, D.P. Toxic and micronutrient elements in organic, brown and polished rice in Brazil. *Food Addit. Contam. Part B Surveill. Commun.* **2014**, *7*, 63–69. [CrossRef]
39. Rekha; Naik, S.N.; Prasad, R. Pesticide residue in organic and conventional food-risk analysis. *J. Chem. Health Saf.* **2006**, *13*, 12–19. [CrossRef]
40. Segura, F.R.; de Oliveira Souza, J.M.; De Paula, E.S.; Jrda, C.M.; Paulelli, A.C.C.; Barbosa, F., Jr.; Batista, B.L. Arsenic speciation in Brazilian rice grains organically and traditionally cultivated: Is there any difference in arsenic content? *Food Res Int.* **2016**, *89*, 169–176. [CrossRef] [PubMed]
41. Hunnell, J.W.; Yasumatsu, K.; Moritaka, S. Iron Enrichment of Rice—Chapter 8. In *Iron Fortification of Foods*; Clydesdale, F.M., Wiemer, K.L., Eds.; Academic Press: Cambridge, MA, USA, 1985; pp. 121–132. [CrossRef]
42. Campbell, K.G.; Thompson, Y.M.; Guy, S.O.; McIntosh, M.; Glaz, B. "Is, or is not, the two great ends of Fate": Errors in agronomic research. *Agron. J.* **2015**, *107*, 718–729. [CrossRef]

43. Fitzgerald, M. Chapter 12—Rice: Grain-Quality Characteristics and Management of Quality Requirements. In *Cereal Grains*, 2nd ed.; Woodhead Publishing: Sawston, UK, 2017; pp. 291–315. [CrossRef]
44. Lisle, A.J.; Martin, M.; Fitzgerald, M.A. Chalky and translucent rice grains differ in starch composition and structure and cooking properties. *Cereal Chem.* **2000**, *77*, 627–632. [CrossRef]
45. Ishimaru, T.; Parween, S.; Saito, Y.; Shigemitsu, T.; Yamakawa, H.; Nakazono, M.; Masumura, T.; Nishizawa, N.K.; Kondo, M.; Sreenivasulu, N. Laser microdissection-based tissue-specific transcriptome analysis reveals a novel regulatory network of genes involved in heat-induced grain chalk in rice endosperm. *Plant Cell Physiol.* **2019**, *60*, 626–642. [CrossRef]
46. Martin, M.; Fitzgerald, M.A. Proteins in rice grains influence cooking properties. *J. Cereal Sci.* **2002**, *36*, 285–294. [CrossRef]
47. Balet, S.; Guelpa, A.; Fox, G.; Manley, M. Rapid Visco Analyser (RVA) as a tool. for measuring starch-related physiochemical properties in cereals: A review. *Food Anal. Methods* **2019**, *12*, 2344–2360. Available online: https://escholarship.org/uc/item/0cf6r30w (accessed on 1 June 2022). [CrossRef]
48. Champagne, E.T.; Bett, K.L.; Vinyard, B.T.; McClung, A.M.; Barton, F.E.I.; Moldenhauer, K.; Linscombe, S.; McKenzie, K. Correlation between cooked rice texture and rapid visco analyser measurements. *Cereal Chem.* **1999**, *76*, 764–771. [CrossRef]
49. Calingacion, M.; Laborte, A.; Nelson, A.; Resurreccion, A.; Concepcion, J.C.; Daygon, V.D.; Mumm, R.; Reinke, R.; Dipti, S.; Bassinello, P.Z.; et al. Diversity of global rice markets and the science required for consumer-targeted rice breeding. *PLoS ONE* **2014**, *9*, e85106. [CrossRef] [PubMed]
50. Fitzgerald, M.A.; McCouch, S.R.; Hall, R.D. Not just a grain of rice: The quest for quality. *Trends Plant Sci.* **2009**, *14*, 133–139. [CrossRef] [PubMed]
51. Champagne, E.T. Rice aroma and flavor: A literature review. *Cereal Chem.* **2008**, *85*, 445–454. [CrossRef]
52. Bhattacharjee, P.; Singhal, R.S.; Kulkarni, P.R. Basmati rice: A review. *Int. J. Food Sci. Technol.* **2002**, *37*, 1–12. [CrossRef]
53. Bailey, R.L.; West Jr, K.P.; Black, R.E. The epidemiology of global micronutrient deficiencies. *Ann. Nutr. Metab.* **2015**, *66*, 22–33. [CrossRef]
54. Black, R.E.; Allen, L.H.; Bhutta, Z.; Bhutta, A.; Caulfi, L.E.; De Onis, M.; Ezzati, M.; Mathers, C.; Rivera, J. Maternal and child undernutrition: Global and regional exposures and health consequences. *Lancet* **2008**, *371*, 243–260. [CrossRef]
55. Maganti, S.; Swaminathan, R.; Parida, A. Variation in iron and zinc content in traditional rice genotypes. *Agric. Res.* **2020**, *9*, 316–328. [CrossRef]
56. Bergman, C.J. Rice: Prevention and Management of Type 2 Diabetes and Coronary Heart Disease. In *The Future of Rice Demand: Quality Beyond Productivity*; de Oliveira, A.C., Pegoraro, C., Viana, V.E., Eds.; Springer International Publishing: New York, NY, USA, 2020; pp. 205–223. [CrossRef]
57. Bergman, C.J.; Xu, Z. Genotype and environment effects on tocopherol, tocotrienol, and gamma-oryzanol. contents of Southern U.S. rice. *Cereal Chem.* **2003**, *80*, 446–449. [CrossRef]
58. Facchinelli, A.; Sacchi, E.; Mallen, L. Multivariate statistical and GIS-based approach to identify heavy metal sources in soils. *Environ. Pollut.* **2001**, *114*, 313. [CrossRef]
59. Wei, B.; Yang, L. A review of heavy metal contaminations in urban soils, urban road dusts and agricultural soils from China. *Microchem. J.* **2010**, *94*, 99–107. [CrossRef]
60. Shimbo, S.; Zhang, Z.; Watanabe, T.; Nakatsuka, H.; Matsuda-Inoguchi, N.; Higashikawa, K.; Ikeda, M. Cadmium and lead contents in rice and other cereal products in Japan in 1998–2000. *Sci. Total Environ.* **2001**, *281*, 165–175. [CrossRef] [PubMed]
61. Meharg, A.A.; Williams, P.N.; Adomako, E.; Lawgali, Y.Y.; Deacon, C.; Villada, A.; Cambell, R.C.J.; Sun, G.; Zhu, Y.; Feldmann, J.; et al. Geographical variation in total and inorganic arsenic content of polished (white) rice. *Environ. Sci. Technol.* **2009**, *43*, 1612–1617. [CrossRef]
62. Zarcinas, B.A.; Pongsakul, P.; McLaughlin, M.J.; Cozens, G. Heavy metals in soils and crops in Southeast Asia 2. Thailand. *Environ. Geochem Health* **2004**, *26*, 359–371. [CrossRef]
63. Gimeno-García, E.; Andreu, V.; Boluda, R. Heavy metals incidence in the application of inorganic fertilizers and pesticides to rice farming soils. *Environ. Pollut.* **1996**, *92*, 19–25. [CrossRef] [PubMed]
64. Zeng, F.; Ali, S.; Zhang, H.; Ouyang, Y.; Qiu, B.; Wu, F.; Zhang, G. The influence of pH and organic matter content in paddy soil on heavy metal availability and their uptake by rice plants. *Environ. Pollut.* **2011**, *159*, 84–91. [CrossRef]
65. Oerke, E. Crop losses to pests. *J. Agric. Sci.* **2006**, *144*, 31–43. [CrossRef]
66. Mostafalou, S.; Abdollahi, M. Pesticides and human chronic diseases: Evidences, mechanisms, and perspectives. *Toxicol. Appl. Pharm.* **2013**, *268*, 157–177. [CrossRef]
67. Kohler, H.R.; Triebskorn, R. Wildlife ecotoxicology of pesticides: Can we track effects to the population level and beyond? *Sci. Mag.* **2013**, *341*, 759–765. [CrossRef]
68. Gill, R.J.; Ramos-Rodriguez, O.; Raine, N.E. Combined pesticide exposure severely affects individual and colony-level traits in bees. *Nature* **2012**, *491*, 105–108. [CrossRef]
69. Anonymous. Pesticides Industry Sales and Usage 2006 and 2007 Market Estimates. Available online: https://www.epa.gov/sites/default/files/2015-10/documents/market_estimates2007.pdf (accessed on 1 June 2022).
70. Ragnarsdottir, K.V. Environmental fate and toxicology of organophosphate pesticides. *J. Geol. Soc.* **2000**, *157*, 859–876. [CrossRef]
71. Marroquin-Cardona, A.G.; Johnson, N.M.; Phillips, T.D.; Hayes, A.W. Mycotoxins in a changing global environment—A review. *Food Chem. Toxicol.* **2014**, *69*, 220–230. [CrossRef] [PubMed]

72. Hussein, H.S.; Brasel, J.M. Toxicity, metabolism, and impact of mycotoxins on humans and animals. *Toxicology* **2001**, *167*, 101–134. [CrossRef] [PubMed]
73. Phan, L.T.K.; Tran, T.M.; Audenaert, K.; Jacxsens, L.; Eeckhout, M. Contamination of Fusarium proliferatum and Aspergillus flavus in the rice chain linked to crop seasons, cultivation regions, and traditional agricultural practices in Mekong Delta, Vietnam. *Foods* **2021**, *10*, 2064. [CrossRef]
74. Nielsen, K.F.; Mogensen, J.M.; Johanson, M.; Larsen, T.O.; Frisvad, J.C. Review of secondary metabolites and mycotoxins from the Aspergillus niger group. *Anal. Bioanal. Chem.* **2009**, *395*, 1225–1242. [CrossRef]

foods

MDPI

Article

Effects of Hard Water Boiling on Chalky Rice in Terms of Texture Improvement and Ca Fortification

Sumiko Nakamura and Ken'ichi Ohtsubo *

Faculty of Applied Life Sciences, Niigata University of Pharmacy and Applied Life Sciences, 265-1, Higashijima, Akiha-ku, Niigata 956-8603, Japan; snaka@nupals.ac.jp
* Correspondence: ohtsubok@nupals.ac.jp; Tel./Fax: +81-250-25-1218

Abstract: In the present paper, we investigated the characteristics of chalky rice grains generated by ripening under high temperature and compared them with whole grains. We evaluated 14 unpolished *Japonica* rice grains harvested in Japan in 2021, and these samples (original grains) were divided into two groups (a whole grain group and a chalky grain one). We found that not only activities of endogenous amylase and proteinase, but also cell wall-degrading enzymes, such as xylanase and cellulase, changed markedly between chalky grains and whole grains. Using rice grains blended with 30% of chalky grains as the material, we compared the sugar and mineral contents and textural properties of the rice grains soaked and boiled in either ordinary water or hard water, such as Evian or Contrex. It was shown that xylanase, in addition to amylase and proteinase, may play an important role in changing the texture of the boiled chalky rice grains. For the sake of preventing the above-mentioned deterioration in the texture of boiled grains of chalky rice, we tried to use hard water, such as Evian or Contrex, to soak and cook the chalky rice grains. It was shown that the hard water was useful for the prevention of texture deterioration of the boiled rice grains due to inhibition of the activities of endogenous hydrolytic enzymes, such as α-amylase, β-amylase, proteinase, and xylanase. Furthermore, we found that the hard water was useful in increasing the calcium absorption through the meal by 2.6 to 16.5 times.

Keywords: chalky rice; xylanase activity; cellulase activity; hard water; hardness

Citation: Nakamura, S.; Ohtsubo, K. Effects of Hard Water Boiling on Chalky Rice in Terms of Texture Improvement and Ca Fortification. *Foods* **2023**, *12*, 2510. https://doi.org/10.3390/foods12132510

Academic Editor: Saroat Rawdkuen

Received: 3 June 2023
Revised: 19 June 2023
Accepted: 26 June 2023
Published: 28 June 2023

1. Introduction

Rice (*Oryza sativa* L.) is one of the three most important cereal crops in the world, along with corn and wheat, and supplies calories to about half of the world's population. Rice is widely grown in over 100 countries [1]. Global warming is the most serious environmental issue, and high temperature stress during the rice ripening period causes deterioration not only in the grain yield, but also in quality, by generating chalky grains [2,3]. Global warming has serious implications for our future. Mitsui et al. [4], Asaoka et al. [5], and Nakata et al. [6] have reported that high temperatures cause the inhibition of starch synthases and the activation of α-amylase.

The endosperm starch of rice grains is damaged by high-temperature ripening, which leads to the lowering of the food quality by factors such as the hardness and stickiness of boiled rice grains and the pasting properties of rice flours [7,8]. In our previous paper, we reported that there are marked differences in properties between chalky grains and whole grains using 54 *Japonica* unpolished rice grains damaged severely or lightly by high-temperature ripening [9]. We reported that the α-amylase activity, proteinase activity, and n-6/n-3 ratio of polyunsaturated fatty acid [10,11] of whole grains were significantly lower than those of chalky rice grains. Furthermore, the chalky rice grains, after boiling, were different from whole grains not only in their physical properties, but also in taste components such as oligo saccharides and amino acids, which affect the eating quality of boiled rice [9]. We also made it possible to estimate the degree of damage to rice grains ripened under high temperatures using the pasting properties measured by an RVA as explanatory variables [9,12].

The texture of boiled rice grains is markedly affected by the cell wall of endosperm [13]. Boiled rice grains become softer with the degradation of the cell wall, which suggests the important role of cell wall-degrading enzymes. For examples, softening of the texture by the addition of cellulase [14] and xylanase [15] on boiling were reported in [16]. We formerly reported that endogenous xylanase and cellulase play important roles in determining the texture of the cooked rice grains, similarly to the amylose content [17]. Furthermore, it seems necessary to measure the amounts of tasty substances, such as oligo saccharides and free amino acids, in addition to the physical properties, such as the hardness and stickiness of the boiled rice grains [18–21]. Therefore, it is necessary to compare the texture- and taste-related substances in boiled rice for the purpose of elucidating the difference in palatability between whole grains and chalky grains [10,22,23]. There are few scientific reports about the relationship between high-temperature damage of rice grains and changes in proteinase activity, amino acids, and fatty acid composition [24,25].

Recently, rice industries, such as rice catering services, rice cracker makers, and rice wine brewers, have suffered from the high-temperature ripening of rice grains, as high-temperature-damaged rice grains have textures too soft and too sticky for the rice processing procedure [20,26]. Therefore, it is necessary to improve the texture of the high-temperature-damaged rice grains.

Furthermore, rice consumers request not only palatable rice, but also "healthy rice", such as brown rice, pigmented rice, and pre-germinated brown rice in order to prevent the lifestyle-related diseases by supplying dietary fiber, vitamins, minerals, gamma-amino butylic acid (GABA), etc. [27]. The elderly population has increased markedly in north-eastern Asia, and various kinds of health disorder, such as osteoporosis, have become extremely important problems [28].

In this study, we attempted to elucidate how high-temperature ripening affects the deterioration in quality of chalky grains. Although we reported that not only α-amylase, but also protease, activities are activated by high-temperature-damaged chalky rice grains, we searched for another cause of the quality deteriorations of the chalky rice grains. Cell wall-degrading enzymes, such as cellulase and xylanase, were our novel targets as candidates as the causes of damage to the quality of chalky rice grains. Another aim of this investigation was the development of a method to improve the physical properties of boiled chalky rice grains.

To achieve the above-mentioned objectives, we collected and analyzed 14 *Japonica* rice samples (original grains), which were divided to two groups (a whole grain group and a chalky grain one). A novel method was used to maintain good texture by inhibiting glycolytic enzymes and cell wall-degrading enzymes during the boiling process. We measured the contents of sugars and the textural properties of boiled rice grains boiled after soaking in two types of hard water. Using rice grains consisting of either 100% whole grains or rice blended with 30% chalky grains, we measured the contents of sugars and the textural properties of boiled rice after soaking in two types of hard water.

2. Materials and Methods

2.1. Materials

The unpolished rice samples were purchased in 2021 at a local market, and were subjected to measurement in 2022 (*Japonica* subspecies) (n = 16). These original rice samples were divided manually, based on their apparent chalkiness, into two groups (whole grain and chalky grain): The ordinary *Japonica* rice included *Kumasannokagayaki* (Kumamoto prefecture), *Tsukiakari* (Iwate), *Sasashigure* (Miyagi A), *Sasashigure* (Miyagi B), *Hitomebore* (Miyagi), *Yosakoibijin* (Kochi), *Morinokumasan* (Kumamoto), *Yumeshizuku* (Saga), and *Tsubusuke* (Chiba)(n = 9). The high-quality premium *Japonica* rice included *Koshihikari* (Niigata A), *Koshihikari* (Niigata B), and *Koshihikari* (Hyogo C) (n = 3). The high-amylose *Japonica–Indica hybrid* rice was *Koshinomenjiman* (Niigata), and the *Indica* rice was *Koshinokaori* (Niigata) ($n = 2$). The types of low-amylose *Japonica* rice were *Milky queen* (Kyoto) and *Yumepirika* (Hokkaido) ($n = 2$). Each sample was stored at 10 °C in a rice storage chamber.

The Evian (hardness: 304 mg/L, pH: 7.2) and Contrex (hardness: 1468 mg/L, pH: 7.2) for cooking were purchased at a local market in Niigata City. We used purified water (hardness: 17 mg/L, pH: 7.0) as a control.

2.2. *Measurement of the Moisture Content of Rice Flour*

The moisture contents of the polished and unpolished rice flours were measured using an oven-drying method by drying flour samples of about 2 g for 1 h at 135 °C. An aluminum cup (Wt) and a cup containing the flour sample (Ws), were compared before heating (Wsb) and after heating (Wsa). The moisture content was calculated as follows.

$$\text{Moisture Content (\%)} = 100 \times ((\text{Wt} + \text{Wsb}) - (\text{Wt} + \text{Wsa}))/\text{Wsb}$$

2.3. *Preparation of Two Kinds of Unpolished Rice Flours*

Whole or chalky unpolished rice grains of 14 rice samples—*Kumasannokagayaki* (Kumamoto), *Tsukiakari* (Iwate), *Sasashigure* (Miyagi A), *Hitomebore* (Miyagi), *Yosakoibijin* (Kochi), *Morinokumasan* (Kumamoto), *Yumeshizuku* (Saga), *Tsubusuke* (Chiba), *Koshihikari* (Niigata A), *Koshihikari* (Hyogo C), *Koshinomenjiman* (Niigata), *Koshinokaori* (Niigata), *Milky queen* (kyoto), and *Yumepirika* (Hokkaido)—were used as materials for rice flour. These whole or chalky rice grains were pulverized into rice flours using a cyclone mill (SFC-S1; UDY, Corp., Fort Collins, CO, USA).

2.4. *Soaking of Polished or Unpolished Rice Flours in 2 Types of Hard Water*

Two kinds of rice cultivars, *Koshihikari* (Hyogo, C) and *Tsubusuke* (Chiba), were used for the polished rice samples, and *Koshihikari* (Niigata, A) was used for the unpolished rice. The whole or chalky rice grains of polished rice were soaked in Evian, Contrex, or purified water (tap water treated with a water purifier) at 8 °C for 3 h in refrigerator, respectively, and the unpolished rice was soaked similarly for 48 h. These polished or unpolished rice flours were prepared by pulverizing rice grains after lyophilization (freeze dryer: FD-1, Eyela, Tokyo, Japan). Pulverization was carried out using a cyclone mill (SFC-S1; UDY, Corp., Fort Collins, CO, USA).

2.5. *α-Amylase Activity*

The α-amylase activity of the whole or chalky grains of unpolished rice flour ($n = 14$), as well as those of polished rice flours ($n = 2$) and unpolished rice flour ($n = 1$) soaked in Evian, Contrex, or purified water were determined using an enzyme assay kit (Megazyme International Ireland, Ltd., Wicklow, Ireland), as described in our former report [12].

2.6. *β-Amylase Activity*

The β-Amylase activity of the whole or chalky grains of unpolished rice flour ($n = 14$) and those of polished rice flours ($n = 2$) and unpolished rice flour ($n = 1$) soaked in Evian, Contrex, or purified water were determined using an enzyme assay kit (Megazyme International Ireland, Ltd.). For measurement of the β-amylase activity, rice flour (0.1 g) was extracted with 0.5 mL of extraction buffer, pH 8.0, at 20 °C for 60 min, and was thereafter centrifuged for 10 min at 2000× *g*. The extraction solution (0.6 mL; 6-fold dilution) and substrate (0.1 mL) were preincubated at 40 °C for 5 min. Thereafter, each sample solution was incubated at 40 °C for exactly 10 min, followed by the addition of a stopping reagent (3.0 mL). The absorbance was measured at 400 nm.

2.7. *Protease Activity*

The protease activity of the whole or chalky grains of the unpolished rice flour ($n = 14$), as well as that of the polished rice flours ($n = 2$) and unpolished rice flour ($n = 1$) soaked in two types of hard water or purified water were measured using casein as substrate. Protease activity was measured using the same method described in our former report [12].

2.8. Xylanase Activity

The xylanase activity of the whole or chalky grains of unpolished rice flour (n = 14), as well as that of polished rice flour (n = 2) and unpolished rice flour (n = 1) soaked in two types of hard water and purified water were determined using a kit (Megazyme International Ireland Ltd., Wicklow, Ireland). For measurement of the xylanase activity, the rice flour (0.1 g) was suspended in 0.1 M $C_2H_3NaO_2$ (pH 4.5) buffer containing BSA (0.5 mg/mL) and NaN_3 (0.02% (w/v)) at room temperature for 15 min, then centrifuged for 10 min at 1000× g. Afterward, the supernatants of the extraction solution (0.05 mL; 2-fold dilution) and substrate solution (0.05 mL) were preincubated at 40 °C for 3 min. Thereafter, each sample solution was incubated at 40 °C for exactly 10 min, followed by the addition of a stopping reagent (Tris-HCl buffer 2% (w/v)) (pH 10.0) (1.5 mL). The absorbance was measured at 400 nm.

2.9. Cellulase Activity

The cellulase activity of the whole or chalky grains of unpolished rice flour (n = 14) and that of polished rice flour (n = 2) and unpolished rice flour (n = 1) soaked in two types of hard water and purified water were determined by a kit (Megazyme International Ireland, Ltd.). For measurement of the cellulase activity, the rice flour (0.1 g) was suspended in 0.1 M $C_2H_3NaO_2$ (pH 4.5) buffer containing BSA (0.5 mg/mL) and NaN_3 (0.02% (w/v)) at room temperature for 15 min, then centrifuged for 10 min at 1000× g. After that, the supernatants of the extraction solution (0.2 mL; 2-fold dilution) and substrate (0.05 mL) were preincubated at 40 °C for 3 min. Thereafter, each sample solution was incubated at 40 °C for exactly 10 min, followed by the addition of a stopping reagent (Tris-HCl buffer 2% (w/v)) (pH 10.0) (3.0 mL). The absorbance was measured at 400 nm.

2.10. Polishing and Boiling of Rice Samples

We prepared polished rice (milling yield of 90–91%) using an experimental friction-type rice milling machine (Yamamotoseisakusyo Co., Yamagata, Japan). Samples of 10 g of the blended polished rice grains (blending ratio: whole rice grains:chalky rice grains = 7:3) were added to 16 g (1.6 times, w/w) of purified water in an aluminum cup as control samples, and another set of rice grains samples (10 g) was added to 16 g (1.6 times, w/w) of Evian or Contrex [13]. After soaking for 1 h, the samples were boiled in an electric rice cooker (SR-SW 182 National, Japan). The cooked rice samples were kept in the vessel for 2 h at 25 °C, then used for the measurements [13]. The rice samples which we utilized were *Kumasannokagayaki* (Kumamoto), *Tsukiakari* (Iwate), *Sasashigure* (Miyagi A), *Sasashigure* (Miyagi B), *Hitomebore* (Miyagi), *Yosakoibijin* (Kochi), *Morinokumasan* (Kumamoto), *Yumeshizuku* (Saga), *Koshihikari* (Niigata A), and *Koshihikari* (Niigata B) (n = 10).

In contrast, the polished rice grains of 100% whole or chalky rice were boiled similarly [13]. The rice samples which we utilized were *Koshihikari* (Hyogo C) and *Tsubusuke* (Chiba).

2.11. Measurements of Textural Properties of Boiled Rice Grains

The physical properties of boiled rice grains were measured based on bulk measurements (10 g) and single-grain measurements (the high-compression/low-compression method), which included low compression (compression ratio = 23%: twice), intermediate compression (compression ratio = 46%: twice), and high compression (compression ratio = 92%: twice), according to the 2 × 3 bite method for blended rice. A low compression test (compression ratio = 25%) and a high compression test (compression ratio = 90%) were utilized for 100% whole and chalky rice using a Tensipresser (My Boy System, Taketomo Electric Co., Tokyo, Japan), according to the method described by Okadome et al. [29]. The bulk measurements were repeated 5 times, and single-grain measurements were calculated by measuring 20 individual grains. We used the following parameters for the physical properties of the boiled rice grains for bulk measurement: hardness, toughness, adhesion, stickiness, and cohesiveness [30]. The cohesiveness was measured as A2/A1 for the ratio of hardness with low compression, A4/A3 for the ratio of hardness with intermediate compression, and A6/A5 for the ratio of hardness with high compression. of the parameters used for single-grain measurement were H1 for surface hardness, H2 for overall hardness, S1 for surface stickiness, S2 for overall stickiness, L3 for surface adhesion, S1/H1 (balance H1) for the ratio of stickiness to hardness of the surface layer, S2/H2 (balance H2) for the ratio of stickiness to hardness of the overall layer, A3/A1 (balance A1) for the ratio of adhesiveness to hardness of the surface layer, and A6/A4 (balance A2) for the ratio of adhesiveness to hardness of the overall layer.

2.12. Measurement of D-Glucose, Maltose, and Saccharose Contents

The cooked rice flour sample was prepared by pulverization after lyophilization.

D-glucose, maltose, and saccharose (0.1 g) were extracted from each sample by shaking with 1 mL of 60% ethanol at room temperature for 1 h, and then were measured by the Sucrose/D-Glucose/D-Fructose content assay method (F-kit, Roche/ R-Biopharm AG., Darmstadt, Germany).

2.13. Measurement of Color Difference of Boiled Rice after Soaking in 2 Types of Hard Water

The color differences of the boiled blended rice (blend rice: whole rice grains: chalky rice grains = 7:3) soaked in Evian, Contrex, or purified water were measured using a color difference meter (Color Meter NW-11, Nippon Denshoku Co., Tokyo, Japan).

2.14. Analysis of Calcium and Magnesium Contents

The calcium and magnesium of the whole and chalky *Koshihikari* (Hyogo C) grains of polished or unpolished boiled rice after being soaked in 2 types of hard water or purified water of whole grains of polished rice of were analyzed by an ICP (inductively coupled plasma) emission spectrometry. The absorbance values were measured as 423 nm for calcium and 285 nm for magnesium. Moreover, the tests on boiled whole and chalky polished *Tsubusuke* (Chiba) rice were analyzed similarly. The measurements of the microminerals in the rice samples were carried out at Japan Food Research Laboratories.

2.15. Statistical Analyses

We used Excel Statics (ver. 2006; Microsoft Corp., Tokyo, Japan) for the statistical analysis of the significance of regression coefficients using Student's t-test, one-way analysis of variance, and Tukey's test. The method of Tukey's multiple comparison was statistically analyzed using Excel NAG Statistics 2.0 (The Numerical Algorithms Group, Ltd., Tokyo, Japan).

3. Results and Discussion

3.1. Activities of Various Hydrolytic Enzymes in Whole or Chalky Unpolished Rice Grains

In addition to amylose content, the cell wall of the rice endosperm also affects the texture of boiled rice grains. When the cellulose of the cell wall is hydrolyzed by cellulase (endo-1,4-beta-D-glucanase), boiled rice grains become softer and stickier [31].

As shown in Figure 1, the endo-xylanase activities of chalky unpolished rice grains of premium *Japonica* rice Koshihikari ($n = 2$), ordinary *Japonica* rice ($n = 6$), low-amylose *Japonica* rice ($n = 2$), and high-amylose *Japonica–Indica* hybrid rice ($n = 1$) were significantly higher than those of whole rice grains. Two samples of ordinary *Japonica* rice and *Indica* rice showed similar tendencies. As a result, whole unpolished rice grains were shown to have significantly lower endo-xylanase activities than chalky unpolished rice grains.

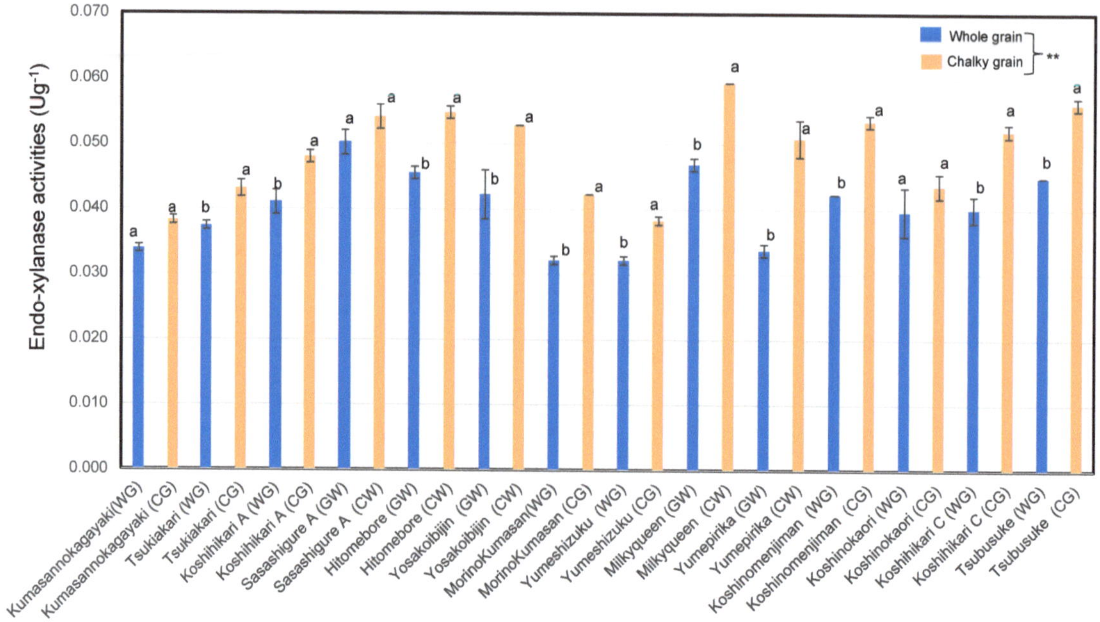

Figure 1. End-xylanase activities of whole and chalky rice grains of 14 *Japonica* unpolished rice samples in 2021. Different letters (a, b) indicate that whole and chalky grains of each same rice samples are significantly different. ** Correlation is significant at 1% according to the method of Tukey's multiple comparison.

The ratio of the endo-xylanase activities of chalky unpolished rice grains to those of whole unpolished rice grains in low-amylose rice (1.38 $\pm$ 0.2) was extraordinary higher than those of premium Koshihikari (1.24 $\pm$ 0.1) rice and ordinary *Japonica* rice cultivars (1.20 $\pm$ 0.1). Chalky unpolished rice grains showed markedly higher endo-xylanase activities, by 1.1–1.5 times, compared to whole unpolished rice grains in 12 *Japonica* rice variants in 2021.

It was found that the endo-xylanase activity of whole unpolished rice grains is lower than that of chalky unpolished rice grains. We also found that the tendency was stronger in the low-amylose rice group.

As shown in Figure 2, the endo-cellulase activity of chalky unpolished rice grains of premium *Japonica* Koshihikari (n = 1) rice, ordinary *Japonica* rice (n = 7), and low-amylose *Japonica* rice (n = 2) were significantly lower than that of whole rice grains. The premium *Japonica* Koshihikari (n = 1) rice and Indica rice (n = 1) showed similarly tendencies. As a result, whole unpolished rice grains were shown to have significantly higher endo-cellulase activities than chalky unpolished rice grains, showing an opposite trend to the endo-xylanase activities. The whole unpolished rice grains showed markedly higher endo-cellulase activities, by 0.8–1.6 times, than chalky unpolished rice grains in 12 *Japonica* rice variants in 2021.

As a result of the statistical treatments using all the rice samples, the xylanase activities of chalky rice grains were shown to be higher than those of whole rice grains ($p < 0.01$) by Tukey's one-way ANOVA. On the contrary, the cellulase activity of the chalky grains was shown to be lower than that of the whole grains ($p < 0.05$).

It seems very interesting that the activities of not only starch-related enzymes, but also cell wall-degrading enzyme activities, change along with the high-temperature ripening of rice grains.

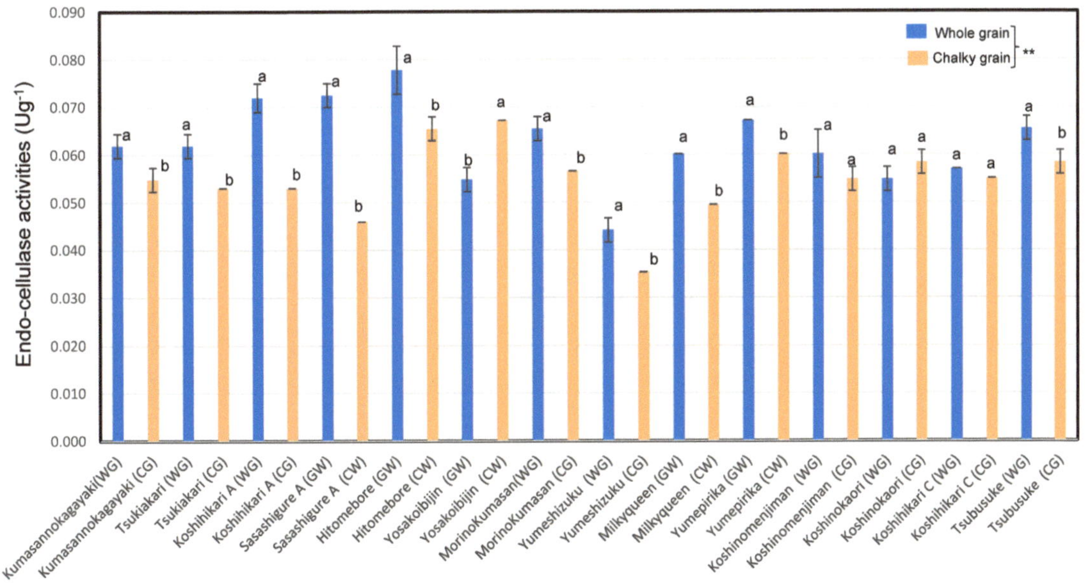

Figure 2. Cellulase activities of whole and chalky rice grains in 14 unpolished *Japonica* rice samples in 2021. Different letters (a, b) indicate that whole and chalky grains in each rice sample are significantly different. ** Correlation is significant at 1% according to the method of Tukey's multiple comparison.

Tsujii et al. reported that endogenous poly-galacturonase activities showed a positive correlation with pectin contents and a negative correlation with the hardness of cooked rice [13]. In our previous paper, xylanase activity showed positive correlations with amylose content and cellulase activity and a negative correlation with the adhesion of cooked rice [17]. Our results, shown in Figures 1 and 2, are harmonized with the report about the discriminative DNA markers encoding 1,4-beta-xylanase and endo-1,4-beta-glucanase 13 in Indica rice, Indica–*Japonica* hybrid rice, and *Japonica* rice (tropical *Japonica* rice) [17].

Alpha-glucosidase hydrolyzes maltose and soluble starch to glucose, which is reported to affect the eating quality of rice grains [32]. Iwata et al. reported that alpha-glucosidase activity showed a positive correlation with GBSS activity and amylose content [33].

As shown in Supplementary Table S1, the α-amylase activity levels of chalky unpolished rice grains of premium *Japonica* Koshihikari ($n = 1$) rice, ordinary *Japonica* rice ($n = 7$), low-amylose *Japonica* rice ($n = 2$), and high-amylose *Japonica–Indica* hybrid rice ($n = 1$) were significantly higher than those of whole rice grains. These results were consistent with the report by Mitsui et al. [4] and our former reports [9,12].

As shown in Supplementary Table S1, the β-amylase activity levels of chalky unpolished rice grains of premium *Japonica* Koshihikari ($n = 1$) rice, ordinary *Japonica* rice ($n = 3$), low-amylose *Japonica* rice ($n = 1$), and high-amylose *Japonica-Indica* hybrid rice ($n = 1$) were significantly higher than those of whole rice grains. This result is in accordance with our previous report [9,12].

As shown in Supplementary Table S1, the protease activity of chalky unpolished grains of ordinary *Japonica* rice ($n = 2$), low-amylose *Japonica* rice ($n = 1$), and high-amylose *Japonica–Indica* hybrid rice ($n = 1$) were significantly higher than those of the whole rice grains.

As a result of statistical treatment using all the rice samples, the α-amylase and proteinase of the chalky grains showed significant higher activity levels than those of whole grains ($p < 0.01$). Unfortunately, beta-amylase activity did not show a significant difference between the two rice groups.

Our results are consistent with our previous report that the activity levels of not only α-amylase, but also β-amylase and protease, were higher for chalky rice grains [9,12,34]. Sun et al. reported that isoamylase is a factor in grain chalkiness using QTLs studies [35].

It has been reported that the gene expression of gibberellin is closely related with the activation of α-amylase, protease activity, and cell wall-degrading enzymes [36].

Our results may show that the high-temperature ripening has close relationship with the changes in the various enzyme activities affected by plant hormones, such as gibberellin [37].

The degradation of the cell walls of cereal grains during germination has been studied from a physiological viewpoint [25,38–40].

3.2. *Activities of Various Hydrolytic Enzymes of Rice Grains Soaking in Hard Water*

In our previous paper, we reported the different properties of whole grains and chalky grains. These chalky rice grains are characterized by high α- and β-amylase activity levels, high protease activity, low apparent amylose content, and low degrees of hardness and stickiness of boiled rice grains compared to whole grains [9]. In this study, we attempted to improve the physical properties of boiled chalky rice grains by reducing various enzyme activity levels.

As shown in Figure 3 and Supplementary Table S2, the α-amylase activity levels of whole grains soaked in purified water (0.05 ± 0.0 Ug^{-1}) were significantly higher than those soaked in Evian (hardness: 304 mg/L, Ca: 8.0 mg, Mg: 2.6 mg, pH: 7.2) (0.03 ± 0.0 Ug^{-1}) or Contrex (hardness: 1468 mg/L, Ca: 46.8 mg, Mg: 7.45 mg, pH: 7.2) (0.04 ± 0.0 Ug^{-1}). On the contrary, those of two- or fourfold-diluted Evian and Contrex did not show inhibition of α-amylase activities.

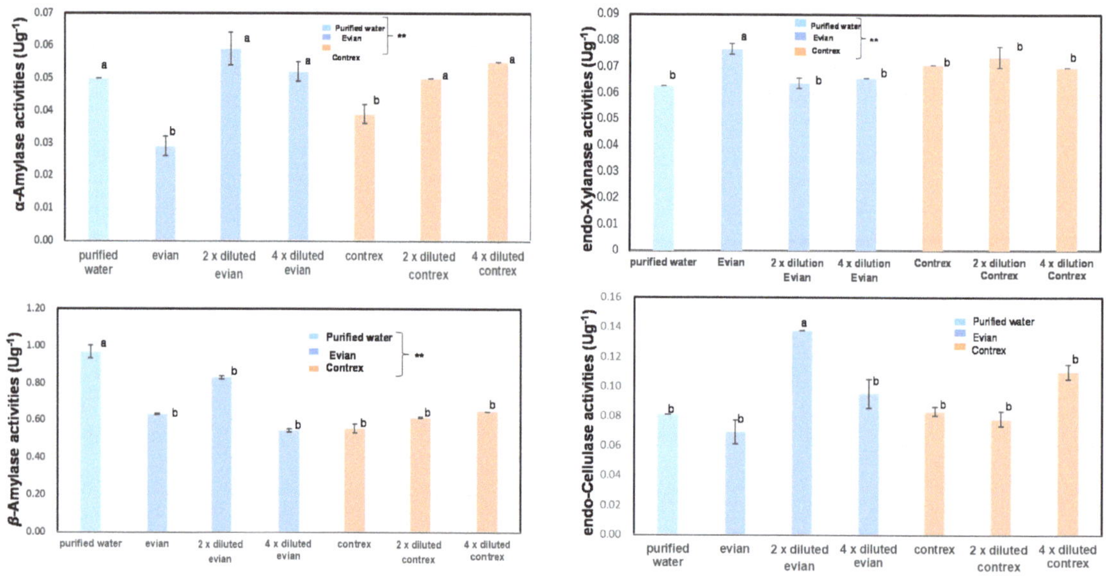

Figure 3. α- and β-amylase activity, proteinase activity, and cell wall-degrading enzyme activity of unpolished rice soaking in Evian, Contrex, or purified water. Within each measurement (α-amylase, β-amylase, proteinase, endo-xylanase, endo-cellulase) in the same column, different letters (a, b) indicate that whole unpolished Koshihikari rice flour, when soaked in Evian, two- or fourfold dilutions of Evian, Contrex, two- or fourfold dilutions of Contrex, or purified water were significantly different. Different letters (a, b) indicate that samples soaked in different types of water were significantly different. ** Correlation was significant at 1% according to the method of Tukey's multiple comparison.

As shown in Figure 3, the β-amylase activity levels of whole grains soaked in purified water (0.967 ± 0.0 Ug^{-1}) were significantly higher than those soaked in Evian (0.630 ± 0.0 Ug^{-1}), two- or fourfold dilutions of Evian (0.830 ± 0.0 Ug^{-1}), (0.543 ± 0.0 Ug^{-1}), Contrex (0.555 ± 0.0 Ug^{-1}), or two- or fourfold dilutions of Contrex (0.612 ± 0.0 Ug^{-1}), (0.648 ± 0.0 Ug^{-1}).

As shown in Figure 3, the proteinase activity levels of the whole grains soaking in purified water (0.072 ± 0.0 Ug^{-1}) were significantly higher than those soaking in Evian (0.065 ± 0.0 Ug^{-1}), fourfold dilutions of Evian (0.066 ± 0.0 Ug^{-1}), or fourfold dilutions of Contrex (0.066 ± 0.0 Ug^{-1}). Those soaked in twofold dilutions of Evian (0.072 ± 0.0 Ug^{-1}), Contrex (0.071 ± 0.0 Ug^{-1}), or twofold dilutions of Contrex (0.072 ± 0.0 Ug^{-1}) showed almost no reduction in proteinase activity.

As shown in Figure 3, the endo-xylanase activity levels of whole grains soaking in purified water (0.063 ± 0.0 Ug^{-1}) were significantly lower than those of the samples soaking in Evian (0.077 ± 0.0 Ug^{-1}). However, those soaked in Contrex (0.071 ± 0.0 Ug^{-1}), two- or fourfold dilutions of Contrex (0.074 ± 0.0 Ug^{-1}), (0.070 ± 0.0 Ug^{-1}), and two- or fourfold dilutions of Evian (0.064 ± 0.0 Ug^{-1}), (0.066 ± 0.0 Ug^{-1}) did not show any significant effects regarding endo-xylanase activity.

As shown in Figure 3, the endo-cellulase activities of whole grains soaked in purified water (0.081 ± 0.0 Ug^{-1}) were significantly lower than those soaking in two- or fourfold dilutions of Evian (0.138 ± 0.0 Ug^{-1}), (0.095 ± 0.0 Ug^{-1}), or fourfold dilutions of Contrex (0.110 ± 0.0 Ug^{-1}). Those of the samples soaked in Evian (0.069 ± 0.0 Ug^{-1}) were lower than those in the purified water group, and those soaked in Contrex (0.083 ± 0.0 Ug^{-1}) and twofold dilutions of Contrex (0.078 ± 0.0 Ug^{-1}) showed similar values to those in the purified water group.

As a result, various enzyme activities were inhibited by soaking in hard water with optimal concentrations. The activity levels of various hydrolytic enzymes in polished rice soaked in hard water showed a similar tendency to those of unpolished rice grains.

3.3. Textural Properties of Boiled Rice Grains

In our previous report, boiled chalky rice grains showed lower hardness and stickiness values and higher retrogradation degrees after boiling compared with the whole grains [12].

In the recent commercial *Japonica* rice market, rice grains containing about 30% of chalky rice are graded as low-class and priced lower than whole rice grains.

In this study, we measured the physical properties of boiled rice of whole and chalky grains after soaking in Evian, Contrex, and purified water using the individual grain method. Both low-compression (25%) and high-compression (90%) tests were conducted using a Tensipresser.

Ogawa et al. [41] reported that water absorption and the swelling of boiled rice, adding calcium, were inhibited compared to soaked in water.

As shown in Supplementary Table S3, H1 (the hardness of the surface layer of the boiled rice grains) and H2 (the hardness of the overall layer) of chalky boiled rice grains were significantly lower than those of whole boiled rice grains, a similarly tendency to that observed in our previous report [9,12]. Compared with other measurements, such as assays of enzyme activities (Supplementary Table S1, Table 1), the measurements of physical properties always show larger standard deviations, as shown in our previous report (references No. 9, No. 12, and No. 34).

As shown in Supplementary Table S3, the textural properties of chalky grains boiled after soaking in purified water change compared with whole grains, such as through an increase in hardness and stickiness.

In polished Koshihikari, H2 (hardness of the overall layer) and S1 (stickiness of the surface layer) of chalky boiled rice grains were significantly higher after soaking in Contrex compared to those of whole rice grains soaking in purified water, and the S2 (stickiness of overall layer) and L3 (the adhered of surface layer) of chalky boiled rice grains showed similar tendencies.

On the other hand, the S1 of chalky boiled rice grains after soaking in Evian was significantly higher compared to the whole rice grains soaked in purified water, while the H2, S2, and L3 of chalky boiled rice grains were slightly lower than those of whole rice grains soaked in purified water.

Table 1. Physical properties of in 10 kinds of boiled *Japonica* rice with 30% chalky grains, blended after soaking in Evian, Contrex, and purified water, in 2021.

	Hardness $\times 10^5$ (N/cm^2)	Toughness $\times 10^5$ (N/cm^2)	Adhesion $\times 10^5$ (N/cm^2)	Stickiness $\times 10^5$ (N/cm^2)	Cohesiveness (A6/A5)
Kumasannokagayaki (purified water)	0.86 ± 0.05 b	17.12 ± 0.53 a	12.41 ± 0.49 b	12.24 ± 1.33 b	0.38 ± 0.01 a
Kumasannokagayaki (Evian)	1.15 ± 0.17 a	17.96 ± 0.46 a	12.67 ± 2.35 b	17.79 ± 5.89 b	0.38 ± 0.02 a
Kumasannokagayaki (Contrex)	1.38 ± 0.05 a	18.33 ± 0.87 a	14.40 ± 1.16 a	26.70 ± 079 a	0.38 ± 0.00 a
Tsukiakari (purified water)	1.63 ± 0.03 a	18.03 ± 0.71 a	12.44 ± 2.02 b	17.94 ± 1.78 a	0.38 ± 0.02 a
Tsukiakari (Evian)	1.23 ± 0.18 b	19.96 ± 0.80 a	12.64 ± 1.20 b	19.13 ± 4.21 a	0.41 ± 0.01 a
Tsukiakari (Contrex)	1.56 ± 0.37 a	19.22 ± 2.47 a	13.48 ± 0.00 a	17.55 ± 1.00 a	0.40 ± 0.00 a
Koshihikari A (purified water)	1.14 ± 0.19 b	16.16 ± 0.41 a	14.51 ± 0.29 a	25.66 ± 1.34 a	0.38 ± 0.00 a
Koshihikari A (Evian)	1.67 ± 0.19 a	17.22 ± 2.56 a	12.33 ± 1.12 a	19.74 ± 4.50 b	0.41 ± 0.04 a
Koshihikari A (Contrex)	1.29 ± 0.09 b	16.89 ± 1.06 a	14.16 ± 1.05 a	25.95 ± 4.67 a	0.41 ± 0.05 a
Sasashigure A (purified water)	1.36 ± 0.13 b	13.32 ± 0.74 b	12.83 ± 1.49 b	22.05 ± 5.50 b	0.41 ± 0.05 a
Sasashigure A (Evian)	1.28 ± 0.46 b	19.35 ± 1.47 a	13.52 ± 2.37 b	24.23 ± 1.52 b	0.42 ± 0.02 a
Sasashigure A (Contrex)	1.70 ± 0.11 a	19.72 ± 1.63 a	15.30 ± 1.66 a	31.40 ± 3.19 a	0.42 ± 0.01 a
Hitomebore (purified water)	1.59 ± 0.25 a	19.92 ± 2.29 a	15.02 ± 0.28 a	22.91 ± 0.63 a	0.37 ± 0.03 a
Hitomebore (Evian)	1.23 ± 0.07 b	17.55 ± 1.25 b	12.78 ± 0.19 b	14.80 ± 0.68 b	0.38 ± 0.00 a
Hitomebore (Contrex)	1.54 ± 0.06 a	17.88 ± 1.79 b	13.35 ± 0.45 b	24.13 ± 6.77 a	0.41 ± 0.02 a
Yosakoibijin (purified water)	1.28 ± 0.10 a	22.07 ± 2.55 a	10.65 ± 0.14 a	16.04 ± 5.47 a	0.39 ± 0.05 a
Yosakoibijin (Evian)	1.40 ± 0.42 a	20.11 ± 3.03 a	13.87 ± 1.56 a	21.20 ± 0.16 a	0.42 ± 0.04 a
Yosakoibijin (Contrex)	1.27 ± 0.46 a	18.17 ± 2.55 b	11.21 ± 0.84 a	18.97 ± 4.47 a	0.40 ± 0.04 a
Koshihikari B (purified water)	1.66 ± 0.27 a	21.43 ± 0.29 a	12.80 ± 1.10 b	30.99 ± 0.74 a	0.39 ± 0.05 a
Koshihikari B (Evian)	1.38 ± 0.11 a	18.65 ± 0.47 b	12.33 ± 1.96 b	23.93 ± 0.15 b	0.42 ± 0.04 a
Koshihikari B (Contrex)	1.42 ± 0.22 a	18.69 ± 1.04 b	15.18 ± 0.96 a	22.87 ± 6.17 b	0.40 ± 0.04 a
Sasashigure B (purified water)	1.46 ± 0.40 b	19.56 ± 1.07 b	11.09 ± 0.26 a	17.89 ± 0.84 a	0.39 ± 0.03 a
Sasashigure B (Evian)	1.45 ± 0.21 b	22.26 ± 0.92 a	11.01 ±0.07 a	17.66 ± 0.24 a	0.40 ± 0.04 a
Sasashigure B (Contrex)	1.88 ± 0.01 a	22.76 ± 0.82 a	11.61 ± 0.31 a	16.70 ± 4.54 a	0.41 ± 0.01 a
Morinokumasan (purified water)	0.92 ± 0.00 a	17.84 ± 0.11 b	14.68 ± 0.33 a	28.51 ± 3.76 a	0.40 ± 0.00 a
Morinokumasan (Evian)	1.16 ± 0.34 a	18.49 ± 2.41 a	12.86 ± 0.52 a	21.81 ± 2.96 b	0.40 ± 0.02 a
Morinokumasan (Contrex)	1.08 ± 0.16 a	17.80 ± 1.67 b	14.38 ± 0.83 a	25.61 ± 1.72 b	0.41 ± 0.00 a
Yumeshizuku (purified water)	1.35 ± 0.25 b	18.90 ± 1.40 a	15.12 ± 1.85 a	18.48 ± 5.02 b	0.37 ± 0.05 b
Yumeshizuku (Evian)	1.47 ± 0.21 b	17.68 ± 1.14 a	12.20 ± 1.87 b	31.78 ± 2.58 a	0.42 ± 0.01 a
Yumeshizuku (Contrex)	1.23 ± 0.01 a	18.69 ± 1.85 a	15.14 ± 0.83 a	31.86 ± 0.85 a	0.42 ± 0.04 a

The physical properties of boiled grains of blended rice were measured using bulk measurements, which included low compression (compression ratio = 23%: twice), intermediate compression (compression ratio = 46%: twice), and high compression (compression ratio = 92%: twice), according to the 2 × 3 bite method. Within each measurement (hardness, toughness, adhesion, stickiness, cohesiveness) in the same column, different letters (a, b) mean that blended boiled rice in each rice sample was significantly different. The value of hardness is indicated by the height and that of toughness by the area of continuous progressive compression in Tensipresser. Cohesiveness (A6/A5) is shown as the ratio of hardness under high compression. Values are shown as mean ± standard deviation.

In polished Tsubusuke, the H2 and S1 of the chalky boiled rice grains after soaking in Contrex were significantly higher than those of whole rice grains soaked in purified water, and the S2 of chalky boiled rice grains showed a similar tendency. On the other hand, the S1 and S2 of chalky boiled rice grains after soaking in Evian were significantly higher than those of whole rice grains soaked in purified water, and the L3 of chalky boiled rice grains showed a similar tendency.

As shown in Table 1 and Supplementary Figure S1, the ratio of the hardness of boiled 30% chalky blended rice after soaking in Evian to the hardness of the rice soaked in purified water was 1.05 ± 0.24 times (n = 5). The ratio of toughness was 1.04 ± 0.17 times (n = 6), the ratio of adhesion was 0.97 ± 0.14 times (n = 5), the ratio of stickiness was 1.06 ± 0.35 times (n = 5), and the ratio of cohesiveness was 1.05 ± 0.04 times (n = 8). The ratio of the hardness of those soaked in Contrex to the hardness of those soaked in purified water was 1.11 ± 0.23 times (n = 5), the ratio of toughness was 1.04 ± 0.19 times (n = 6), the ratio of adhesion was 1.06 ± 0.10 times (n = 7), the ratio of stickiness was 1.21 ± 0.44 times (n = 6), and cohesiveness was 1.05 ± 0.04 times (n = 9).

As a result, the various physical properties of boiled 30% chalky blended rice after soaking in hard water showed higher values than those of the rice soaked in purified water, and the physical properties of the rice soaked in Contrex were slightly higher than those of the rice soaked in Evian.

In the recent commercial market, rice grains containing about 30% chalky grains were classified as low-grade; thus, the prices are lower than those of whole rice grains. In our previous report, the hardness and toughness of boiled 30% chalky blended rice were lower than those of whole boiled rice grains, and the stickiness and adhesion showed a similar tendency.

We found that the boiled 30% chalky grain blended rice, after soaking in hard water, showed slightly higher hardness, toughness, stickiness, and cohesiveness compared to the rice soaked in purified water, which means that the physical properties of boiled 30% chalky blended rice were improved in terms of textural qualities.

3.4. D-Glucose, Maltose, and Saccharose Contents in Boiled Rice Grains

Awazuhara et al. [42] showed that the thermal dependency and stability of enzymes producing reducing sugar are different between outer endosperm and inner endosperm of rice.

As shown in Supplementary Table S4, the D-glucose, maltose, and saccharose in chalky boiled rice grains were significantly higher than those in whole boiled rice grains, which was consistent with our previous reports [9,34].

In Koshihikari, the D-glucose and maltose levels in chalky boiled rice grains after soaking in Evian and Contrex were significantly lower than those in grains soaked in purified water. The sugar content showed similarity to that of whole grains soaking in purified water. The saccharose contents of chalky boiled rice grains after soaking in Evian and Contrex showed a similar tendency.

In Tsubusuke, the D-glucose, maltose, and saccharose levels of chalky boiled rice grains after soaking in Evian and Contrex were lower than those soaked in purified water, and the sugar content was higher than that of whole grains soaked in purified water.

Shibuya et al. [31] demonstrated the isolation of cell walls from different parts of rice grains of a *Japonica* variety, their macromolecule composition, and the sugar linkages contained in these cell walls. Tsujii et al. [16] showed that the extent of decomposition of pectin has a negative correlation with the hardness value of cooked rice.

In two kinds of boiled rice, after soaking in Evian, Contrex, or purified water, the stickiness of the overall layer (S2) of boiled rice showed a positive correlation with D-glucose (r = 0.70, $p < 0.05$), maltose (r = 0.77, $p < 0.01$), and saccharose (r = 0.78, $p < 0.01$). The hardness of the surface layer (H1) of boiled rice showed a negative correlation with D-glucose (r = −0.74, $p < 0.01$), maltose (r = −0.68, $p < 0.05$), and saccharose (r = −0.76, $p < 0.01$), and the endo-cellulase activities showed a negative correlation with D-glucose (r = −0.67, $p < 0.05$), maltose (r = −0.58, $p < 0.05$), and saccharose (r = −0.58, $p < 0.05$). Furthermore, α-amylase activity showed a negative correlation with the stickiness of surface layer (S1) (r = −0.70, $p < 0.05$) of boiled rice, as shown in Supplementary Table S5.

It was shown that rice grains boiled after soaking in Evian or Contrex contained lower amounts of glucose, due to the lower enzyme activity, than of the rice grains boiled after soaking in purified water.

As shown in Table 2, the D-glucose content of the 30% chalky blended boiled grains after soaking in Evian or Contrex was significantly lower than that of blended rice soaked in purified water (n = 10), and the maltose and saccharose contents showed similar tendencies.

Table 2. Oligosaccharides of boiled 30% chalky blended rice after soaking in Evian, Contrex, and purified water in 10 kinds of *Japonica* rice in 2021.

	D-Glucose Content (g/100 g)	Maltose Content (g/100 g)	Saccharose Content (g/100 g)
Kumasannokagayaki (purified water)	0.065 ± 0.004 a	0.095 ± 0.002 a	0.313 ± 0.008 a
Kumasannokagayaki (evian)	0.058 ± 0.001 b	0.091 ± 0.001 a	0.299 ± 0.001 a
Kumasannokagayaki (contrex)	0.055 ± 0.002 b	0.087 ± 0.002 a	0.299 ± 0.000 a
Tsukiakari (purified water)	0.078 ± 0.002 a	0.134 ± 0.001 a	0.302 ± 0.002 a
Tsukiakari (evian)	0.057 ± 0.002 b	0.090 ± 0.000 b	0.260 ± 0.006 b
Tsukiakari (contrex)	0.055 ± 0.002 b	0.088 ± 0.007 b	0.254 ± 0.007 b
Koshihikari A (purified water)	0.074 ± 0.001 a	0.100 ± 0.000 a	0.345 ± 0.000 a
Koshihikari A (evian)	0.058 ± 0.002 b	0.089 ± 0.000 b	0.311 ± 0.006 b
Koshihikari A (contrex)	0.057 ± 0.001 b	0.089 ± 0.001 b	0.306 ± 0.008 b
Sasashigure A (purified water)	0.077 ± 0.002 a	0.097 ± 0.000 a	0.331 ± 0.009 a
Sasashigure A (evian)	0.063 ± 0.002 b	0.087 ± 0.001 b	0.313 ± 0.006 a
Sasashigure A (contrex)	0.066 ± 0.002 b	0.091 ± 0.002 a	0.322 ± 0.009 a
Hitomebore (purified water)	0.062 ± 0.001 a	0.088 ± 0.001 a	0.281 ± 0.008 a
Hitomebore (evian)	0.050 ± 0.001 b	0.073 ± 0.003 b	0.259 ± 0.007 a
Hitomebore (contrex)	0.046 ± 0.001 b	0.072 ± 0.002 b	0.252 ± 0.006 a
Yosakoibijin (purified water)	0.060 ± 0.001 a	0.115 ± 0.003 a	0.514 ± 0.007 a
Yosakoibijin (evian)	0.051 ± 0.001 b	0.107 ± 0.001 a	0.492 ± 0.002 a
Yosakoibijin (contrex)	0.055 ± 0.001 b	0.112 ± 0.001 a	0.500 ± 0.005 a
Koshihikari B (purified water)	0.070 ± 0.001 a	0.092 ± 0.001 a	0.295 ± 0.001 a
Koshihikari B (evian)	0.061 ± 0.001 b	0.083 ± 0.001 b	0.280 ± 0.003 a
Koshihikari B (contrex)	0.063 ± 0.001 b	0.082 ± 0.001 b	0.277 ± 0.001 a
Sasashigure B (purified water)	0.062 ± 0.000 a	0.085 ± 0.000 a	0.295 ± 0.000 a
Sasashigure B (evian)	0.051 ± 0.000 b	0.073 ± 0.003 b	0.270 ± 0.001 a
Sasashigure B (contrex)	0.054 ± 0.002 b	0.073 ± 0.001 b	0.277 ± 0.001 a
Morinokumasan (purified water)	0.075 ± 0.000 a	0.113 ± 0.003 a	0.411 ± 0.001 a
Morinokumasan (evian)	0.063 ± 0.001 b	0.099 ± 0.003 a	0.378 ± 0.003 a
Morinokumasan (contrex)	0.066 ± 0.000 b	0.104 ± 0.001 a	0.385 ± 0.001 a
Yumeshizuku (purified water)	0.074 ± 0.000 a	0.098 ± 0.001 a	0.318 ± 0.002 a
Yumeshizuku (evian)	0.057 ± 0.001 b	0.079 ± 0.001 b	0.275 ± 0.001 b
Yumeshizuku (contrex)	0.060 ± 0.000 b	0.084 ± 0.001 b	0.285 ± 0.001 b

Within each measure (D-glucose content, maltose content, saccharose content) in the same column, different letters (a, b) mean that in each sample, the properties of the rice were significantly different. Values are shown as mean ± standard deviation.

The ratio of D-glucose of the boiled 30% chalky blended rice after soaking in Evian to D-glucose of the rice soaking in purified water was 0.82 ± 0.05 times (n = 10). The ratio of maltose was 0.86 ± 0.08 times (n = 8), that of saccharose was 0.92 ± 0.04 times (n = 10), and of the results for blended boiled grains after soaking in Contrex showed a similar tendency. As a result, the sugar content of the boiled 30% chalky blended rice were lower after soaking in hard water compared to that of the boiled 30% chalky blended rice soaked in purified water.

In our previous report, the sugar contents of the boiled 30% chalky blended rice after soaking in purified water were 1.1 times higher than those of 100% whole grain rice [9]. The reason why blended boiled rice contained a higher sugar content than 100% whole grains is due to the higher activity levels of multiple amylases and lower activity levels of starch-synthesizing enzymes.

In this study, it was shown that the boiled 30% chalky blended rice, after soaking in Evian and Contrex, had a lower sugar content than that of the rice boiled after soaking in purified water due to lower enzyme activity.

Onishi et al. [43] showed that cooked rice boiled in hard water is harder, with less coloring than rice boiled in soft water.

As shown in Supplementary Table S6, a ratio of the color difference ($\Delta E^*(ab)$) of boiled 30% chalky blended rice after soaking in Evian to the color difference of the rice soaked in purified water was 1.06 ± 0.19 times, and those of a ratio of the color difference of rice soaked in Contrex was 0.92 ± 0.18 times.

As a result, the color difference of boiled 30% chalky blended rice after soaking in Evian showed slightly higher values than of the rice soaked in purified water, while the color differences in the rice soaked in Contrex were slightly lower.

In this report, the color difference of boiled 30% chalky blended rice after soaking in hard water showed a similar tendency to the boiled rice after soaking in purified water. Although it is well-known that color differences in boiled rice are affected by amino-carbonyl reactions from sugar and amino acids, in our investigation, various enzyme activities were inhibited by soaking in hard water of an optimal concentration. The reason why blended 30% chalky rice, when boiled after soaking in Evian or and Contrex, did not show a marked color difference was due to the lower sugar and amino acid contents than the rice boiled after soaking in purified water. As multiple amylases and proteinase activity levels were reduced by the hard water, the boiled rice contained lower amounts of mono-or oligosaccharides, as shown in Table S4.

3.5. Calcium and Magnesium Contents in Whole and Chalky Polished or Unpolished Rice, and Those in Boiled Rice, after Soaking

The ash distribution in brown rice is calculated as 51% in bran, 10% in germ, 11% in polish, and 28% in milled rice. Some minerals are also present according to some calculations, which have shown that milled rice retains 63% of the sodium and 74% of the calcium content of brown rice [44].

Calcium deficiency is a global problem, especially in the aging population [45]. Significant impairments in bone mineral density and bone fracture are generally found in low-calcium-intake populations [46]. To supplement calcium deficiency, boiled rice as a staple food is one of the best ways, because many people eat rice almost every day.

As shown in Table 3, unpolished chalky rice contained about 1.3 times more calcium than unpolished whole rice. Similarly, polished chalky rice contained about 1.3 times more calcium than whole rice in the case of Koshihikari.

As a result, it seems that the calcium levels in chalky unpolished or polished rice grains were significantly higher than in whole rice grains, and magnesium showed a similar tendency. Okuda showed that minerals and protein contents, negatively influence the quality of sake in abundance, are distributed more in the outer layer [20,21].

Polished whole rice grains boiled after soaking in Evian contained 3.5 times more calcium than those boiled after soaking in purified water, and furthermore, those boiled after soaking in Contrex contained as much as 16.5 times more than those boiled after soaking in purified water in the case of Koshihikari.

In the case of Tsubusuke, chalky polished rice grains boiled after soaking in Evian or Contrex contained 2.7 times and 13.5 times more calcium than those boiled after soaking in purified water. Additionally, whole polished rice grains boiled after soaking in Evian or Contrex contained 2.6 times and 13.5 times more calcium than those boiled after soaking in purified water.

As shown in Table 3, in polished Koshihikari rice, the magnesium content in whole and chalky grains showed a similar tendency with calcium, and polished chalky Koshihikari rice grains showed a slightly higher magnesium content than whole grains.

Polished whole rice grains boiled after soaking in Evian contained 1.3 times more magnesium than rice boiled after soaking in purified water, and furthermore, rice boiled after soaking in Contrex contained about 1.8 times more than rice boiled after soaking in purified water in the case of Koshihikari.

In polished whole or chalky boiled Tsubusuke rice grains, the ratio of magnesium content in chalky rice boiled after soaking in Evian or Contrex to rice boiled after soaking in purified water was 1.1 or 1.3. The ratio of magnesium content in whole rice boiled after soaking in Evian or Contrex to rice boiled after soaking in purified water was 1.1 or 1.6.

Table 3. Calcium and magnesium contents of whole and chalky unpolished rice, those of polished rice, and those of polished rice boiled after soaking in Evian, Contrex, and purified water using kinds of *Japonica* rice in 2021.

	Calcium (mg/100 g)	Magnesium (mg/100 g)
Koshihikari C (WG) (unpolished rice)	9.6 ± 0.0 b	103.0 ± 0.0 a
Koshihikari C (CG) (unpolished rice)	12.3 ± 0.0 a	102.0 ± 0.0 a
Koshihikari C (WG) (polished rice)	5.4 ± 0.0 b	19.3 ± 0.0 b
Koshihikari C (CG) (polished rice)	6.8 ± 0.0 a	22.5 ± 0.0 a
Koshihikari C (WG) (polished boiled rice) (pulified water)	5.5 ± 0.0 b	17.5 ± 0.0 b
Koshihikari C (CG) (polished boiled rice) (pulified water)	6.9 ± 0.0 a	20.4 ± 0.0 a
Koshihikari C (WG) (polished boild rice) (Evian)	19.0 ± 0.0 b	22.1 ± 0.0 b
Koshihikari C (CG) (polished boild rice) (Evian)	23.9 ± 0.0 a	25.8 ± 0.0 a
Koshihikari C (WG) (polished boiled rice) (Contrex)	90.5 ± 0.0 b	31.3 ± 0.0 b
Koshihikari C (CG) (polished boiled rice) (Contrex)	114.0 ± 0.0 a	36.5 ± 0.0 a
Tsubusuke (WG) (polished boiled rice) (pulified water)	6.1 ± 0.0 a	26.6 ± 0.1 b
Tsubusuke (CG) (polished boiled rice) (pulified water)	6.4 ± 0.0 a	29.4 ± 0.1 a
Tsubusuke (WG) (polished boiled rice) (Evian)	15.8 ± 0.0 b	30.4 ± 0.1 a
Tsubusuke (CG) (polished boiled rice) (Evian)	17.5 ± 0.0 a	32.6 ± 0.1 a
Tsubusuke (WG) (polished boiled rice) (Contrex)	82.6 ± 0.2 b	41.3 ± 0.1 a
Tsubusuke (CG) (polished boiled rice) (Contrex)	86.1 ± 0.2 a	39.0 ± 0.1 a

Within each measure (calcium, magnesium) in the same column, different letters (a, b) mean that each rice sample was significantly different. Abbreviation: WG, whole grains; CG, chalky grains. Values are shown as mean ± standard deviation.

As a result, it became possible to improve the quality of the chalky rice grains not only in terms of textural properties, but also in terms of bio-functionality and mineral absorption, by boiling in hard water. According to the Dietary Reference Intakes for Japanese (2020), it is recommended to absorb 650 mg of calcium and 370 mg of magnesium per day. If one eats 300 g of boiled rice soaked in Contrex every day, he or she can ingest 342 mg of calcium and 110 mg of magnesium. These figures mean that about 53% of the calcium and 30% of the magnesium necessary per day can be consumed from only boiled rice.

4. Conclusions

In the present paper, we investigated the characteristics of chalky rice grains generated by ripening under high temperatures, compared with whole grains. We found that the activity levels of not only endogenous amylase and proteinase, but also cell wall-degrading enzymes such as xylanase and cellulase, changed markedly. As we reported in the former study boiled grains of chalky rice become softer and non-stickier compared with boiled grains of whole rice, it was ascertained that the change was intrinsic for the chalky rice grains. It was shown that xylanase, in addition to amylase and proteinase, may take on an important role contributing to the change in texture of the boiled chalky rice grains. In order to prevent the above-mentioned deterioration in the texture of the boiled chalky rice grains, we used hard water, such as Evian or Contrex, to soak and cook the chalky rice grains. It was shown that the hard water was useful for prevention of the texture deterioration of the boiled rice grains due to inhibition of the activities of endogenous hydrolytic enzymes, such as α-amylase, β-amylase, and proteinase. Furthermore, we found that hard water was useful for a 2.6- to 16.5-fold increase in calcium absorption through consumption of boiled rice grains soaked and cooked using hard water.

Supplementary Materials: The following supporting information can be downloaded at: https://www.mdpi.com/article/10.3390/foods12132510/s1, Table S1: Comparison between α, β-amylase activities and proteinase activities of whole and chalky unpolished rice grains in 14 *Japonica* rice variants in 2021. Table S2: α, β-amylase activities, proteinase activities, and cell wall degrading enzyme activities of unpolished Koshihikari soaked in Evian, Contrex or purified water. Table S3: Physical properties of boiled chalky rice after soaked in Evian, Contrex, and purified water, and those of whole rice grains soaked in purified water, in 2 kinds of *Japonica* rice in 2021. Table S4: Oligosaccharides in boiled whole rice grains after soaking in purified water and in boiled chalky rice grains soaked in purified water or 2 kinds of hard water. Table S5: Correlation between endogenous amylase, proteinase, cell wall-degrading enzymes, physical properties, and reducing sugars of boiled rice after soaking in Evian, Contrex, or purified water using whole and chalky *Japonica* rice grains (n = 2). Table S6: Comparison between color difference (ΔE*(ab)) of boiled 30% chalky blended rice after soaking in Evian, Contrex, or purified water in 10 kinds of *Japonica* rice in 2021. Supplement Figure S1: Comparison between the ratio of the physical properties of boiled rice after soaking in hard water to the physical properties of rice soaking in purified water, using 10 kinds of boiled blended rice with 30% chalky grains, in 2021. Different letters (a, b) mean that each rice sample was significantly different in Tables S1–S6, and Figure S1.

Author Contributions: S.N. and K.O., designed this research, S.N. performed the experiments; S.N. and K.O. wrote the manuscript. All authors have read and agreed to the published version of the manuscript.

Funding: A part of this research was supported by Grant in Aid Scientific Research 22K05604, Japan.

Data Availability Statement: The datasets generated for this study are available upon request to the corresponding author.

Conflicts of Interest: The authors declare no conflict of interest.

Abbreviations

WG: whole grains; CG: chalky grains. Blend rice: 30% chalky grains blended rice.

References

1. Childs, N.W. Chapter 1: Production and utilization of rice. In *RICE: Chemistry and Technology*; Scientific Societies: St. Paul, MN, USA, 2004; pp. 1–23.
2. Morita, S.; Shiratsuch, I.H.; Takahashi, J.; Fujita, K. Effect of temperature on grain ripening in rice plants. *Jpn. Crop Sci.* **2004**, *73*, 77–83. [CrossRef]
3. Liao, J.-L.; Zhou, H.-W.; Zhang, H.-Y.; Zhong, P.-A.; Huang, Y.-J. Comparative proteomic analysis of differentially expressed proteins in the early milky stage of rice grains during high temperature stress. *J. Exp. Bot.* **2014**, *65*, 655–671. [CrossRef]
4. Mitsui, T.; Shiraya, S.; Kaneko, K.; Wada, K. Proteomics of rice grain under high temperature stress. *Front. Plant Sci.* **2013**, *4*, 36. [CrossRef]
5. Asaoka, M.; Okuno, K.; Sugimoto, Y.; Kawakami, J.; Fuwa, H. Effect of environmental temperature during development of rice plants on some properties of endosperm starch. *Starch-Stärke* **1984**, *31*, 189–193. [CrossRef]
6. Nakata, M.; Fukamatsu, Y.; Miyashita, T.; Hakata, M.; Kimura, R.; Nakata, Y.; Kuroda, M.; Yamaguchi, T.; Yamakawa, H. High temperature-induced expression of rice α- amylase in developing endosperm produces chalky grains. *Front. Plant Sci.* **2017**, *8*, 2089. [CrossRef] [PubMed]
7. Nevam, A.Y.M.; Emon, R.M.; Malek, M.A.; Hasan, M.M.; Alam, M.A.; Muharam, M.; Aslani, F.; Rafii, M.Y.; Ismail, M.R. Relationship between high temperature and formation of chalkiness and their effects on quality of rice. *Biomed. Res. Int.* **2018**, *2018*, 1653721. [CrossRef] [PubMed]
8. Igarashi, T.; Yanagihara, T.; Kanda, H.; Kawamoto, K.; Masaki, K. Development of new eating quality evaluation method based on iodine adsorption multispectral analysis of rice flour. *J. Crop Sci.* **2009**, *78*, 66–73. [CrossRef]
9. Nakamura, S.; Hasegawa, M.; Kobayashi, Y.; Komata, C.; Katsura, J.; Maruyama, Y.; Ohtsubo, K. Palatability and bio-functionality of chalky grains generated by high-temperature ripening and development of formulae for estimating the degree of damage using a rapid visco analyzer of *Japonica* unpolished rice. *Foods* **2022**, *11*, 3422. [CrossRef]

10. Simopoulos, A.P. The important of the ratio of omega-6/omega-3 essential fatty acids. *Biomed. Pharmacother.* **2002**, *56*, 365–379. [CrossRef] [PubMed]
11. Nakamura, S.; Katsura, J.; Maruyama, Y.; Ohtsubo, K. Relationship between fatty acid composition and starch properties of 30 japonica rice cultivars. *Cereal Chem.* **2019**, *96*, 228–242. [CrossRef]
12. Nakamura, S.; Satoh, A.; Aizawa, M.; Ohtsubo, K. Characteristics of physicochemical properties of chalky grains of *Japonica* rice generated by high temperature during ripening. *Foods* **2022**, *11*, 97. [CrossRef] [PubMed]
13. Tsujii, Y.; Kiyose, N.; Tatsuda, N.; Yaguchi, Y.; Uchino, M.; Takano, K. Change in the endospermous cell wall of cooked rice and its effects on the palatability of cooked rice. *Food Preserv. Sci.* **2009**, *35*, 127–134. [CrossRef]
14. Shibuya, N.; Iwasaki, T. Polysaccharides and glycoproteins in the rice endosperm cell wall. *Agric. Biol. Chem.* **1978**, *42*, 2259–2266.
15. Shibuya, N.; Iwasaki, T. Effect of cell wall degrading enzymes on the cooking properties of milled rice and the texture of cooked rice. *Nippon. Shokuhin Kogakukaishi* **1984**, *31*, 656–660. [CrossRef] [PubMed]
16. Tsujii, Y.; Uwaya, M.; Uchino, M.; Takano, K. Effect of pectin contents and polygalacturonase activity on cooked rice texture. *Food Preserv. Sci.* **2010**, *36*, 177–182. [CrossRef]
17. Nakamura, S.; Machida, K.; Ohtsubo, K. Search for cell-wall-degrading enzymes of world-wide rice grains by PCR and their effects on the palatability of rice. *Biosci. Biotechnol. Biochem.* **2012**, *76*, 1645–1654. [CrossRef] [PubMed]
18. Kasai, M.; Ishiguro, K.; Kyoda, H.; Hamazono, T.; Hatae, K.; Shimada, A. Change in the amounts of reducing sugars and free amino acids in rice during the cooking processes. *JSHE* **2000**, *51*, 579–585.
19. Matsuzaki, A.; Takano, T.; Sakamoto, S.; Kuboyama, T. Relation between eating quality and chemical components in milled rice and amino acid contents in cooked rice. *J. Crop Sci.* **1992**, *61*, 561–567. [CrossRef]
20. Okuda, M. Rice used for Japanese sake making. *Biosci. Biotechnol. Biochem.* **2019**, *83*, 1428–1441. [CrossRef]
21. Koseki, T.; Okuda, M.; Yonehara, Y.; Hachida, K.; Iwata, H. Effects of high temperature during grain-filling on characteristics of rice for brewing. *J. Brew. Soc. Jpn.* **2004**, *99*, 591–596. [CrossRef]
22. Yamakawa, H.; Hakata, M. Atlas of rice grain filling-related metabolism under high temperature joint analysis of metabolome and transcriptome demonstrated inhibition of starch accumulation and induction of amino acid accumulation. *Plant Cell Physiol.* **2010**, *51*, 796–809. [CrossRef]
23. Yamakawa, H.; Hirai-Kimura, R.; Nakata, Y.; Nakata, M.; Kuroda, M.; Yamaguchi, T. An activity-staining method on filtration paper enables high-throughput screening of temperature-sensitive and inactive mutations of rice α- amylase for improvement of rice grain quality. *Plant Cell Physiol.* **2017**, *58*, 658–667. [CrossRef]
24. Mitsui, T.; Yamakawa, H.; Kobata, T. Molecular physiological aspects of chalking mechanism in rice grains under high-temperature stress. *Plant Prod. Sci.* **2016**, *19*, 22–29. [CrossRef]
25. Tashiro, M.; Kurata, A.; Maki, Z. Occurrence of Cysteine Proteinase Inhibitors in Cereals. *Sci. Rep. Kyoto Prefect. Univ. Nat. Sci. Living Sci.* **1990**, *41*, 15–21.
26. Kitadume, R.; Nakamura, S.; Kumagai, T.; Takahashi, H.; Ohtsuno, K. Characteristics of chalky rice grains and their influence on rice cracker processing. *Nippon. Shokuhin Kogakukaishi* **2012**, *59*, 621–627. [CrossRef]
27. Nakamura, S.; Ikeuch, T.; Araki, A.; Kasuga, K.; Watanabe, K.; Hirayama, M.; Ito, M.; Ohtsubo, K. Possibility for prevention of type 2 diabetes mellitus and dementia using three kinds of brown rice blends after high-pressure treatment. *Foods* **2022**, *11*, 818. [CrossRef] [PubMed]
28. Hu, J.F.; Zhao, X.H.; Jia, J.B.; Parpia, B.; Campbell, T.C. Dietary calcium and bone density among middle-aged and elderly women in China. *Am. J. Clin. Nutr.* **1993**, *58*, 219–227. [CrossRef]
29. Okadome, H.; Toyoshima, H.; Sudo, M.; Ando, I.; Numaguchi, K.; Ohtsubo, K. Palatability evaluation for *Japonica* rice grains based on multiple physical measurements of individual cooked rice grain. *J. Jpn. Soc. Food Sci. Technol.* **1998**, *45*, 398–407. (In Japanese) [CrossRef]
30. Odahara, M.; Sokooshi, H.; Takahashi, T.; Okadome, H.; Ohtsubo, K. The effect of sushi vinegar on texture of sushi rice before and after storage under low temperature. *Nippon. Shokuhin Kagaku Kogaku Kaishi* **2004**, *51*, 620–625. [CrossRef]
31. Shibuya, N.; Nakane, R.; Yasui, A.; Tanaka, K.; Iwasaki, T. Comparative studies on cell wall preparations from rice bran, germ, and endosperm. *Cereal Chem.* **1985**, *62*, 252–258.
32. Ohtsubo, K.; Nakamura, S. Evaluation of palatability of cooked rice. In *Adovances in International Rice Research*; Li, J., Ed.; INTECH: Zagred, Croatia, 2017; pp. 91–110.
33. Iwata, H.; Iwase, S.; Takahama, K.; Matuura, H.; Itani, T.; Aramaki, I. Relation between α-glucosidase activity and physical and chemical properties of rice. *Nippon. Shokuhin Kagaku Kogaku Kaishi* **2001**, *48*, 482–490. [CrossRef]
34. Nakamura, S.; Yamaguchi, H.; Benitani, Y.; Ohtsubo, K. Development of a novel formula for estimating the amylose content of starch using *Japonica* milled rice flours based on the iodine absorption curve. *Biosci. Biotechnol. Biochem.* **2020**, *84*, 2347–2359. [CrossRef]
35. Wenqian, S.; Qiaoling, Z.; Yue, Y.; Xianjin, Q.; Kun, X.; Sibin, Y. Identification of genomic regions and the isoamylase gene for reduced grain chalkiness in rice. *PLoS ONE* **2015**, *10*, e0122013.
36. Ikeda, A.; Ueguchi-Tanaka, M.; Sonoda, H.; Kitano, M.; Koshioka, Y.; Futsuhara, M.; Matsuoka, M.; Yamaguchi, I. Slender rice, a constitutive gibberellin response mutant, is caused by a null mutation of the SLR_1, an ortholog of the height-regulating gene GAI/RGA/D8. *Plant Cell* **2001**, *13*, 999–1010. [CrossRef] [PubMed]

37. Ueguchi-Tanaka, M.; Ashikari, M.; Nakajima, M.; Itoh, H.; Katoh, E.; Kobayashi, M.; Chow, T.Y.; Hsing, Y.C.; Kitano, H.; Yamaguchi, I.; et al. *GIBBERELLIN INSENSITIVE DWARF1* encodes a soluble receptor for gibberellin. *Nature* **2005**, *437*, 693–698. [CrossRef] [PubMed]
38. Fincher, G.B.; Stone, B.A. Some chemical and morphological changes induced by gibberellic acid in embryo-free wheat grain. Aust. *J. Plant Physiol.* **1974**, *1*, 297.
39. Glennie, C.W. Endosperm cell wall modification in sorghum grain during germination. *Cereal Chem.* **1984**, *61*, 285.
40. Gram, N.H.; Juliano, B.O. Studies on alkali-soluble rice bran hemicelluloses. *Carbohydr. Res.* **1970**, *12*, 273.
41. Ogawa, N.; Inagaki, A.; Yamanaka, N.; Shimomura, M. Effect of calcium and sodium in rice cooking water on the properties of cooked rice (Part 1). *Jpn. Soc. Home Econ.* **2006**, *57*, 669–675.
42. Awazuhara, M.; Nakagawa, A.; Yamaguch, J.; Fujiwara, T.; Hayashi, H.; Hatae, K.; Chino, M.; Shimada, A. Distribution and characterization of enzymes responsible for starch degradation in rice. (Oryza sativa Cv. Koshihikari). *J. Agric. Food Chem.* **2000**, *48*, 245–252. [CrossRef]
43. Onishi, M.; Shoji, I.; Ogawa, N.; Katoh, Y.; Nagaoka, S.; Shimomura, M. Histological study of rice grains cooked in water containing calcium ions. *Jpn. Soc. Home Econ.* **2002**, *53*, 1087–1096.
44. Champagne, E.T.; Wood, D.F.; Juliano, B.O.; Bechtel, D.B. The rice grain and its gross composition. In *Rice-Chemistry and Technology*, 3rd ed.; Champagne, E.T., Ed.; American Association Cereal Chemistry Int.: St. Paul, MN, USA, 2004; pp. 77–107.
45. Fujita, T. Calcium paradox: Consequences of calcium deficiency manifested by a wide variety of diseases. *J. Bone Miner. Metab.* **2000**, *18*, 234–236. [CrossRef] [PubMed]
46. Kanis, J.A.; Johanansson, H.; Oden, A.; Laet, C.D.; Johnell, O.; Eisman, J.A.; Closkey, E.M.; Mellstrom, D.; Pols, H.; Reeve, J.; et al. A meta-analysis of milk intake and fracture risk: Low utility for case finding. *Osteoporos. Int.* **2005**, *16*, 799–804. [CrossRef] [PubMed]

foods

MDPI

Article

Salinity Treatments Promote the Accumulations of Momilactones and Phenolic Compounds in Germinated Brown Rice

Mehedi Hasan [1], Nguyen Van Quan [1], La Hoang Anh [1,2], Tran Dang Khanh [3,4] and Tran Dang Xuan [1,2,*]

[1] Graduate School of Advanced Science and Engineering, Hiroshima University, 1-5-1 Kagamiyama, Higashi-Hiroshima 739-8529, Japan; mehedihasanjony110@gmail.com (M.H.); nvquan@hiroshima-u.ac.jp (N.V.Q.); hoanganh6920@gmail.com (L.H.A.)

[2] Center for the Planetary Health and Innovation Science (PHIS), The IDEC Institute, Hiroshima University, 1-5-1 Kagamiyama, Higashi-Hiroshima 739-8529, Japan

[3] Agricultural Genetics Institute, Pham Van Dong Street, Hanoi 122000, Vietnam; tdkhanh@vaas.vn or tdkhanh@vnua.edu.vn

[4] Center for Agricultural Innovation, Vietnam National University of Agriculture, Hanoi 131000, Vietnam

* Correspondence: tdxuan@hiroshima-u.ac.jp; Tel.: +81-82-424-6927

Abstract: This is the first investigation, conducted in a completely randomized design (CRD), to determine the effects of different salinity levels (75 and 150 mM) and germination periods (3, 4, and 5 days) on momilactone and phenolic accumulations in germinated brown rice (GBR) var. Koshihikari. Particularly, the identification of bioactive compounds was confirmed using electrospray ionization-mass spectrometry (ESI-MS) and nuclear magnetic resonance (NMR) spectroscopy (^{1}H and ^{13}C). Momilactone A (MA) and momilactone B (MB) amounts were determined by ultra-performance liquid chromatography–electrospray ionization-mass spectrometry (UPLC–ESI-MS), whereas other compounds were quantified by spectrophotometry and high-performance liquid chromatography (HPLC). Accordingly, GBR under B2 treatment (75 mM salinity for 4 days) showed the greatest total phenolic and flavonoid contents (14.50 mg gallic acid and 11.06 mg rutin equivalents, respectively, per g dry weight). GBR treated with B2 also accumulated the highest quantities of MA, MB, ρ-coumaric, ferulic, cinnamic, salicylic acids, and tricin (18.94, 41.00, 93.77, 139.03, 46.05, 596.26, and 107.63 µg/g DW, respectively), which were consistent with the strongest antiradical activities in DPPH and ABTS assays (IC_{50} = 1.58 and 1.78 mg/mL, respectively). These findings have implications for promoting the value of GBR consumption and rice-based products that benefit human health.

Keywords: momilactones; phenolics; antioxidants; bioactive compounds; germinated brown rice; salinity

Citation: Hasan, M.; Quan, N.V.; Anh, L.H.; Khanh, T.D.; Xuan, T.D. Salinity Treatments Promote the Accumulations of Momilactones and Phenolic Compounds in Germinated Brown Rice. *Foods* **2023**, *12*, 2501. https://doi.org/10.3390/foods12132501

Academic Editor: Costantino Fadda

Received: 28 May 2023
Revised: 13 June 2023
Accepted: 23 June 2023
Published: 27 June 2023

1. Introduction

Rice (*Oryza sativa* L.) provides about 20% of the world's dietary energy, which is comparatively higher than wheat (19%) and maize (5%) [1]. Rice contains various secondary metabolites, including phenolic acids, flavonoids, terpenoids, steroids, and alkaloids [2]. Nowadays, rice not only plays a vital role as an indispensable food source but has also been demonstrated to possess certain health benefits for human consumption. Notably, despite brown rice (BR) possessing a higher nutritional and bioactive composition in its bran and embryo, it is less popular than white rice (WR) [1,3]. BR includes around 2% of the total dietary fiber and serves as a vital source of γ-oryzanol, vitamin E, minerals, phenolic compounds, phytosterols, and phytic acid. Therefore, the utilization of brown rice as a nutritional and functional food has emerged as a recent trend. However, due to the compact structure of its outer bran layer, brown rice tends to have a firmer texture, making it more challenging to process and less digestible compared to white rice.

As an inevitable consequence, germinated brown rice (GBR) has been found to have effective alternative features while still maintaining its inherent nutritional value. The quality of GBR is enhanced through increased water absorption on the outer kernel, resulting in a softened texture. Additionally, enzymatic activities during seed germination modify the bioactive substances through interactions between proteins and carbohydrates in the grain endosperm [4,5]. Accordingly, GBR has been reported to have a proliferation of bio-functional constituents such as γ-aminobutyric acid (GABA) [1,6], vitamins, and amino acids [7], as well as a reduction of sugar [8], compared to non-GBR. On the other hand, previous studies indicated that subjecting germinated brown rice (GBR) to abiotic stresses and various germination conditions can lead to improved nutritional profiles and an elevated accumulation of bioactive compounds and antioxidant properties. Different soaking and germination periods revealed stimulatory effects on the growth of sprouts and increased contents of total phenolics, total flavonoids, and GABA in GBR [9–11]. Meanwhile, abiotic stresses such as salt and cold conditions may improve the contents of GABA, polyphenols, and antioxidant activity [6,12]. Therefore, the utilization of abiotic stresses and diverse germination conditions for GBR presents a promising strategy to promote the consumption value of brown rice. In a recent study, Choe et al. [12] reported that GBR accumulated a greater content of polyphenols and flavonoids, which was in line with the motivated antioxidant activity during the treatments with calcium chloride ($CaCl_2$). Nevertheless, none of the researchers focused on the effects of salinity treatment on the accumulation of bioactive compounds and antioxidant capacity of GBR.

In rice, although present in relatively small quantities, secondary metabolites such as phenolics, terpenes, and lactones play significant roles in both nutritional value and physiological processes, including metabolism, synthesis, and responses to environmental factors. For example, tricin, an important flavonoid, can be isolated from various rice plant organs (grains, leaves, brans, and husks). Tricin has been reported to have potentials for antioxidants [13], anti-skin aging [13], and anticancer [14–16] in numerous studies. Additionally, in rice, more prevalent are phenolic acids such as ρ-coumaric, ferulic, cinnamic, and salicylic acids, which have been recognized for their bioactive properties, including antioxidant, anti-inflammatory, and anticancer activities [17]. Notably, these phenolic and flavonoid compounds are accumulated with dominant contents in the bran layer [3], so they are generally found in greater amounts in BR compared to WR [18,19]. Moreover, the quantities of these phenolic compounds in GBR are up to twice higher than those in BR [3]. On the other hand, momilactones A (MA) and B (MB) have been acknowledged as valuable diterpene lactones derived from rice, which have recently exhibited antioxidant [13], anticancer (leukemia [20], lymphoma [21], and colon cancer [22]), anti-diabetes [23,24], anti-obesity [24], and anti-skin aging [13] properties. Though MA and MB have shown high potential for medicinal and pharmaceutical purposes, their contents in GBR have never been elucidated [23]. Alongside the mentioned valuable compounds, antioxidant property is also an integral criterion determining the value of rice consumption [25]. Notably, 2,2-diphenyl-1-1-picrylhydrazyl (DPPH) and 2,2′-azino-bis(3-ethylbenzothiazoline-6-sulfonic acid) (ABTS) assays are the most popular and convenient protocols to examine the antioxidant capacity of rice samples [26]. Of which, the ABTS assay relies on the production of a blue/green $ABTS^{\bullet+}$ radical cation, while the DPPH assay involves the reduction of the purple-colored $DPPH^{\bullet}$ radical to 1,1-diphenyl-2-picryl hydrazine [26]. In addition, ABTS and DPPH radicals also exhibit differences in molecular weight, stability, affinity, solubility, absorption ability, and pH requirements [26,27]. Generally, both DPPH and ABTS assays have their advantages and limitations, which should be applied in combination to obtain a more comprehensive understanding of the antioxidant properties of target products.

Among rice varieties, Koshihikari is a famous Japonica model rice cultivar that is widely distributed throughout Japan [28]. Koshihikari rice grains are small, plump, relatively lightweight, and have a rounded shape [28]. They exhibit a light brown or tan color and a smooth, glossy texture [28]. Meanwhile, the rice husks are typically thin, light, and have a pale brown color [28]. Owing to its favorable physical attributes, accompanied by its well-established aroma and taste, Koshihikari rice has garnered extensive popularity and preference among consumers [28]. However, the predominant cultivation of this particular variety contributes annually to a substantial production of rice by-products, including brans and husks, which have historically been subjected to inadequate utilization or wastage [13,24]. Conversely, scientific investigations have revealed the presence of valuable bioactive compounds within these by-products, exhibiting significant health-promoting effects [13,23,24]. Accordingly, the objective of this research endeavor was to procure Koshihikari rice husks for the purpose of isolating bioactive compounds, with a specific focus on phenolics and momilactones. Moreover, the study aimed to assess the variations in these compounds within Koshihikari BR seeds that were subjected to different salt treatments (0, 75, and 150 mM) and varying durations of germination (3, 4, and 5 days). Furthermore, an investigation was conducted to explore the correlation between the levels of bioactive compounds and the antioxidant capacity exhibited during exposure to salt conditions.

2. Materials and Methods

2.1. Materials

Rice (*Oryza sativa* var. Koshihikari) husks were collected from rice mills allocated near Hiroshima University, Higashi-Hiroshima Campus, Japan, in September 2019. In the specification, mature and healthy rice grains were selected for milling to ensure the quality of the husks. After that, the obtained husks were thoroughly cleaned with water to remove dust and impurities. The husk samples were then dried and preserved (voucher specimen: KOS-MOMI 19HJ) at the laboratory of Plant Physiology and Biochemistry, Graduate School of Advanced Science and Engineering, Hiroshima University, Japan. Brown rice of the Koshihikari was purchased from a Japan Agriculture (JA) shop in Hiroshima, Japan, to prepare germinated brown rice (GBR).

For extraction and isolation processes, methanol, hexane, and ethyl acetate were purchased from Junsei Chemical Co., Ltd. (Tokyo, Japan), while silica gel was bought from Sigma-Aldrich (St. Louis, MO, USA). Isolated compounds were dissolved in deuterated dimethyl sulfoxide (DMSO-d_6) and deuterated chloroform ($CDCl_3$) procured from Sigma-Aldrich (St. Louis, MO, USA). Standards comprising ferulic acid, cinnamic acid, and salicylic acid and chemicals including sodium acetate (CH_3COONa), sodium carbonate (Na_2CO_3), sodium hypochlorite (NaClO), aluminum chloride ($AlCl_3$), Folin–Ciocalteu's reagent, 2,2-diphenyl-1-picrylhydrazyl (DPPH), potassium persulfate ($K_2S_2O_8$), and 2,2′-azinobis(3-ethylbenzothiazoline-6-sulfonic acid) (ABTS) were acquired from Kanto Chemical Co., Inc. (Tokyo, Japan). Formic acid, trifluoroacetic acid, acetonitrile, methanol plus, and distilled water used for HPLC and UPLC analyses were obtained from Sigma-Aldrich (St. Louis, MO, USA), EMD Millipore Corporation (Billerica, MA, USA), Fisher Chemical (Hampton, VA, USA), Kanto Chemical Co., Inc. (Tokyo, Japan), and Nacalai Tesque (Kyoto, Japan), respectively.

2.2. Isolation of Tricin, ρ-Coumaric Acid, and Momilactones A (MA) and B (MB)

The isolation process for tricin, ρ-coumaric acid, and momilactones A (MA) and B (MB) followed the methods described in the previous study [13]. Briefly, 30 kg of rice husks were dried in an oven at 50 °C for six days and then extracted with 100% MeOH for two weeks at room temperature. The MeOH crude extract was then mixed with an appropriate amount of distilled water and partitioned sequentially with hexane and EtOAc. Next, the obtained EtOAc extract was subjected to column chromatography using silica gel as the stationary phase and a hexane:EtOAc (v/v) mixture as the mobile phase. MA and MB were isolated from the eluate of hexane:EtOAc (8:2, v/v), while tricin and ρ-coumaric acid were purified from the eluate of hexane:EtOAc (7:3, v/v).

2.3. Confirmation of Isolated Tricin, ρ-Coumaric Acid, and Momilactones A (MA) and B (MB) by ^{1}H- and ^{13}C-Nuclear Magnetic Resonance (NMR) and Electrospray Ionization-Mass Spectrometry (ESI-MS)

The identification of isolated tricin, ρ-coumaric acid, MA, and MB was confirmed by ^{1}H- and ^{13}C-nuclear magnetic resonance (NMR) spectra. Of which, ^{1}H- and ^{13}C NMR spectra of ρ-coumaric acid (in DMSO-d_6) were received on an NMR spectrometer (Bruker Ascend 400, BRUKER BioSpin, Fällanden, Switzerland) at 400 and 101 MHz, respectively. Meanwhile, ^{1}H- and ^{13}C NMR spectra of tricin (in DMSO-d_6) and MA and MB (in $CDCl_3$) were acquired on an NMR spectrometer (JNM-ECA600, JEOL Ltd., Tokyo, Japan) at 600 and 151 MHz, respectively. Coupling constants (J) and chemical shifts (δ) were indicated in Hz and parts per million (ppm), respectively. The shorthand notations s, d, t, q, dd, and dt represent the resonance multiplicities singlet, doublet, triplet, quartet, doublet of doublets, and doublet of triplets, respectively.

Additionally, tricin, ρ-coumaric acid, MA, and MB were confirmed using electrospray ionization-mass spectrometry (ESI-MS) (LTQ Orbitrap XL, Thermo Fisher Scientific, Waltham, MA, USA). The compounds (10 μg/mL) were dissolved in a MeOH:ACN mixture (8:2, v/v) and injected with a volume of 3 μL into the ESI system (positive ion mode) using an auto-sampler. The flow rate was 0.2 mL/min. The ESI conditions were set up as follows: ion source and capillary voltages were 4.5 kV and 50 V, respectively. Tube lens offset was 80 V. Capillary temperature was 330 °C. Gas carrier was nitrogen, and the sheath and aux flow rates were 50 arb and 10 arb, respectively. The mass spectra were recorded at 60,000 resolution with a scan range of 100–2000 m/z. To identify the MS/MS spectra, the PubChem online database (National Center for Biotechnology Information, U.S. National Library of Medicine, Bethesda, MD, USA) and literature were used as references.

2.4. Preparation for Germination

The germination process was generated following the method described by Cáceres et al. [29], with several modifications. Nine treatments were applied during the germination stage. The experimental conditions were as follows: soaking time of 36 h, temperature of 30 °C, and different salt (NaCl) concentrations, including 0, 75, and 150 mM. All treatments are presented in Table 1. Germination was conducted for 3, 4, and 5 days in darkness for all treatments. First, 100 g of brown rice was individually measured for nine plastic pots. The rice was soaked in 0.1% NaOCl at a ratio of 1:2 (w/v) for 30 min to remove or eliminate surface bacteria and fungi without damaging the internal organs [29]. Furthermore, it was also washed five times with clean tap water and dried for 5 min to remove residual NaClO. Then, 75 mM and 150 mM aqueous solutions of NaCl were prepared with distilled water for different treatments. All rice samples were soaked with salinity solution (grain:solution ratio, 1:2 w/v) and kept in an incubator for various periods at 30 °C (Table 1). Following the soaking period, the seeds were washed with distilled water to remove salinity. All the trays containing brown rice seeds were placed in an incubator at 30 °C for 3, 4, and 5 days in the dark for germination (Table 1). Relative humidity was around 65% in a closed system. The seeds were washed every four hours with distilled water to avoid bacterial and fungal invasions.

2.5. Extracted Phytochemicals from GBR

After germination, GBR was washed twice with distilled water and drained for 5 min. The samples were then dried in an oven for 7 days at 40 °C. For extraction, 50 g of GBR powder were saturated in 80% methanol for one week with two replications at room temperature. The extractions were then filtered after centrifugation (10,000 rpm) for 10 min at 4 °C. Subsequently, all methanolic extracts were evaporated at 50 °C to obtain methanol crude extract. Finally, the crude extracts were dissolved in methanol to achieve stock solutions of 20 mg/mL for further experiments.

Table 1. Description of different treatments.

Treatments Code	NaCl Concentration (mM)	Germination Time (Day)
A1	0	
A2	75	3
A3	150	
B1	0	
B2	75	4
B3	150	
C1	0	
C2	75	5
C3	150	

2.6. Total Phenolic Content (TPC) and Total Flavonoid Content (TFC) in GBR Extracts

Total phenolic content (TPC) of the GBR extracts was quantified based on the Folin–Ciocalteu method described by Mohammadabadi et al. [30] with several modifications. Briefly, a mixture of GBR sample, 10% Folin–Ciocalteu's reagent, and 7.5% Na_2CO_3 with volumes of 20, 100, and 80 μL, respectively, was generated and incubated for 30 min at 25 °C in darkness. The results were scanned at 765 nm. Total flavonoid content (TFC) was quantified following the aluminum chloride colorimetric method described in the research of Bueno-Costa et al. [31]. Concretely, a total volume of 100 μL of mixture (1:1, *v*/*v*) comprising GBR sample and 2% $AlCl_3$ was incubated for 15 min at 25 °C in darkness. The absorbance was measured at 430 nm. The calibration curves of TPC (0.0052x + 0.0645, r^2 = 0.9969) and TFC (0.009x + 0.0644; r^2 = 0.9998) established by applying the standards of gallic acid and rutin with concentrations ranging from 6.25 to 100 μg/mL were employed for the estimation of TPC and TFC in GBR samples. Of which, TPC and TFC were indicated in milligrams of gallic acid equivalent (GAE) and rutin equivalent (RE), respectively, per one gram of sample dry weight (DW).

2.7. Antioxidant Activities of GBR

Radical scavenging activities of GBR extracts were determined via 2,2-diphenyl-1-1-picrylhydrazyl (DPPH) and 2,2′-azino-bis(3-ethylbenzthiazoline-6-sulfonic acid) (ABTS) assays based on the methods presented by Anh et al. [32]. For the DPPH assay, a mixture of GBR sample, DPPH working solution (0.5 mM), and acetate buffer (0.1 mM, pH 5.5) with volumes of 80, 40, and 80 μL, respectively, was incubated for 20 min at 25 °C in darkness. In the ABTS assay, 200 μL of a combination (1:9, *v*/*v*) of GBR sample and ABTS working solution was incubated for 30 min at 25 °C in darkness. The radical scavenging activities (%) were determined as the reduced absorbance at 517 and 734 for DPPH and ABTS assays, respectively, compared to the control (MeOH).

$$\text{Radical scavenging activity (\%)} = (A_c - (A_s - A_b)/A_c) \times 100 \quad (1)$$

where A_c is the absorbance of the control (MeOH), A_s is the absorbance of the sample, and A_b is the absorbance of the blank (without radical solution).

2.8. Identification and Quantification of Momilactones A (MA) and B (MB) in GBR by Ultra-Performance Liquid Chromatography–Electrospray Ionization-Mass Spectrometry (UPLC–ESI-MS)

MA and MB in GBR samples were identified and quantified by ultra-performance liquid chromatography–electrospray ionization-mass spectrometry (UPLC–ESI-MS). In particular, the UPLC–ESI-MS system consisted of a mass spectrometer (LTQ Orbitrap XL, Thermo Fisher Scientific, Waltham, MA, USA) and an electrospray ionization (ESI) source. A volume of 3.0 μL of GBR sample (in MeOH) was injected by an autosampler (Vanquish autosampler, Thermo Fisher Scientific, Waltham, MA, USA) into a column (1.7 μm, 50 × 2.1 mm i.d.) (Acquity UPLC® BEH C18, Waters Cooperation, Milford, MA,

USA) at 25 °C. A mobile phase gradient was applied, of which solvents A and B were 0.1% trifluoroacetic acid in water and 0.1% trifluoroacetic acid in acetonitrile, respectively. The gradient program was established following the same procedure published by Anh et al. [33]. MS analysis was conducted with a positive Fourier transform mass spectrometer (FTMS) mode with 60,000 resolution and 100–1000 m/z of scan range. By using various MA and MB standard concentrations (0.5, 1, 5, and 10 µg/mL), the calibration curves for MA and MB were created. Using standard curves, MA and MB quantities were determined by applying the MA and MB peak areas detected in each sample.

2.9. Identification and Quantification of Tricin, ρ-Coumaric Acid, Ferulic Acid, Cinnamic Acid, and Salicylic Acid by High-Performance Liquid Chromatography (HPLC)

The presence and quantification of tricin, ρ-coumaric acid, ferulic acid, cinnamic acid, and salicylic acid by high-performance liquid chromatography (HPLC) analyses were compared with the standards attained by the method presented by Anh et al. [33]. In brief, the HPLC system consisted of a pump (PU-4180 RHPLC, Jasco, Tokyo, Japan), a controller (LC-Net II/ADC, Jasco, Japan), and a detector (UV-4075 UV/VIS, Jasco, Tokyo, Japan). A column (130 Å, 5 µm, 2.1 × 100 mm) (XBridge BEH Shield RP18, Waters Cooperation, Milford, MA, USA) was used as a stationary phase. Solvent A (0.1% formic acid in water) and solvent B (acetonitrile) were applied as mobile phases, which were fixed in the same gradient program reported by Anh et al. [33]. Each operation was continued for 35 min at room temperature. Every sample was identified by a corresponding peak scanned at 350 nm for tricin and 280 nm for ρ-coumaric acid, ferulic acid, cinnamic acid, and salicylic acid. The peak area was used to quantify these compounds.

2.10. Statistical Analysis

All experiments were conducted in a completely randomized design (CRD) with three replications. The analyses were performed using Minitab software (Minitab 16.2.3, Minitab Inc., State College, PA, USA) through one-way and two-way ANOVA. The outcomes were presented as means ± standard deviations (SD) (n = 3). The same software was used for Pearson's correlation coefficients among different parameters.

3. Results and Discussion

3.1. Confirmation of Isolated Tricin, ρ-Coumaric Acid, and Momilactones A (MA) and B (MB)

Isolated tricin, ρ-coumaric acid, and momilactones A (MA) and B (MB) were identified and confirmed using electrospray ionization-mass spectrometry (ESI-MS) and ^{1}H- and ^{13}C-nuclear magnetic resonance (NMR) methods. The mass spectra of these compounds are shown in Figure 1.

Tricin: ESI-MS (m/z): 331.08139 $[M + H]^+$ ($C_{17}H_{15}O_7$) (Figure 1). The mass spectrum of tricin was compared with that in published data by Quan et al. [13]. The ^{1}H NMR (600 MHz, DMSO-d_6) δ 12.90 (1H, s, 5-OH), 10.84 (d, J = 124.2, 7-OH), 9.26 (s, 1H, 4-OH), 7.26 (s, H-60 and H-20), 6.92 (s, H-3), 6.49 (d, J = 2.1, H-8), 6.13 (d, J = 2.1, H-6), 3.82 (s, 2OCH$_3$), 3.26 (s, 220H), 2.43 (dt, J = 3.6 and 1.8, 171H) (Supplementary Figure S1). The ^{13}C NMR (151 MHz, DMSO-d6) δ 182.35 (C-4), 164.66 (C-2), 164.19 (C-7), 161.94 (C-5), 157.87 (C-9), 148.72 (C-30 and C-50), 140.38 (C-40), 56.90 (2OCH$_3$), 40.05 (dp, J = 42.0, 21.0 Hz) (Supplementary Figure S1). The NMR results of tricin are matched with reference data from a previous study [34].

ρ-Coumaric acid: ESI-MS (m/z): 165.05424 $[M + H]^+$ ($C_9H_9O_3$) (Figure 1). The ^{1}H NMR (400 MHz, DMSO-d_6) δ 7.51 (dd, J = 16.0 and 8.6, H-7, H-2, H-6), 6.79 (d, J = 8.6, H-3, H-5), and 6.29 (d, J = 16.0, H-8) (Supplementary Figure S1). The ^{13}C NMR (101 MHz, DMSO-d_6) δ 168.42 (COOH), 160.05 (C-4), 144.64 (C-7), 130.53 (C-2, C-6), 125.75 (C-1), 116.22 (C-3, C-5), and 115.80 (C-8) (Supplementary Figure S1). The NMR and ESI-MS results are entirely similar to those in the literature [35].

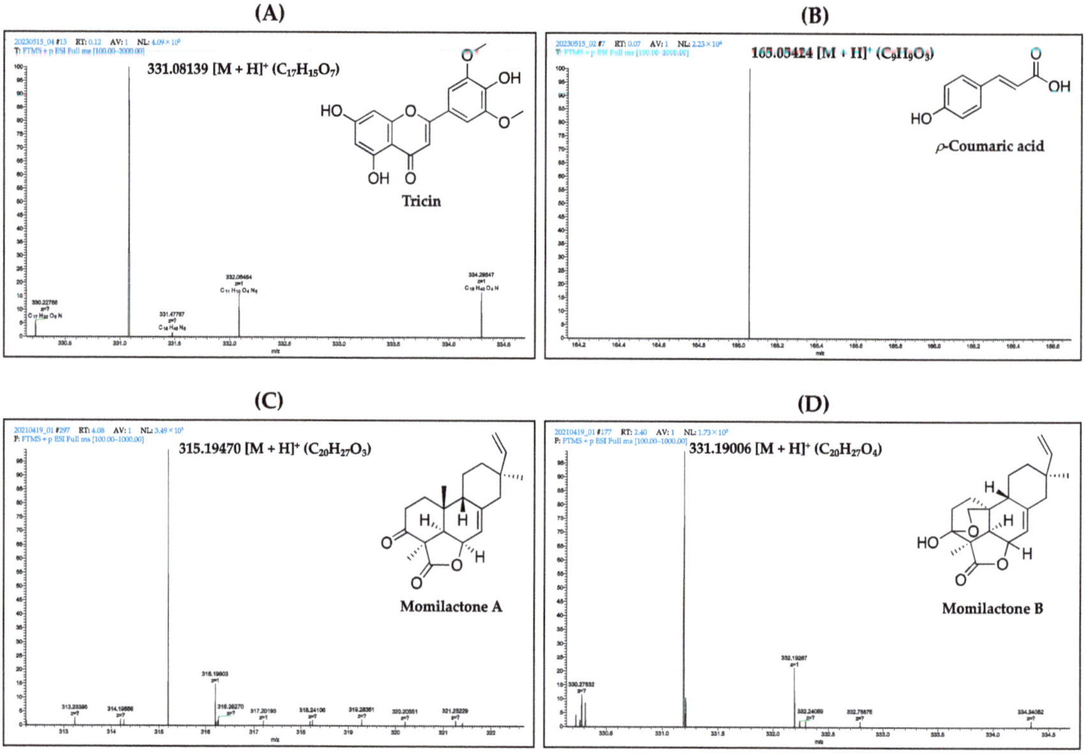

Figure 1. Mass spectra of isolated (**A**) tricin, (**B**) *p*-coumaric acid, (**C**) momilactone A (MA), and (**D**) momilactone B (MB) in this study by ESI-MS.

MA: ESI-MS (m/z): 315.19470 $[M + H]^+$ ($C_{20}H_{27}O_3$) (Figure 1). The mass spectrum of MA was confirmed based on reference data reported by Quan et al. [24]. The ^{1}H-NMR (600 MHz, $CDCl_3$) δ 5.87 (s, 1H), 5.86 (s, 1H), 5.85 (s, 1H), 5.83 (d, J = 17.0, 11.0, H-15), 5.71 (d, *J* = 5.0, H-7), 5.00 (d, *J* = 1.0, 1H), 4.97 (d, *J* = 1.0, 1H), 4.95 (d, *J* = 1.0, 1H), 4.93 (d, *J* = 1.0, 1H), 4.84 (t, *J* = 5.1, H-6), 4.10–4.06 (m, 1H), 3.97 (s, 1H), 3.95 (s, 1H), 3.30 (s, 1H), 3.28 (s, 1H), 3.21 (s, 1H), 3.18 (s, 1H), 2.67–2.56 (m, H-2), 2.32 (d, *J* = 5.1, H-5), 2.21 (d, *J* = 12.0, 2H-14), 2.08–2.04 (m, 1H), 1.93–1.86 (m, 1H), 1.79 (dd, *J* = 12.9, 3.9, 1H), 1.77–1.72 (m, H-9, H-11α), 1.63–1.55 (m, H2-1β, H2-12), 1.53 (s, H-18), 1.00 (s, H-20), 0.89 (s, H-17) (Supplementary Figure S1). The ^{13}C NMR (151 MHz, $CDCl_3$) δ 205.24 (C-3), 174.37 (C-19), 149.03 (C-8), 148.10 (C-15), 114.12 (C-7), 110.25 (C-16), 73.23 (C-6), 53.64 (C-4), 50.26 (C-9), 47.60 (C-14), 46.54 (C-5), 40.20 (C-13), 37.31 (C-12), 34.95 (C-1), 32.53 (C-10), 31.29 (C-2), 24.06 (C-11), 22.03 (C-20), 21.87 (C-17), 21.54 (C-18) (Supplementary Figure S1). The NMR spectrum is matched with published data in the report of Quan et al. [23].

MB: ESI-MS (m/z): 331.19006 $[M + H]^+$ ($C_{20}H_{27}O_4$) (Figure 1). The obtained results were confirmed by comparing them with those in the preceding report [24]. The ^{1}H NMR (600 MHz, $CDCl_3$) δ 5.82 (dd, *J* = 17.5, 10.7, H-15), 5.69 (d, *J* = 4.8, H-7), 4.98–4.92 (m, 1H), 4.13 (s, 1H), 4.08 (dd, J = 9.2, 3.4, 1H), 3.58 (dd, J = 9.2, 2.1, 1H), 2.20 (dd, J = 6.8, 2.0, H-5), 2.14–2.07 (m, H-2, H-14), 2.04–1.98 (m, 1H), 1.75–1.64 (m, H-9, H-11α), 1.59–1.57 (m, 1H), 1.57–1.51 (m, H-1β, H-12), 1.50 (d, *J* = 4.2, 1H), 1.48–1.43 (m, 1H), 1.41 (s, H-18), 1.26–1.19 (m, 1H), 0.87 (s, H-17) (Supplementary Figure S1). The ^{13}C NMR (151 MHz, $CDCl_3$) δ 180.53 (C-19), 148.91 (C-15), 146.76 (C-8), 114.09 (C-7), 110.30 (C-16), 96.67 (C-3), 73.81 (C-6), 72.79 (C-20), 50.41 (C-4), 47.49 (C-14), 44.76 (C-9), 43.05 (C-5), 40.06 (C-13), 37.29 (C-12), 30.81 (C-10), 28.89 (C-1), 26.51 (C-2), 24.86 (C-11), 21.94 (C-17), and 19.06 (C-18) (Supplementary Figure S1). NMR data were compared to published results by Quan et al. [23].

3.2. Phenolic and Momilactone Contents in GBR

3.2.1. Total Phenolic (TPC) and Flavonoid (TFC) Contents

In essence, the chemical composition of natural products determines their biological activity [32]. Additionally, methods for the identification and quantification of natural compounds have been rapidly advancing [36]. Therefore, exploring the phytochemical profiles of targeted products is necessary for studies concerning their potential bioactivity. In our research, the initial assessment focused on the compound groups of phenolics and flavonoids, which may contribute to the pharmaceutical and medicinal properties (e.g., antioxidant, antibacterial, anticancer, cardioprotective, immune system-promoting and anti-inflammatory, and skin-protective effects) of the targeted products [37]. The total phenolic (TPC) and flavonoid (TFC) contents of GBR are shown in Figure 2. There were significant differences in TPCs and TFCs among different treatments. Particularly, the highest TPCs were found in treatments B2 and C2 (14.50 and 14.36 mg GAE/g DW, respectively). Whereas the lowest TPC was observed in A2 (6.17 mg GAE/g DW). On the other hand, the highest TFC (11.06 mg RE/g DW) was detected in B2, while the lowest TFC (2.54 mg RE/g DW) was found in A2. Previous studies indicated that TPC and TFC increased in rice seedlings under salinity effects [38–40], which might be due to the upregulation of genes encoding the major biosynthetic enzymes (e.g., phenylalanine ammonia lyase and chalcone synthase) in plant responses to biotic stresses [33,41,42]. In this study, TPC and TFC were remarkably stimulated in treatments with 75 mM salinity. However, they were remarkably decreased when increasing to an extreme salt level of 150 mM. Our findings suggest that a moderate salinity level of 75 mM and 4-day germination are the most appropriate conditions for proliferating TPC and TFC in GBR.

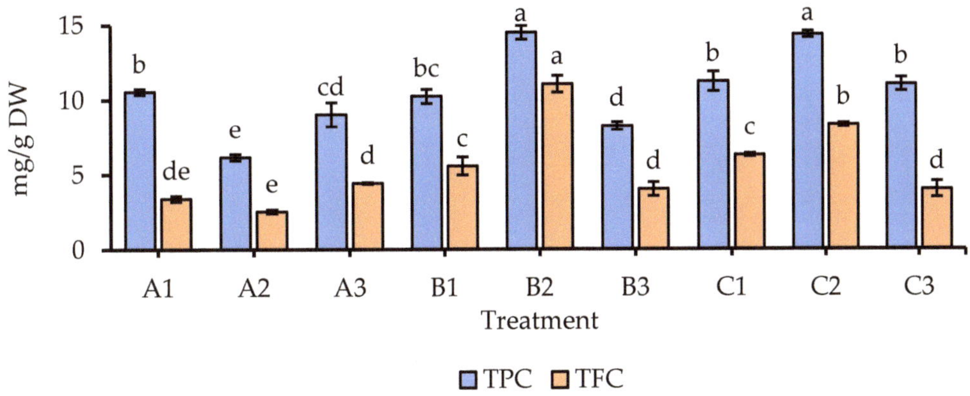

Figure 2. Total phenolic (TPC) and total flavonoid (TFC) contents of GBR extracts. TPC and TFC outcomes are expressed as mg gallic acid equivalent per g dry weight (mg GAE/g DW) and mg rutin equivalent per g dry weight (mg RE/g DW), respectively. Whiskers enclosed in a column express the standard deviation (SD). Different letters attached to a column (same color) indicate significant differences at $p < 0.05$. A1: 0 mM NaCl and 3-day germination; A2: 75 mM NaCl and 3-day germination; A3: 150 mM NaCl and 3-day germination; B1: 0 mM NaCl and 4-day germination; B2: 75 mM NaCl and 4-day germination; B3: 150 mM NaCl and 4-day germination; C1: 0 mM NaCl and 5-day germination; C2: 75 mM NaCl and 5-day germination; C3: 150 mM NaCl and 5-day germination.

Considering TPC analysis, the Folin–Ciocalteu method is widely used because it provides a quick and cost-effective estimation, allowing for comparative analyses between samples [43]. However, the method reveals limitations in specifically quantifying TPC due to its reactivity with other components such as amino acids, peptides, and reducing sugars [43]. For a more accurate determination of TPC, an improved method such as solid-phase extraction using the Sep-Pak C18 column cartridges is required to purify the extract and eliminate unwanted components [30]. Though the Folin–Ciocalteu method serves as a valuable screening tool to assess the relative phenolic content in various samples, it may not provide precise quantification of individual phenolic compounds [43]. Therefore, in further investigation, we employed HPLC analysis to determine the profiles of specific phenolics found in GBR in this study.

3.2.2. Contents of Tricin, ρ-Coumaric, Ferulic, Cinnamic, and Salicylic Acids in GBR

Tricin, ρ-coumaric acid, ferulic acid, cinnamic acid, and salicylic acid play multifunctional roles benefiting human health, including antioxidants, anticancer, and anti-chronic diseases [44]. In the present study, these phenolic compounds were identified (Supplementary Figure S2) and quantified (Table 2). Accordingly, tricin, ρ-coumaric acid, ferulic acid, salicylic acid, and cinnamic acid were found in increasing quantities in B2 treatment (75 mM salinity and 4-day germination), which accounted for 107.63, 93.77, 139.03, 46.05, and 596.26 μg/g DW, respectively. The elevated contents of ρ-coumaric, salicylic, and ferulic acids are in line with those in rice seedlings subjected to salinity (100 mM) in a previous study [38]. Meanwhile, cinnamic acid and tricin amounts decreased [38], which does not align with our findings. This might be due to the differences in genetic diversity among tested rice varieties. In fact, different rice varieties (tolerant and susceptible cultivars) exhibit dissimilar mechanisms in phenolic accumulation to cope with stress conditions [33,38–40]. On the other hand, the present study revealed a notable decrease in the quantities of these phenolics in GBR when exposed to extreme levels of salinity (150 mM). This finding implies that the most optimal conditions for the proliferation of bioactive phenolics in GBR are a salinity of 75 mM maintained for 4 days during the germination process.

Table 2. Quantities of momilactones, tricin, ρ-coumaric acid, ferulic acid, cinnamic acid, and salicylic acid (μg/g DW) in GBR.

Treatment Code	MA	MB	Tricin	ρ-Coumaric Acid	Ferulic Acid	Cinnamic Acid	Salicylic Acid
A1	7.33 ± 0.39^{c}	18.68 ± 0.89^{c}	44.43 ± 8.92^{cd}	46.43 ± 3.37^{de}	64.92 ± 3.34^{bc}	28.16 ± 0.64^{b}	290.27 ± 68.05^{bc}
A2	2.92 ± 0.06^{ef}	9.30 ± 0.09^{e}	41.12 ± 5.57^{cde}	39.88 ± 0.63^{ef}	53.23 ± 1.88^{c}	21.86 ± 1.09^{c}	349.04 ± 83.3^{b}
A3	1.93 ± 0.09^{f}	7.27 ± 0.12^{f}	29.55 ± 2.04^{ef}	36.29 ± 1.66^{f}	49.05 ± 2.72^{c}	22.58 ± 1.10^{c}	194.16 ± 14.77^{cd}
B1	5.68 ± 1.38^{cd}	18.88 ± 0.57^{c}	33.93 ± 1.38^{def}	61.77 ± 1.96^{b}	57.22 ± 7.63^{bc}	11.82 ± 0.82^{d}	88.49 ± 22.89^{de}
B2	18.94 ± 0.47^{a}	41.00 ± 0.51^{a}	107.63 ± 6.75^{a}	93.77 ± 4.35^{a}	139.03 ± 5.16^{a}	46.05 ± 0.88^{a}	596.26 ± 1.14^{a}
B3	4.19 ± 0.03^{de}	12.70 ± 0.75^{d}	25.51 ± 0.94^{f}	44.99 ± 1.44^{def}	61.98 ± 2.36^{bc}	22.01 ± 0.52^{c}	52.86 ± 3.2^{e}
C1	4.90 ± 0.17^{de}	11.97 ± 0.05^{d}	49.54 ± 0.34^{c}	44.48 ± 2.90^{def}	52.55 ± 4.77^{c}	-	-
C2	10.17 ± 0.49^{b}	24.79 ± 0.55^{b}	65.13 ± 3.06^{b}	59.95 ± 5.51^{bc}	76.55 ± 9.07^{bc}	-	-
C3	1.70 ± 0.01^{f}	7.20 ± 0.29^{f}	31.67 ± 0.59^{ef}	51.52 ± 3.19^{cd}	60.81 ± 5.97^{bc}	-	-
ANOVA							
Period	***	***	***	***	***	***	***
Treatment	***	***	***	***	***	***	***
Period × Treatment	***	***	***	***	***	***	***

Data are expressed as means ± SD (standard deviation). Different superscript letters (^{a,b,c,d,e,f}) in a column indicate significant differences at $p < 0.05$; *** denotes a significant difference at $p < 0.001$. MA: momilactone A; MB: momilactone B; DW: dry weight; -: not detected; A1: 0 mM NaCl and 3-day germination; A2: 75 mM NaCl and 3-day germination; A3: 150 mM NaCl and 3-day germination; B1: 0 mM NaCl and 4-day germination; B2: 75 mM NaCl and 4-day germination; B3: 150 mM NaCl and 4-day germination; C1: 0 mM NaCl and 5-day germination; C2: 75 mM NaCl and 5-day germination; C3: 150 mM NaCl and 5-day germination.

3.2.3. Contents of Momilactones A (MA) and B (MB) in GBR

Momilactones A (MA) and B (MB) have been known as valuable bioactive compounds from rice with various health-related benefits, including antioxidant [13], anticancer (leukemia [20], lymphoma [21], and colon cancer [22]), anti-diabetes [23,24], anti-obesity [24], and anti-skin aging activities [13,45]. Recently, based on an improved technique for sample preparation and quantification, momilactones can also be detected with high detection sensitivity in different rice plant parts (e.g., leaves, roots, husks, etc.) [23,45,46]. However, their exploitation from rice sources has still faced many limitations due to the lack of commercial availability and difficulties in the isolation process [23,45]. A few reports about momilactone isolation and purification have been published, and in those studies, a minor amount of momilactones can be isolated from rice sources [45–48]. Additionally, the published artificial syntheses of MA were also challenging since they included multiple complicated steps, required high costs, resulted in low yields (40–50%), and were environmentally unfriendly [49]. On the other hand, the synthetic methods of MB have never been reported. In fact, due to their limited availability, research on momilactones has been relatively scarce and underdeveloped during the last half-century. Recently, only two studies focused on optimizing the extraction conditions of MA and MB from rice husks [48,50], while no research has been conducted to enhance the momilactone contents of rice grains to increase their consumption value. Therefore, this study investigated, for the first time, the effects of different conditions (salinity and germination periods) on the accumulation of MA and MB in GBR. In Supplementary Figure S3, the presence of MA and MB in GBR is confirmed by comparing their retention times and mass spectra with those of the standards. Numerous studies indicated that the antioxidant, anti-diabetic, and anticancer potentials of MB were greater than those of MA [23,51,52]. However, Chung et al. [53] announced that the endogenous quantity of MA was generally higher than that of MB in different 99 rice varieties. In contrast to previous reports, our findings demonstrate that Koshihikari GBR exhibited a greater amount of MB than MA in all treatments (Table 2). Significantly, the highest accumulation of MA (18.94 μg/g DW) and MB (41.00 μg/g DW) was recorded in the B2 treatment. The amounts of MA and MB in GBR under B2 were significantly higher than those of preceding studies, in which MA and MB quantities ranged from 2.07 to 16.44 μg/g DW and 1.06 to 12.73 μg/g DW, respectively [13,23,24]. The increased accumulation of MA and MB in GBR in B2 treatment may be caused by the elevated expression of related genes to momilactone biosynthesis, including OsCPS4, OsKSL4, CYP99A3, OsMAS, and OsMAS2 [33]. However, at a strong salinity level of 150 mM, both MA and MB accumulations in GBR were significantly reduced compared to non-salinity treatment. The results suggest that a moderate salt concentration of 75 mM and 4-day germination are the most ideal conditions to stimulate MA and MB contents, which may contribute to the pharmaceutical and medicinal values of GBR.

3.3. *Antioxidant Activity of GBR by the DPPH and ABTS Radical Scavenging Assays*

In humans, oxidative stress is closely associated with inflammation, which is considered a key physiological process in the development of various chronic diseases such as diabetes, aging, and cancer [54,55]. Specifically, inflammation can worsen oxidative stress and vice versa [54,55]. Numerous experimental findings have demonstrated the presence and impact of oxidative stress in several chronic diseases, which result in higher rates of morbidity and mortality [54,55]. Based on that, assessing the antioxidant properties of the samples is an essential step in our study. According to Figure 3, changes in salinity levels and germination periods might promote the antioxidation activity of GBR. Among all treatments, B2 showed the highest antiradical activities against DPPH and ABTS (IC_{50} = 1.58 and 1.78 mg/mL, respectively) compared to others. Several studies have consistently demonstrated that different germination conditions and salt stress enhanced the antioxidant capacity of GBR [6,56,57]. This result may be due to the upregulation of relevant enzymes such as superoxide dismutase, catalase, glutathione peroxidase, and ascorbate peroxidase in rice under abiotic stresses [58]. Therefore, salinity effects could potentially contribute to enhancing the antioxidant capacities of GBR in this research. However, the salt concentration should be carefully considered since extreme levels may lead to reduced antioxidant capacities of GBR [59]. In agreement with Falcinelli et al. [59], we indicate that the antioxidant activity of GBR significantly decreased at a high NaCl concentration of 150 mM. Meanwhile, a moderate salinity level of 75 mM may be the most effective condition to elevate the antioxidant activity of GBR.

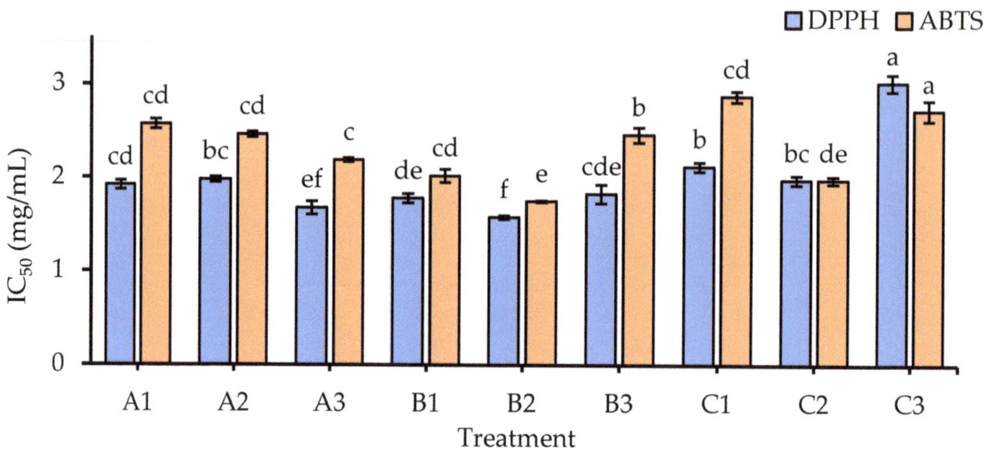

Figure 3. Antioxidant activities of GBR extracts. IC_{50} is the required concentration (mg/mL) for scavenging 50% of radicals. Whiskers enclosed in a column express the standard deviation (SD). Different letters attached to a column (same color) indicate significant differences at $p < 0.05$. DPPH: 2,2-diphenyl-1-picrylhydrazyl assay; ABTS: 2,2′-azino-bis(3-ethylbenzothiazoline-6-sulfonic acid assay; A1: 0 mM NaCl and 3-day germination; A2: 75 mM NaCl and 3-day germination; A3: 150 mM NaCl and 3-day germination; B1: 0 mM NaCl and 4-day germination; B2: 75 mM NaCl and 4-day germination; B3: 150 mM NaCl and 4-day germination; C1: 0 mM NaCl and 5-day germination; C2: 75 mM NaCl and 5-day germination; C3: 150 mM NaCl and 5-day germination.

3.4. *Correlation between Antioxidant Activities and Phytochemicals of GBR*

Pearson's correlation coefficients between antioxidant activities and phytochemicals are displayed in Table 3. Accordingly, a concomitant accumulation of MA, MB, tricin, *ρ*-coumaric, ferulic, cinnamic, and salicylic acids is recorded, which was strongly correlated with the antioxidant activities of GBR. Previous studies have extensively reported on the roles of tricin, *ρ*-coumaric, ferulic, cinnamic, and salicylic acids in antioxidant abilities [60]. Particularly, these compounds can detoxify free radicals by donating hydrogen ions, thus strengthening antioxidant capacities [60]. Conversely, while the antioxidant activities of MA and MB have been mentioned in several publications [13,52], their underlying mechanisms remain unclear. In another consideration, Anh et al. [33] hypothesized that MA and MB might not directly contribute to the antioxidant responses of rice against adverse stresses, but they might play a role in signaling the production of antioxidant compounds such as phenolics [33]. Based on that, in this study, the upregulated contents of MA and MB in GBR under moderate salinity (75 mM NaCl) and 4-day germination might lead to the proliferation of tricin, *ρ*-coumaric, ferulic, cinnamic, and salicylic acids, thereby increasing the antioxidant capacity of GBR, which requires further validation.

Additionally, due to the well-established correlation between oxidative stress and chronic diseases [54,55], extensive research has been undertaken to explore the use of antioxidant substances for the treatment of such disorders [61]. However, the failures have been documented through clinical evaluations, which might be attributed to the single use of antioxidant agents to target specific diseases [61]. Furthermore, interactions among compounds may hold greater significance than individual ones, leading to enhanced therapeutic efficiency [13,32,62]. Therefore, the simultaneous proliferation of bioactive compounds and antioxidant activity of GBR in this study might potentially lead to a synergistic effect that benefits human health. Our findings may support the promotion of rice consumption values as well as the development of pharmaceuticals, functional foods and supplements. For example, GBR treated with B2 (75 mM salinity for 4 days) can be applied to produce a fermented functional beverage known as kombucha, which has become increasingly popular because of its health benefits [63]. On the other hand, considering the impacts of human digestion on targeted products, their bioaccessibility and bioavailability during the digestion process should be thoroughly examined in future studies [64].

Table 3. Pearson's correlation coefficients between phytochemicals and antioxidant activities of GBR.

	MA	MB	ρ-Cou	Tri	Fer	Sal	Cin	DPPH	ABTS	TFC
MB	0.984 ***									
ρ-Cou	0.888 ***	0.915 ***								
Tri	0.940 ***	0.898 ***	0.837 ***							
Fer	0.908 ***	0.901 ***	0.905 ***	0.886 ***						
Sal	0.610 **	0.572 **	0.497 *	0.613 **	0.653 **					
Cin	0.548 **	0.526 **	0.419 *	0.458 *	0.594 **	0.900 ***				
DPPH	0.510 **	0.540 **	0.355 *	0.397 *	0.431 *	0.577 **	0.722 **			
ABTS	0.053	0.043	0.063	0.003	0.259	0.18	0.347 *	0.041		
TFC	0.860 ***	0.858 ***	0.851 ***	0.864 ***	0.809 ***	0.304 *	0.216	0.410 *	−0.07	
TPC	0.743 **	0.744 **	0.733 **	0.727 **	0.670 **	0.068	−0.029	0.09	−0.113	0.861 ***

*, **, and *** indicate significances at $p < 0.05$, 0.01, and 0.001, respectively; MA: momilactone A; MB: momilactone B; ρ-Cou: ρ-coumaric acid; Tri: tricin; Fer: ferulic acid; Sal: salicylic acid; Cin: cinnamic acid; TPC: total phenolic content; TFC: total flavonoid content. ABTS: 2,2′-azino-bis-(3-ethylbenzothiazoline-6-sulfonic acid) assay; DPPH: 2,2-diphenyl-1-picrylhydrazyl assay.

4. Conclusions

This research, for the first time, has identified an optimized treatment (B2: 75 mM NaCl and 4-day germination) that significantly advanced the accumulation of valuable bioactive compounds, including phenolics and momilactones A (MA) and B (MB), in germinated brown rice (GBR, Koshihikari var.). In particular, GBR treated by B2 contained the highest amounts of total phenolics and total flavonoids. Moreover, the isolated bioactive compounds were identified and confirmed by electrospray ionization-mass spectrometry (ESI-MS) and nuclear magnetic resonance (NMR) spectroscopy (^{1}H and ^{13}C). Additionally, the quantification results indicated that GBR under B2 treatment accumulated the greatest quantities of MA, MB, tricin, ρ-coumaric acid, ferulic acid, cinnamic acid, and salicylic acid. The B2 treatment also significantly enhanced the antioxidant activities of GBR, as demonstrated in the antiradical assays (DPPH and ABTS). In the context that BR has been less favored, leading to improper utilization or waste of this source, the outcomes of our study hold promising prospects for enhancing the nutritional value of BR and fostering the advancement of rice-derived products that contribute to human well-being. Consequently, this research serves to incentivize the consumption of BR by highlighting its intrinsic value and potential benefits for human health. Furthermore, the present findings are expected to contribute to the attainment of the Sustainable Development Goals (SDGs) by promoting the overall welfare of individuals, eradicating poverty, and ensuring global food security, particularly in countries reliant on rice cultivation.

Supplementary Materials: The following supporting information can be downloaded at: https://www.mdpi.com/article/10.3390/foods12132501/s1, Figure S1: ^{1}H- and ^{13}C-NMR spectrum of isolated (A) tricin, (B) ρ-coumaric acid, (C) momilactone A (MA), and (D) momilactone B (MB) in this study; Figure S2: High-performance liquid chromatography (HPLC) chromatograms of standards (A) tricin; (B) ρ-coumaric acid; (C) ferulic acid; (D) salicylic acid; (E) cinnamic acid, and (F) detected phenolic compounds in B2 (75 mM NaCl and 4-day germination); Figure S3: UPLC–ESI-MS chromatograms of (A) standard momilactones A (MA) and B (MB), and (B) MA and MB detected in GBR in B2 treatment (75 mM NaCl and 4-day germination).

Author Contributions: Conceptualization, T.D.X., T.D.K. and N.V.Q.; methodology, N.V.Q. and L.H.A.; investigation, M.H. and L.H.A.; data curation, M.H. and L.H.A.; writing—original draft preparation, M.H. and L.H.A.; writing—review and editing, M.H., L.H.A., T.D.X., T.D.K. and N.V.Q.; supervision, T.D.X. and N.V.Q.; project administration, T.D.X., T.D.K. and N.V.Q. All authors have read and agreed to the published version of the manuscript.

Funding: This research received no external funding.

Data Availability Statement: The data used to support the findings of this study can be made available by the corresponding author upon request.

Acknowledgments: We are indebted to Nguyen Phuong Mai, Mohammad Amir Hossen, and Jesmin Akter for their contributive support in sample preparation and analysis.

Conflicts of Interest: The authors declare no conflict of interest.

References

1. Cho, D.H.; Lim, S.T. Germinated brown rice and its bio-functional compounds. *Food Chem.* **2016**, *196*, 259–271. [CrossRef] [PubMed]
2. Wang, W.; Li, Y.; Dang, P.; Zhao, S.; Lai, D.; Zhou, L. Rice secondary metabolites: Structures, roles, biosynthesis, and metabolic regulation. *Molecules* **2018**, *23*, 3098. [CrossRef] [PubMed]
3. Ravichanthiran, K.; Ma, Z.F.; Zhang, H.; Cao, Y.; Wang, C.W.; Muhammad, S.; Aglago, E.K.; Zhang, Y.; Jin, Y.; Pan, B. Phytochemical profile of brown rice and its nutrigenomic implications. *Antioxidants* **2018**, *7*, 71. [CrossRef] [PubMed]
4. Ding, J.; Yang, T.; Feng, H.; Dong, M.; Slavin, M.; Xiong, S.; Zhao, S. Enhancing contents of γ-aminobutyric acid (GABA) and other micronutrients in dehulled rice during germination under normoxic and hypoxic conditions. *J. Agric. Food Chem.* **2016**, *64*, 1094–1102. [CrossRef]
5. Ng, L.T.; Huang, S.H.; Chen, Y.T.; Su, C.H. Changes of tocopherols, tocotrienols, γ-oryzanol, and γ-aminobutyric acid levels in the germinated brown rice of pigmented and nonpigmented cultivars. *J. Agric. Food Chem.* **2013**, *61*, 12604–12611. [CrossRef]
6. Nascimento, L.Á.D.; Avila, B.P.; Colussi, R.; Elias, M.C. Effect of abiotic stress on bioactive compound production in germinated brown rice. *Cereal Chem.* **2020**, *97*, 868–876. [CrossRef]
7. Trachoo, N.; Boudreaux, C.; Moongngarm, A.; Samappito, S.; Gaensakoo, R. Effect of germinated rough rice media on growth of selected probiotic bacteria. *Pak. J. Biol. Sci.* **2006**, *9*, 2657–2661. [CrossRef]
8. Choi, I.D.; Kim, D.S.; Son, J.R.; Yang, C.I.; Chun, J.Y.; Kim, K.J. Physico-chemical properties of giant embryo brown rice (Keunnunbyeo). *J. Appl. Biol. Chem.* **2006**, *49*, 95–100.
9. Owolabi, I.O.; Chakree, K.; Takahashi, Y.C. Bioactive components, antioxidative and anti-inflammatory properties (on RAW 264.7 macrophage cells) of soaked and germinated purple rice extracts. *Int. J. Food Sci. Technol.* **2019**, *54*, 2374–2386. [CrossRef]
10. Munarko, H.; Sitanggang, A.B.; Kusnandar, F.; Budijanto, S. Effect of different soaking and germination methods on bioactive compounds of germinated brown rice. *Int. J. Food Sci. Technol.* **2021**, *56*, 4540–4548. [CrossRef]
11. Islam, M.Z.; Shim, M.J.; Jeong, S.Y.; Lee, Y.T. Effects of soaking and sprouting on bioactive compounds of black and red pigmented rice cultivars. *Int. J. Food Sci. Technol.* **2022**, *57*, 201–209. [CrossRef]
12. Choe, H.; Sung, J.; Lee, J.; Kim, Y. Effects of calcium chloride treatment on bioactive compound accumulation and antioxidant capacity in germinated brown rice. *J. Cereal Sci.* **2021**, *101*, 103294. [CrossRef]
13. Quan, N.V.; Thien, D.D.; Khanh, T.D.; Tran, H.D.; Xuan, T.D. Momilactones A, B, and tricin in rice grain and by-products are potential skin aging inhibitors. *Foods* **2019**, *8*, 602. [CrossRef]
14. Shalini, V.; Bhaskar, S.; Kumar, K.S.; Mohanlal, S.; Jayalekshmy, A.; Helen, A. Molecular mechanisms of anti-inflammatory action of the flavonoid, tricin from Njavara rice (*Oryza sativa* L.) in human peripheral blood mononuclear cells: Possible role in the inflammatory signaling. *Int. Immunopharmacol.* **2012**, *14*, 32–38. [CrossRef] [PubMed]
15. Shalini, V.; Pushpan, C.K.; Sindhu, G.; Jayalekshmy, A.; Helen, A. Tricin, flavonoid from Njavara reduces inflammatory responses in hPBMCs by modulating the p38MAPK and PI3K/Akt pathways and prevents inflammation associated endothelial dysfunction in HUVECs. *Immunobiology* **2016**, *221*, 137–144. [CrossRef]
16. Cai, H.; Hudson, E.A.; Mann, P.; Verschoyle, R.D.; Greaves, P.; Manson, M.M.; Steward, W.P.; Gescher, A.J. Growth-inhibitory and cell cycle-arresting properties of the rice bran constituent tricin in human-derived breast cancer cells in vitro and in nude mice in vivo. *Br. J. Cancer* **2004**, *91*, 1364–1371. [CrossRef] [PubMed]
17. Wongsa, P. Phenolic compounds and potential health benefits of pigmented rice. In *Recent Advances in Rice Research*; Ansari, M.U.R., Ed.; IntechOpen: London, UK, 2021.
18. Ghasemzadeh, A.; Karbalaii, M.T.; Jaafar, H.Z.; Rahmat, A. Phytochemical constituents, antioxidant activity, and antiproliferative properties of black, red, and brown rice bran. *Chem. Cent. J.* **2018**, *12*, 17. [CrossRef]
19. Gong, E.S.; Luo, S.; Li, T.; Liu, C.; Zhang, G.; Chen, J.; Zeng, Z.; Liu, R.H. Phytochemical profiles and antioxidant activity of processed brown rice products. *Food Chem.* **2017**, *232*, 67–78. [CrossRef]
20. Park, C.; Jeong, N.Y.; Kim, G.Y.; Han, M.H.; Chung, I.M.; Kim, W.J.; Yoo, Y.H.; Choi, Y.H. Momilactone B induces apoptosis and G1 arrest of the cell cycle in human monocytic leukemia U937 cells through downregulation of pRB phosphorylation and induction of the cyclin-dependent kinase inhibitor $p21^{Waf1/Cip1}$. *Oncol. Rep.* **2014**, *31*, 1653–1660. [CrossRef]
21. Lee, S.C.; Chung, I.M.; Jin, Y.J.; Song, Y.S.; Seo, S.Y.; Park, B.S.; Cho, K.H.; Yoo, K.S.; Kim, T.H.; Yee, S.B.; et al. Momilactone B, an allelochemical of rice hulls, induces apoptosis on human lymphoma cells (Jurkat) in a micromolar concentration. *Nutr. Cancer* **2008**, *60*, 542–551. [CrossRef]
22. Kim, S.J.; Park, H.R.; Park, E.; Lee, S.C. Cytotoxic and antitumor activity of momilactone B from rice hulls. *J. Agric. Food Chem.* **2007**, *55*, 1702–1706. [CrossRef]
23. Quan, N.V.; Tran, H.D.; Xuan, T.D.; Ahmad, A.; Dat, T.D.; Khanh, T.D.; Teschke, R. Momilactones A and B are α-amylase and α-glucosidase inhibitors. *Molecules* **2019**, *24*, 482. [CrossRef]

24. Quan, N.V.; Xuan, T.D.; Tran, H.D.; Ahmad, A.; Khanh, T.D.; Dat, T.D. Contribution of momilactones A and B to diabetes inhibitory potential of rice bran: Evidence from in vitro assays. *Saudi Pharm. J.* **2019**, *27*, 643–649. [CrossRef] [PubMed]
25. Goufo, P.; Trindade, H. Rice antioxidants: Phenolic acids, flavonoids, anthocyanins, proanthocyanidins, tocopherols, tocotrienols, γ-oryzanol, and phytic acid. *Food Sci. Nutr.* **2014**, *2*, 75–104. [CrossRef]
26. Floegel, A.; Kim, D.O.; Chung, S.J.; Koo, S.I.; Chun, O.K. Comparison of ABTS/DPPH assays to measure antioxidant capacity in popular antioxidant-rich US foods. *J. Food Compos. Anal.* **2011**, *24*, 1043–1048. [CrossRef]
27. Olszowy, M.; Dawidowicz, A.L. Is it possible to use the DPPH and ABTS methods for reliable estimation of antioxidant power of colored compounds? *Chem. Pap.* **2018**, *72*, 393–400. [CrossRef]
28. Kobayashi, A.; Hori, K.; Yamamoto, T.; Yano, M. Koshihikari: A premium short-grain rice cultivar–its expansion and breeding in Japan. *Rice* **2018**, *11*, 15. [CrossRef] [PubMed]
29. Cáceres, P.J.; Peñas, E.; Martinez-Villaluenga, C.; Amigo, L.; Frias, J. Enhancement of biologically active compounds in germinated brown rice and the effect of sun-drying. *J. Cereal Sci.* **2017**, *73*, 1–9. [CrossRef]
30. Mohammadabadi, S.S.; Goli, M.; Tabasi, S.N. Optimization of bioactive compound extraction from eggplant peel by response surface methodology: Ultrasound-assisted solvent qualitative and quantitative effect. *Foods* **2022**, *11*, 3263. [CrossRef]
31. Bueno-Costa, F.M.; Zambiazi, R.C.; Bohmer, B.W.; Chaves, F.C.; Silva, W.P.D.; Zanusso, J.T.; Dutra, I. Antibacterial and antioxidant activity of honeys from the state of Rio Grande do Sul, Brazil. *LWT-Food Sci. Technol.* **2016**, *65*, 333–340. [CrossRef]
32. Anh, L.H.; Quan, N.V.; Lam, V.Q.; Iuchi, Y.; Takami, A.; Teschke, R.; Xuan, T.D. Antioxidant, anti-tyrosinase, anti-α-amylase, and cytotoxic potentials of the invasive weed *Andropogon virginicus*. *Plants* **2020**, *10*, 69. [CrossRef] [PubMed]
33. Anh, L.H.; Quan, N.V.; Lam, V.Q.; Takami, A.; Khanh, T.D.; Xuan, T.D. Rice momilactones and phenolics: Expression of relevant biosynthetic genes in response to UV and chilling stresses. *Agronomy* **2022**, *12*, 1731.
34. Wang, Y.; Shen, J.Z.; Chan, Y.W.; Ho, W.S. Identification and growth inhibitory activity of the chemical constituents from *Imperata cylindrica* aerial part ethyl acetate extract. *Molecules* **2018**, *23*, 1807. [CrossRef] [PubMed]
35. Nguyen, D.H.; Zhao, B.T.; Le, D.D.; Kim, K.Y.; Kim, Y.H.; Yoon, Y.H.; Ko, J.Y.; Woo, K.S.; Woo, M.H. Phenolic constituents and their anti-inflammatory activity from *Echinochloa utilis* grains. *Nat. Prod. Sci.* **2016**, *22*, 140–145. [CrossRef]
36. Anh, N.T.H.; Anh, L.H.; Mai, N.P.; Quan, N.V.; Xuan, T.D. Identification and isolation techniques for plant growth inhibitors in rice. *Separations* **2023**, *10*, 105. [CrossRef]
37. Tungmunnithum, D.; Thongboonyou, A.; Pholboon, A.; Yangsabai, A. Flavonoids and other phenolic compounds from medicinal plants for pharmaceutical and medical aspects: An overview. *Medicines* **2018**, *5*, 93. [CrossRef]
38. Xuan, T.D.; Huong, C.T.; Quan, N.V.; Anh, L.H.; Khanh, T.D.; Rayee, R. Improvement of salinity tolerance in rice seedlings by exogenous magnesium sulfate application. *Soil Syst.* **2022**, *6*, 69. [CrossRef]
39. Huong, C.T.; Anh, T.T.T.; Dat, T.D.; Khanh, T.D.; Xuan, T.D. Uniparental inheritance of salinity tolerance and beneficial phytochemicals in rice. *Agronomy* **2020**, *10*, 1032. [CrossRef]
40. Minh, L.T.; Khang, D.T.; Ha, P.T.T.; Tuyen, P.T.; Minh, T.N.; Quan, N.V.; Xuan, T.D. Effects of salinity stress on growth and phenolics of rice (*Oryza sativa* L.). *Int. Lett. Nat. Sci.* **2016**, *57*, 1–10. [CrossRef]
41. Sharma, A.; Thakur, S.; Kumar, V.; Kanwar, M.K.; Kesavan, A.K.; Thukral, A.K.; Bhardwaj, R.; Alam, P.; Ahmad, P. Pre-sowing seed treatment with 24-epibrassinolide ameliorates pesticide stress in *Brassica juncea* L. Through the modulation of stress markers. *Front. Plant Sci.* **2016**, *7*, 1569. [CrossRef]
42. Anh, L.H.; Quan, N.V.; Nghia, L.T.; Xuan, T.D. Phenolic allelochemicals: Achievements, limitations, and prospective approaches in weed management. *Weed Biol. Manag.* **2021**, *21*, 37–67.
43. Prior, R.L.; Wu, X.; Schaich, K. Standardized methods for the determination of antioxidant capacity and phenolics in foods and dietary supplements. *J. Agric. Food Chem.* **2005**, *53*, 4290–4302. [CrossRef]
44. Shahidi, F.; Ambigaipalan, P. Phenolics and polyphenolics in foods, beverages and spices: Antioxidant activity and health effects—A review. *J. Funct. Foods* **2015**, *18*, 820–897. [CrossRef]
45. Kato, T.; Kabuto, C.; Sasaki, N.; Tsunagawa, M.; Aizawa, H.; Fujita, K.; Kato, Y.; Kitahara, Y.; Takahashi, N. Momilactones, growth inhibitors from rice, *Oryza sativa* L. *Tetrahedron Lett.* **1973**, *14*, 3861–3864. [CrossRef]
46. Quan, N.V.; Xuan, T.D.; Tran, H.D.; Thuy, N.T.D. Inhibitory activities of momilactones A, B, E, and 7-ketostigmasterol isolated from rice husk on paddy and invasive weeds. *Plants* **2019**, *8*, 159. [CrossRef]
47. Chung, I.M.; Hahn, S.J.; Ahmad, A. Confirmation of potential herbicidal agents in hulls of rice, *Oryza sativa*. *J. Chem. Ecol.* **2005**, *31*, 1339–1352. [CrossRef]
48. Ahmad, A.; Xuan, T.D.; Minh, T.N.; Siddiqui, N.A.; Quan, N.V. Comparative extraction and simple isolation improvement techniques of active constituents' momilactone A and B from rice husks of *Oryza sativa* by HPLC analysis and column chromatography. *Saudi Pharm. J.* **2019**, *27*, 17–24. [CrossRef]
49. Germain, J.; Deslongchamps, P. Total synthesis of ($\pm$)-momilactone A. *J. Med. Chem.* **2002**, *67*, 5269–5278. [CrossRef]
50. Minh, T.N.; Xuan, T.D.; Ahmad, A.; Elzaawely, A.A.; Teschke, R.; Van, T. Momilactones A and B: Optimization of yields from isolation and purification. *Separations* **2018**, *5*, 28. [CrossRef]
51. Anh, L.H.; Lam, V.Q.; Takami, A.; Khanh, T.D.; Quan, N.V.; Xuan, T.D. Cytotoxic mechanism of momilactones A and B against acute promyelocytic leukemia and multiple myeloma cell lines. *Cancers* **2022**, *14*, 4848. [CrossRef]
52. Fukuta, M.; Xuan, T.D.; Deba, F.; Tawata, S.; Khanh, T.D.; Chung, I.M. Comparative efficacies in vitro of antibacterial, fungicidal, antioxidant, and herbicidal activities of momilactones A and B. *J. Plant Interact.* **2007**, *2*, 245–251. [CrossRef]

53. Chung, I.M.; Jung, T.K.; Kim, S.H. Evaluation of allelopathic potential and quantification of momilactone A, B from rice hull extracts and assessment of inhibitory bioactivity on paddy field weeds. *J. Agric. Food Chem.* **2006**, *54*, 2527–2536. [CrossRef]
54. Ambade, A.; Mandrekar, P. Oxidative stress and inflammation: Essential partners in alcoholic liver disease. *Int. J. Hepatol.* **2012**, *2012*, 853175. [CrossRef] [PubMed]
55. Tucker, P.S.; Scanlan, A.T.; Dalbo, V.J. Chronic kidney disease influences multiple systems: Describing the relationship between oxidative stress, inflammation, kidney damage, and concomitant disease. *Oxidative Med. Cell. Longev.* **2015**, *2015*, 806358. [CrossRef]
56. Lin, Y.T.; Pao, C.C.; Wu, S.T.; Chang, C.Y. Effect of different germination conditions on antioxidative properties and bioactive compounds of germinated brown rice. *BioMed. Res. Int.* **2015**, *2015*, 608761. [CrossRef] [PubMed]
57. Ti, H.; Zhang, R.; Zhang, M.; Li, Q.; Wei, Z.; Zhang, Y.; Tang, X.; Deng, Y.; Liu, L.; Ma, Y. Dynamic changes in the free and bound phenolic compounds and antioxidant activity of brown rice at different germination stages. *Food Chem.* **2014**, *161*, 337–344. [CrossRef] [PubMed]
58. Xuan, T.D.; Khang, D.T. Effects of exogenous application of protocatechuic acid and vanillic acid to chlorophylls, phenolics and antioxidant enzymes of rice (*Oryza sativa* L.) in submergence. *Molecules* **2018**, *23*, 620. [CrossRef]
59. Falcinelli, B.; Sileoni, V.; Marconi, O.; Perretti, G.; Quinet, M.; Lutts, S.; Benincasa, P. Germination under moderate salinity increases phenolic content and antioxidant activity in rapeseed (*Brassica napus* var oleifera Del.) sprouts. *Molecules* **2017**, *22*, 1377. [CrossRef]
60. Soong, Y.Y.; Barlow, P.J. Antioxidant activity and phenolic content of selected fruit seeds. *Food Chem.* **2004**, *88*, 411–417. [CrossRef]
61. Biswas, S.K. Does the interdependence between oxidative stress and inflammation explain the antioxidant paradox? *Oxidative Med. Cell. Longev.* **2016**, *2016*, 5698931. [CrossRef]
62. Quan, N.V.; Xuan, T.D.; Tran, H.D.; Thuy, N.T.D.; Trang, L.T.; Huong, C.T.; Andriana, Y.; Tuyen, P.T. Antioxidant, α-amylase and α-glucosidase inhibitory activities and potential constituents of *Canarium tramdenum* bark. *Molecules* **2019**, *24*, 605. [CrossRef] [PubMed]
63. Quan, N.V.; Iuchi, Y.; Anh, L.H.; Hasan, M.; Xuan, T.D. Simple isolation of cordycepin from *Cordyceps militaris* by dual-normal phase column chromatography and its potential for making kombucha functional products. *Separations* **2022**, *9*, 290. [CrossRef]
64. Un, S.; Quan, N.V.; Anh, L.H.; Lam, V.Q.; Takami, A.; Khanh, T.D.; Xuan, T.D. Effects of in vitro digestion on anti-α-amylase and cytotoxic potentials of *Sargassum* spp. *Molecules* **2022**, *27*, 2307. [CrossRef] [PubMed]

foods

Article

Improving γ-Oryzanol and γ-Aminobutyric Acid Contents in Rice Beverage *Amazake* Produced with Brown, Milled and Germinated Rices

Ana Castanho [1,2], Cristiana Pereira [2], Manuela Lageiro [2,3], Jorge C. Oliveira [4], Luís M. Cunha [1] and Carla Brites [2,5,*]

1 GreenUPorto/Inov4Agro, DGAOT, Faculty of Sciences, University of Porto, 4169-007 Porto, Portugal; anavargascastanho@gmail.com (A.C.); lmcunha@fc.up.pt (L.M.C.)
2 INIAV—Instituto Nacional de Investigação Agrária e Veterinária, 2780-157 Oeiras, Portugal; cristianapereirags@gmail.com (C.P.); manuela.lageiro@iniav.pt (M.L.)
3 GeoBioTec Research Center, Faculdade de Ciências e Tecnologia, Universidade Nova de Lisboa, 2829-516 Caparica, Portugal
4 School of Engineering and Architecture, University College Cork, T12 HW58 Cork, Ireland; j.oliveira@ucc.ie
5 GREEN-IT, ITQB NOVA, 2780-157 Oeiras, Portugal
* Correspondence: carla.brites@iniav.pt

Abstract: Rice is an important source of γ-oryzanol (GO) and γ-aminobutyric acid (GABA), which are bioactive compounds that may benefit blood lipid and pressure control. Both GO and GABA can be improved by germination and fermentation. Fermentation with *A. oryzae* produces *Koji*, a rice-based starter for *Amazake*, a naturally sweet beverage. Germinated rice (brown and milled rice), were tested to improve those bioactive compounds during the fermentation process. The resulting *Koji* was optimised to GO and GABA through a response surface methodology; α-amylase activity and starch content were also assessed. The different rice matrix resulting from the germination largely impacted the biosynthesis of GABA, α-amylase and starch contents. *Amazake*, obtained by germinated rice, has increased GO and GABA contents when compared to the one obtained from milled rice (from a non-detectable value to 27.65 ± 0.23 mg/100 g for GO and from 163.95 ± 24.7 to 271.53 ± 5.7 mg/100 g for GABA). A panel of 136 Portuguese consumers tasted the beverage in a blind overall tasting test followed by an informed test, using 9-point scales. The consumer scores had a mean value of 4.67 ± 1.9 and 4.9 ± 1.8, meaning that cultural differences may play an important role with regard to liking and accepting *Amazake*.

Keywords: γ-oryzanol; γ-aminobutyric acid; α-amylase; *Koji*; *Amazake*; rice bioactives; response surface methodology

Citation: Castanho, A.; Pereira, C.; Lageiro, M.; Oliveira, J.C.; Cunha, L.M.; Brites, C. Improving γ-Oryzanol and γ-Aminobutyric Acid Contents in Rice Beverage *Amazake* Produced with Brown, Milled and Germinated Rices. *Foods* **2023**, *12*, 1476. https://doi.org/10.3390/foods12071476

Academic Editor: Antonio Cilla

Received: 10 January 2023
Revised: 17 March 2023
Accepted: 29 March 2023
Published: 31 March 2023

1. Introduction

Rice is considered a staple food in many countries, particularly in Asia. Despite the low consumption rate in Europe, Portugal is the largest rice consumer, with innumerable ways of consuming rice in its gastronomic tradition. However, rice is not consumed at its full nutritional potential, since the bran and germ are discarded with the industrial processing, and rice is consumed after milling [1]. In recent years, the nutritional compounds of whole rice and rice bran have been studied to explore natural ways of increasing its content in rice-based foods. One of those compounds is γ-oryzanol (GO) [2,3]. GO is a bioactive compound known for its potential to regulate lipid metabolism, reducing the LDL (Low-Density Lipoprotein) and total cholesterol [4–7]. GO is synthesised during the maturation of the rice grain, being responsible for different regulatory functions, properties and structures; therefore, the GO content is not only affected by genetic factors but also by climate and growing conditions [8–10].

Another important compound in rice bran is γ-aminobutyric acid (GABA), a non-protein amino acid. GABA is an important neurotransmitter responsible for lowering the nervous system's activity, effectively lowering blood pressure [11–14] and treating epilepsy [15]. GABA is mainly synthesised by biochemical processes such as germination and fermentation. During those processes, the α-ketoglutarate (contained in large quantities in the bran) is transaminated to glutamic acid, which is decarboxylated to GABA through glutamic acid decarboxylase (GAD) [12,13]. Thus, increased levels of GABA can be obtained by reinforcing the germination status of the grain. The amount of GABA in rice bran ranges from 10.7–58.0 mg/100 g (before germination) and 90.0–350.0 mg/100 g (after 10–12 h of rice germination) [16].

The germination and fermentation processes have been employed empirically to increase nutritional value and enhance the assimilability and palatability of foods. The most important microorganism employed in rice fermentation is *Aspergillus oryzae*, which was named after its occurrence in rice (*Oryza sativa*) [17]. *A. oryzae* has been used in Asia, mainly in Japan, as a starter for a solid-state fermentation named *Koji*, which is the basis of many traditional products, including soy sauce, miso, sake, and *Amazake* [17,18]. Solid-state fermentation is a low moisture fermentation, providing just enough water to support the growth of filamentous fungi's penetrative and aerial hyphae, producing enzymes that would not typically be produced during other fermentation processes [19,20]. Although *A. oryzae* can ferment other products, rice is the most used food substrate for *Koji* production. The process to obtain *Koji* can be seen as a system which includes mainly (1) time, (2) temperature, and (3) moisture [21]. Ventilation, substrate type, pH, and the initial quantity of the inoculum are also essential factors in *Koji*-making [20,22]. *Amazake* is the simplest way of using *Koji*. *Amazake* is a non-alcoholic, white-coloured, sweet drink that can be made only by mixing *Koji* with water or by adding steamed rice. In the processing of *Amazake*, *Koji* acts as a starter for starch hydrolysis, where their α-amylases break the rice starch into glucose. Therefore, besides *Koji*'s unique flavour, *Amazake* is naturally sweet (without added sugars); contrary to other sweet beverages, which contain sucrose or fructose, *Amazake*'s sweetness is mainly realized through naturally produced glucose [23,24]. This is an advantage, since the final glucose content of *Amazake* can be controlled and adjusted through a longer or shorter rice *Koji* fermentation time [25].

Despite being closely related to the *Aspergillus flavus* species, *A. oryzae* is considered a safe organism by the World Health Organisation [26], if the microbiological safety of *Koji* and its products are produced with approved starter cultures, appropriate rice *Koji* fermentation processes, and storage conditions are used [20]. The safety of *A. oryzae* arises from the fact that it is unable to produce aflatoxins, as the genes responsible for its biosynthesis are absent or dysfunctional [18,20,23,24,27]. *Koji Amazake* can also be considered microbiologically safe, as starch is hydrolysed at a temperature range where harmful bacteria do not grow [24].

Few authors have studied the impact of *Amazake* intake on human health. Kurahashi and Yonei [23] studied the excess intake of *Amazake* (for four consecutive weeks) in 24 healthy subjects with high blood glucose levels. The body mass index (BMI) of the subjects did not change, but LDL cholesterol and glycated haemoglobin (HbA1c), a marker for diabetes, had a significant slight decrease, returning to the typical values after the test; the excess intake of *Amazake* led to a significant decrease in blood pressure. The authors related the hypotensive effect with the presence of GABA. Kikushima et al. [28], in a randomised, double-blind, placebo-controlled trial, evaluated the effects of *Koji Amazake*, finding that the daily intake of *Amazake* decreased systolic blood pressure. Other authors [25,29] studied the impact of *Amazake* on other health issues with positive results.

This work aimed to develop a rice-based beverage (*Amazake*) via a germination and rice *Koji* fermentation process with *A. oryzae*. As both GABA and GO can be improved naturally through biochemical processes, this work explores ways of increasing *Amazake's* GABA and GO contents, improving *Amazake's* nutritional value and evaluating its acceptability by Portuguese consumers.

2. Materials and Methods

A schematic view of the experimental methodology can be observed in Figure 1. Different rice matrices were selected for this study: milled, whole and germinated rice; germination rice was subjected to a pre-selection (through the employment of three different germination times) to obtain the greatest GABA and GO contents.

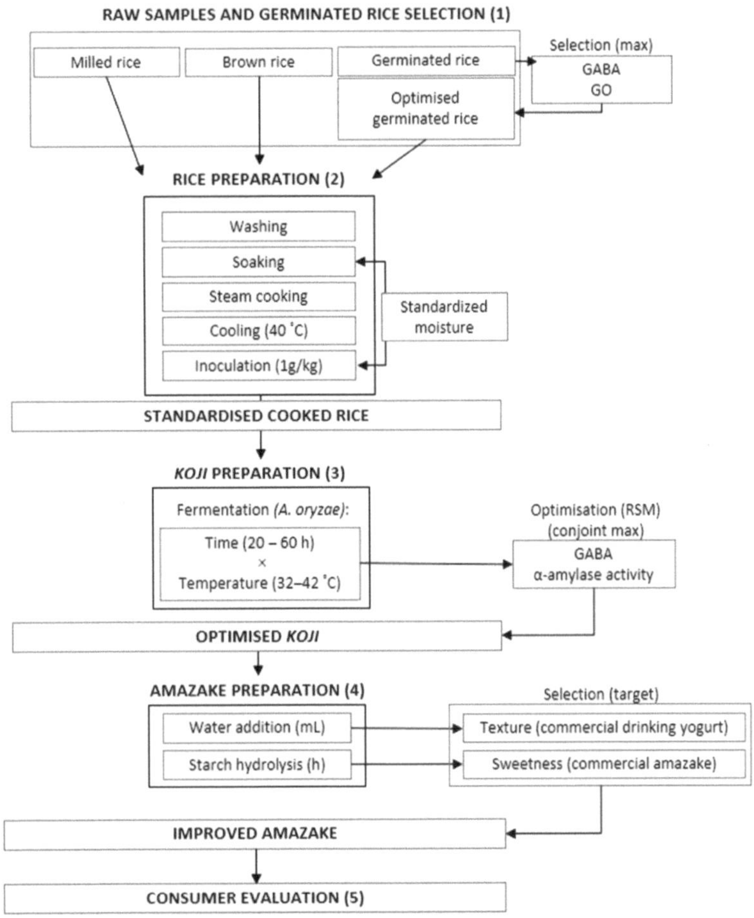

Figure 1. Schematic view of the experimental methodology, with numbers representing materials and methods, as well as discussion sections (e.g., 1 represents Sections 2.1 and 3.1).

The three raw samples were further soaked, cooked and inoculated with *A. oryzae* to produce *Koji*. The GABA content and α-amylase activity were analysed in the *Koji* preparation process to optimise the best inoculation time and temperature.

The optimised *Koji* was subjected to a starch hydrolysis process to obtain *Amazake*. The sweetness and texture parameters were monitored through the starch hydrolysis process to reach similarity with commercial *Amazake* (fermented brown rice *Amazake* from Kenshô, Spain) in terms of sweetness. For texture, the used reference was a commercial yoghurt beverage (plain liquid yoghurt from a major Portuguese retailer brand).

The improved *Amazake* was then presented to consumers to assess its overall likeability, and it was also evaluated for acceptance after elicitation with a final product prototype.

2.1. Sample Selection

Commercial milled and brown rice from the variety Ariete (from Orivarzea SA, Portugal) were acquired in the market and kept sealed at room temperature until use.

Germinated rice was prepared from brown rice, washed three times, and submerged in distilled water for 24 h; the samples were kept at 25 °C for 48 and 72 h, being washed every 4 h to avoid fermentation. Samples of 50 g were collected and freeze-dried. For comparison purposes, the raw brown rice sample was considered the control, and the subsequent samples were identified as 24 h, 48 h and 72 h germinated rice. The GABA and GO data were acquired, and the sample with greater values of those compounds was selected for further processing.

All of the samples were analysed in triplicate.

2.1.1. GABA Extraction and Quantification

GABA extraction and quantification were performed according to Jannoey et al. [30], with modifications. The freeze-dried samples were extracted with 70% ethanol. The samples were dissolved with 2 mL of ethanol, mixed in a vortex, shaken for 30 min at room temperature, and centrifuged at 4 °C for 10 min at 10,000× g (Laborzentrifugen 2K15, Sigma, Osterode am Harz, Germany). The supernatant was collected, and the procedure was repeated two more times, until collecting a total of 6 mL extract. The derivatisation was performed with 2-hydroxynaphthaldehyde (HN), by adding 1 mL of the extracted samples to 0.5 mL HN-methanol solution (0.3%) and 0.5 mL of borate buffer (pH 8), incubating at 80 °C during 15 min in a water-bath and cooling to room temperature. The resulting extract was filtered through a 0.22 µm nylon syringe filter (FilterTECH, Saran, France) before injection.

The quantification of GABA was performed by reverse-phase high-performance liquid chromatography (RP-HPLC) using Waters 2695 series equipment (Waters, Milford, MA, USA) connected to a diode array detector (DAD, Waters 2996, Waters, Milford, MA, USA). The amino acid HN derivative separation was performed in a C18 column (Sunfire 5 µm, 4.6 × 250 mm, Waters, Milford, MA, USA), which remained at a temperature of 40 °C (Waters column thermostat Jetstream 2 plus, Waters, Milford, MA, USA). The mobile phase used was methanol (A): water (B), with the following gradient: 0–4 min: 20–50%(A); 4–6 min: 50–80% (A); 6–8 min: 80–100% (A); 8–10 min: 100% (A) and from 10–14 min: 100–20% (A), at a flow of 0.8 mL min^{-1} and with an injection volume of 20 µL, with a total analysis time of 20 min.

The GABA-HN derivative identification was performed at 231 nm by comparison of the GABA-HN peak retention times from the GABA-HN spectra. Sample fortification with GABA standard was undertaken for all samples prior to HN derivatisation, and the results of the fortified samples were compared with non-fortified samples to validate the GABA identification. The GABA quantification was based on an external calibration curve made with GABA standard aqueous solutions (10 to 1000 mg/L). A linear regression was obtained between the GABA-HN peak area and the GABA content in mg/mL (total GABA-HN area = 144,448,962 [GABA] + 5,176,153) with a determination coefficient of 0.995. GABA quantification was measured in mg of GABA per 100 g of sample, and all samples were extracted in duplicate.

2.1.2. GO Extraction and Quantification

The γ-oryzanol (GO) quantification was performed by RP-HPLC equipped with a DAD using the same HPLC system described for GABA quantification using a Waters Spherisorb ODS 2 separation column (4.6 × 250 mm, 5 µm, Waters, Milford, MA, USA), according to the method reported by Lageiro et al. [8], with minor modifications. The extraction of GO was carried out from freeze-dried *Koji* lipid residue according to the procedures reported by Castanho, et al. [1]. The GO identification was made at 325 nm by retention times and spectra comparisons with GO standards from TCI Europe. The quantification was based on an external calibration curve with GO standard solutions

(10 to 900 mg/L). A linear regression between the total GO area (peak area sum of the different GO compounds' peak areas) and the GO content in mg/mL was obtained (total GO area = 29.10 × [GO] 58.99), with a determination coefficient of 0.9995. GO quantification was measured in mg of GO per 100 g of sample, and all samples were extracted in duplicate.

2.2. Rice Cooking and Inoculation

Rice was cooked according to Saigusa and Ohba [31], with few modifications. The milled rice (250 g) was washed three times and soaked in distilled water for 20 min. It was drained for 15 min, and the water content was adjusted to achieve 118% of the initial raw rice weight with distilled water. The rice was then steamed for 40 min in a rice cooker steamer (CKSTRC4723-050, Oster 600 mL, Sunbeam Products, Inc., Boca Raton, FL, USA), wrapped in 20 × 20 cm non-woven pads (Wells, Portugal). The cooked rice was allowed to rest for 10 min while wrapped. After removing the cloth, the rice was allowed to cool in a tray until the temperature decreased to 40 °C. For brown rice, the cooking method was slightly modified: the raw brown rice was soaked for 5 h to adjust the water content to 118% of the initial raw rice weight and cooked for 60 min. The procedure was the same for germinated rice apart from the initial water correction, as the water uptake was given by the germination time.

The water content of all samples was adjusted (133% raw weight) with distilled water to ensure process standardisation before inoculation. The cooked rice was then inoculated with *A. oryzae* spores obtained from Kenshô (Spain) according to the producer's instructions (1 g/kg of raw rice).

2.3. Koji Preparation and Optimisation

A circumscribed central composite design using rice *Koji* fermentation time and temperature as independent variables was used to obtain a surface response, with the values presented in Table 1. Ten experimental combinations were determined with two levels (−1, +1), two repetitions of the central point (0), and two levels of axial points (−α, +α) [32]. The experimental conditions were defined as between 20 and 60 h and 32 and 42 °C, according to Narahara [33]. The selected points were used for GABA, α-amylase activity, and starch content evaluation.

Table 1. Definition of the circumscribed central composite design independent variable values. The α value related to ±1 is equal to ±1.414.

	Time (h)	Temperature (°C)
Min (−α)	20	32
Max (+α)	60	42
Centre (0)	40	37
Factorial (−1)	26	34
Factorial (+1)	54	41

Before analysis, the preparation of *Koji* samples was based on Hong and Kim [34] and previous pilot studies from the research group. Fresh collected *Koji*, 20 g, was mixed with 60 mL of distilled water and homogenised in an Ultraturax (T25, IKA, Germany). From that mixture, 6 mL were diluted in 16 mL of distilled water and centrifuged at 10,000× *g*; the remaining *Koji* mixture was freeze-dried, ground to flour, passed through a 120 μm sieve, and stored in dry conditions for subsequent analysis, as *Koji* extracts. All the analyses were treated as independent triplicates.

2.3.1. α-Amylase Activity Assessment

The α-amylase activity of *Koji* extracts was measured using the Ceralpha® assay procedure (AOAC Official Method 2002.01) by using benzylidene-blocked *p*-nitrophenyl maltoheptaoside in the presence of thermostable α-glucosidase. The activity was calculated according to the Ceralpha® assay procedure using the Megazyme kit calculator for Microsoft Excel, and expressed as Ceralpha U per g *Koji*, where one Unit of enzyme activity is the amount of enzyme releasing one μmole of *p*-nitrophenol per minute under the defined assay conditions.

2.3.2. Starch Measurement

Starch data was collected using NIR transflection MPA equipment (Bruker Optics, Germany), according to Sampaio, et al. [35]. The data was obtained using the B-FING cereals calibration model (Bruker Optics, Germany).

2.4. *Amazake Selection and Preparation Processes*

Amazake was produced by hydrolysing the optimised *Koji* by adding water at a specific temperature and time. The suspension viscosity was measured, following each water addition, until the target value (plain liquid yoghurt, 0.285 Pa.s) was reached. As the target texture was achieved, the sweetness was determined by °Brix, through time, to achieve commercial *Amazake* sweetness (fermented brown rice *Amazake*, 21%).

The *Amazake* was mixed in a cooking robot (Thermomix TM6, Vorwerk, Germany) at the highest speed in order to homogenise the beverage.

2.4.1. Starch Hydrolysis and Viscosity Measurement

Starch hydrolysis was performed at a small scale in a Rapid Visco Analyser (RVA 4800, Perten Instruments, Sweden) at 50 °C at the constant speed of 1000 rpm for 30 min at different *Koji*:water proportions (1:1; 1:1.5; 1.75 and 1:2). The samples were cooled down to 4 °C, and the viscosity after cooling was measured at the constant speed of 1000 rpm for 2 min.

2.4.2. Brix Measurement

The sweetness of the *Amazake* was measured by its Brix (%) in a refractometer (PR-201, Atago, Japan) after filtering the suspension through a 45 μm syringe filter. The measurements were performed after 10, 15, 25 and 35 min.

2.5. *Consumer Evaluation*

A total of 111 consumers evaluated the *Amazake* beverage at an open science fair, which included students and families, representing a broad demographic group. In the first stage, the consumers tested the product without any information other than that it was a new and experimental rice beverage. The overall liking was evaluated using a 9-point scale, ranging from 1—"dislike extremely" to 9—"like extremely" [36].

In the second stage, after the first questionnaire, a prototype of the product was created, including a bottle full of "DIS! *Amasake*" with a label containing nutritional information, due date, ingredients, and logotype, and a flyer with information about the potential health benefits as well as information about the objectives of the project. The consumers filled out the Food Action Rating Scale (FACT) developed by Schutz [37], that was adapted for the product. The 9-point FACT scale was: 1—"I would drink this only if forced", 2—"I would drink this if there were no other food choice", 3—"I would hardly ever drink this", 4—"I do not like this but would drink this on an occasion", 5—"I would drink this if available but would not go out of my way", 6—"I like this and would drink it now and then", 7—"I would frequently drink this", 8—"I would drink this very often", 9—"I would drink this every opportunity that I have". Both the liking and FACT scales were translated into Portuguese, according to Ribeiro, et al. [38]. Demographic data (age, gender and education level) and free comments about the product were also collected with the overall liking data. Free comments have been analysed by grouping terms into categories and sub-categories derived from the content analysis [39] following a triangulation procedure [40]. All participants were aged 16 years old or above, and were willing to participate. Following the Helsinki statement, an informed consent was given, and participants were assured that all private information would remain anonymous. All minors (<18 years old) were accompanied by their consenting parents or educators. The research team has enforced procedures to guarantee adherence to the European General Data Protection Regulations.

2.6. Statistical Analysis

To analyse data from the rice *Koji* fermentation process, following the Central Composite Design, a quadratic model (see Equation (1)) was fitted to GABA, α-amylase activity, and starch content data, where b_0 represents the constant term, b_1 and b_2 the linear terms, b_3 and b_5 the quadratic terms, and b_4 the interaction term. x_1 represents the rice *Koji* fermentation temperature (°C), and x_2 represents rice *Koji* fermentation time (h).

$$y = b_0 + b_1x_1 + b_2x_2 + b_3{x_1}^2 + b_4x_1x_2 + b_5{x_2}^2 \quad (1)$$

Raw data inspection was performed in advance, and outliers were identified among triplicates at each sampling point, following the Grubbs method (Table S1). A linear model building approach was implemented through a stepwise regression approach, minimising Akaike's AIC parameter [41]. Overall model fitting was assessed by the adjusted coefficient of determination ${R_{adj}}^2$ [32].

The comparison of processing conditions at the additional steps, such as the impact of the germination time on GO and GABA, was performed using a one-way ANOVA, followed by Tukey's post hoc multiple comparison test, if applicable. All tests were performed at a 95% confidence level using XLStat for Microsoft Excel.

3. Results and Discussion

3.1. Sample Selection

The rice bran proteins, lipids, and mineral content are considered undesirable for that specific fermentation process, and traditional rice *Koji* is produced with milled rice [21], so a milled rice sample was included as a control in the improvement of GO and GABA contents; a brown rice sample was also included for presenting a naturally richer content of those compounds due to the presence of the bran [1,42].

As GO and GABA contents also increase with germination, a previous study was conducted to select the best germination times. Figure 2 shows the contents of GO and GABA over germination time.

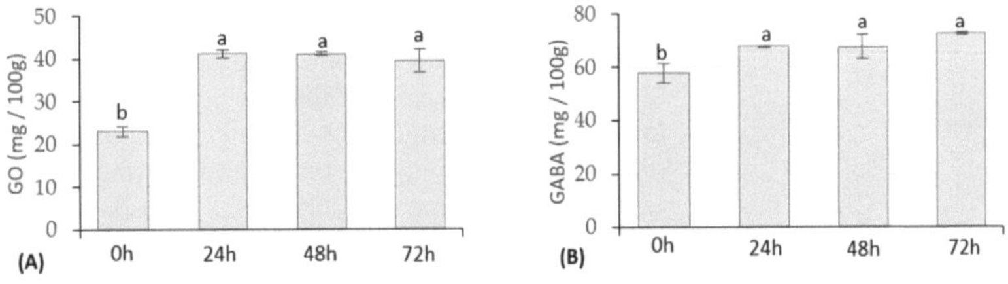

Figure 2. Effect of the germination time on the GO (**A**) and GABA (**B**) contents of brown Ariete variety rice, expressed in mg/100 g (DW). a, b—homogeneous groups according to Tukey's post hoc test at a 95% confidence level.

Both GO and GABA presented an increase in their content after 24 h of immersed germination when compared to raw rice, from 23.09 ± 1.193 to 41.17 ± 0.835 mg/100 g and 57.783 ± 3.60 to 67.73 ± 0.22 mg/100 g, respectively (Figure 2). After 24 h, the GO content lowered, and the GABA content increased; however, those changes were not statistically significant ($p < 0.05$). Wu, et al. [42] also reported an increase of the GO in the first germination stage, followed by a slight decrease; however, regarding GABA, the authors denoted a constant increase during germination. Wu, et al. [42] attributed the enhancement of GO in the first stages to the natural presence of lipases, promoting the decomposition of various lipid substances, including GO. Munarko, et al. [43] tested the effect of the germination process on the GABA content in four different rice varieties, reporting differences in the increase rate of GABA content with the germination time; the same authors also reported differences between GO content changing rates between the four varieties along germination time. The increase of GO content with germination is also reported by other authors [44].

Considering the results obtained, a germinated brown rice sample, after 24 h of submersion, was selected for the fermentation trials to obtain rice *Koji*, in addition to the milled and brown rice samples.

3.2. Rice Preparation

As reported previously, time, temperature, and moisture are the most critical factors influencing *Koji* making [21]. GO and GABA content can be affected by the processes carried out during *Koji* making. Besides temperature and time, the system's moisture is also essential. The system's moisture combines the media's humidity and cooked rice's moisture. While initially the moisture content is about 95–98%, during the rice *Koji* fermentation process, the moisture is lost due to fungal metabolism [24], therefore being essential to cover the *Koji* with a cloth. The system's moisture was standardised by carefully monitoring the water content based on the raw weight of the rice.

The gelatinisation stage of starch, from β-starch to α-starch, is important in *Koji* making, as complete gelatinization is needed to allow the tip of the *A. oryzae* hyphae to extend forward into the rice grain and not only over the surface, thus producing more desirable compounds [24]. The gelatinisation state of the rice was also verified before inoculation.

According to Gomi [17], the optimum temperature for *Koji* making is between 32–40 °C. As *A. oryzae* cannot grow above 44 °C, the cooked rice was cooled to 40–42 °C before inoculation.

3.3. Optimisation of Koji According to GABA and α-Amylase Activity

Temperature and time were the selected variables to optimise the *Koji* making:

(1) Temperature can affect the production of metabolic compounds such as GABA, and can also dictate the type of compounds produced (e.g., temperatures up to 37 °C are reported to increase the production of protease, while higher temperatures increase amylase activity), depending on the strain specificities [21,22]; however, fermentation temperature depends not only on the external specified parameters, but also on the exothermic reaction that occurs during that stage (rising the temperature to 40–42 °C); therefore, it is necessary to mix *Koji* and to cool it and ventilate it, especially when working with large batches [24].

(2) Time is also crucial in *Koji* making, as it defines the stage of growth;

(a) spores start germinating three to five hours after attachment to the matrix;

(b) at 20 h the spores grow on the grain surface; and

(c) the hyphae grow, and the mould extends its hyphae into rice grains, spreading over the matrix at 44 h, secreting high molecular weight hydrolytic enzymes (e.g., amylase and proteases) through the process [18,24].

During this stage, GABA and α-amylase activity were selected as the target parameters to measure, as GO content is related to lipid metabolism and is not expected to change during rice *Koji* fermentation, and α-amylase will break the starch in the following hydrolysis process, therefore being essential to *Amazake* production. Starch, the main compound of rice, was measured to observe the changes occurring during the germination and fermentation processes.

Table 2 and Figure 3 present the regression models' parameters and model fit for GABA, α-amylase activity and starch content, and the contour plots for those models.

Table 2. Significant stepwise regression coefficients (±Std Error) ($p < 0.05$) and R^2_{adj}, from the fitting of Equation (1) to GABA concentration, α-amylase activity and starch content. Values for temperature (x_1) were expressed in °C and incubation time (x_2) was expressed in hours. Non-significant values ($p > 0.05$) are presented as n.s.

	b_0	b_1	b_2	b_3	b_4	b_5	AIC	R^2_{adj}
			GABA (mg/100 g)					
Brown Rice	−3120 ± 777	171 ± 39.5	n.s.	−2.47 ± 0.524	0.307 ± 0.143	−0.255 ± 0.033	197	0.679
Milled Rice	−1811 ± 387	96.7 ± 20.2	8.92 ± 1.38	−1.27 ± 0.271	n.s.	−0.102 ± 0.017	151	0.681
Germinated Rice	1900 ± 971	−98.1 ± 50.8	9.12 ± 3.48	1.21 ± 0.681	n.s.	−0.073 ± 0.043	212	0.715
			α-amylase activity (CU/g)					
Brown Rice	−3061 ± 221	153 ± 11.6	13.1 ± 0.815	−2.06 ± 0.155	n.s.	−0.159 ± 0.01	115	0.922
Milled Rice	−3490 ± 237	173 ± 12.5	16.2 ± 0.88	−2.33 ± 0.168	n.s.	−0.199 ± 0.01	114	0.939
Germinated Rice	44.0 ± 15.2	−2.31 ± 0.826	0.049 ± 0.008	0.03 ± 0.011	n.s.	n.s.	−29.3	0.618
			Starch (g/100 g)					
Brown Rice	350 ± 105	−11.9 ± 5.3	−3.66 ± 0.80	n.s.	0.048 ± 0.019	0.021 ± 0.004	74.4	0.577
Milled Rice	360 ± 75.9	−14.7 ± 3.97	−1.76 ± 0.273	0.202 ± 0.053	n.s.	0.021 ± 0.003	57.3	0.622
Germinated Rice	153 ± 16.8	−1.75 ± 0.442	−2.38 ± 0.441	n.s.	0.038 ± 0.01	0.007 ± 0.002	38.1	0.876

Germination is a process that occurs in three phases [45]:

(1) The seeds become fully hydrated;

(2) Activation of the metabolism to mobilise nutrients to grow the radicle, which emerges at the end of this phase;

(3) The seed absorbs even more water, and there is a significant mobilisation of reserve material (i.e., carbohydrates which are hydrolysed and metabolised, leading to the growing of the seed).

During the germination process, the rice matrix suffers irreversible changes: α-amylase and α-glucosidases are synthesised; thus, the starch content decreases due to the partial hydrolysis of starch [46]; the peptidase activity levels increase [47], leading to protein hydrolysis, which is not only degraded into peptides but also into free amino acids [45]. Given these changes, the GABA synthesis in the germinated rice during fermentation is carried out in a completely new matrix; if, on the one hand, the free glutamic acid present in raw rice bran was consumed as a substrate for producing GABA during the germination, on the other hand, during rice *Koji* fermentation, the enzymes produced by *A. oryzae* can break proteins leading to the existence of more free glutamic acid that is latterly synthesised to GABA. However, as *A. oryzae* produces proteases at a lower temperature, there is a need to decrease the fermentation temperature; at the same time, the time factor is also important to allow the action of the proteases, which may explain the differences between the predicted time to achieve the maximum values of GABA regarding brown and milled rice, and the germinated rice in Figure 3.

The α-amylase predicted time and temperature for maximum α-amylase activity also follows the pattern of GABA synthesis (Figure 3). In milled rice and brown rice samples, α-amylase is formed at the time/temperature suggested in the literature [21,22]; in the germinated rice matrix, α-amylase reaches its more significant activity at the lower temperature and maximum time. Here, the hypothesis presented before may also apply, and as the matrix changes, the proteases formed at lower temperatures may also change, freeing some compounds that promote the increase of α-amylase activity. Despite the differences in the initial starch amounts of milled and brown rice (82.77 ± 1.410 and 73.05 ± 0.351 g/100 g, respectively), the starch that remained after fermentation was similar in both matrices. Germinated rice predicted starch content shows a lower amount of starch at lower temperatures and maximum time, meaning that the starch structure may be responsible for those differences (Figure 3).

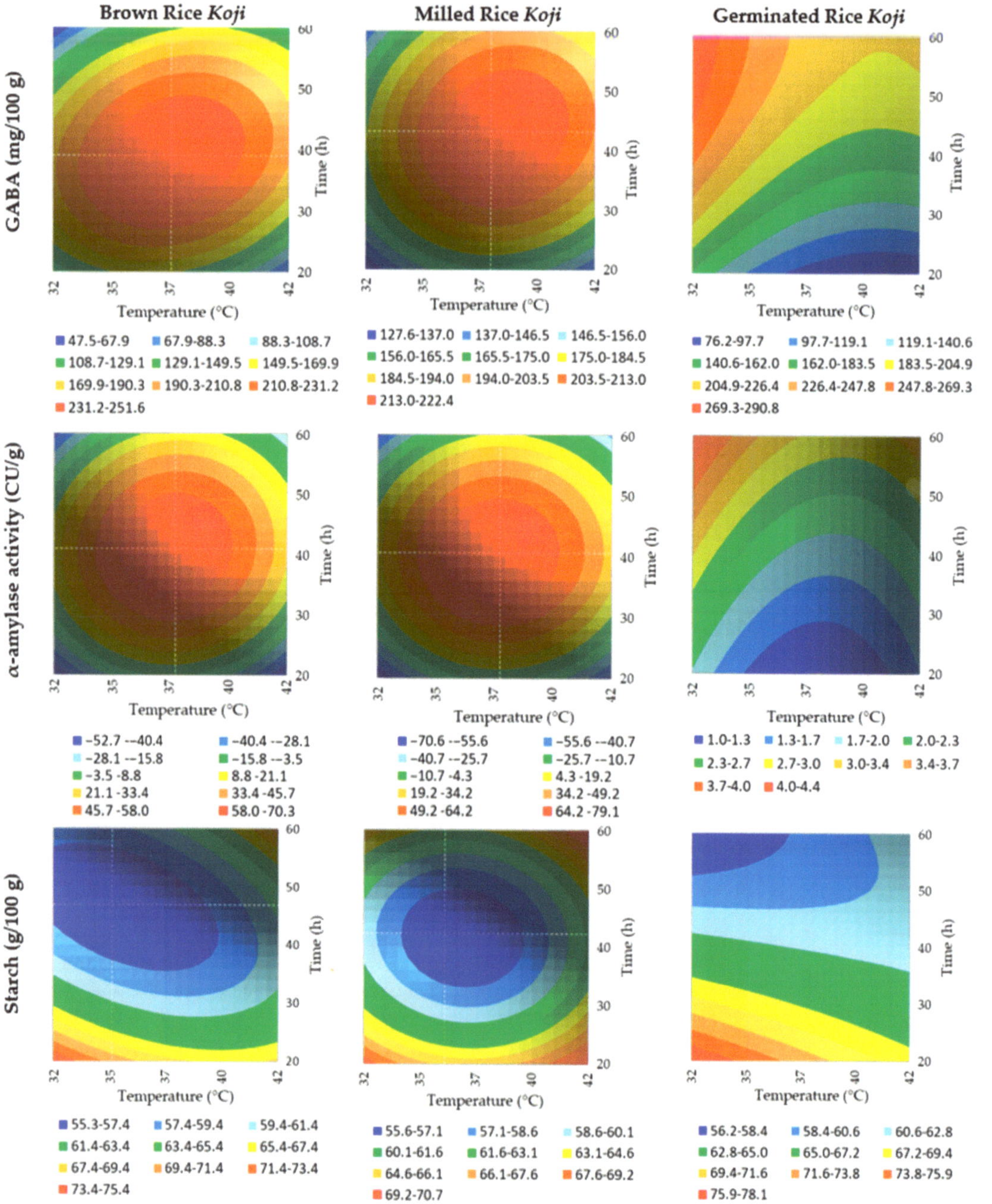

Figure 3. Contour plots of the predicted model for GABA, α-amylase activity and starch on brown, milled and germinated rice *Koji*. In GABA and α-amylase activity plots, white lines represent the temperature (°C) and time (h) targets to obtain the maximum response for GABA and α-amylase, and the minimum response for starch contents.

Table 3 shows the maximum concentrations of GABA and α-amylase activity, according to the simultaneous linear optimisation of both the GABA and α-amylase activity models. The results show the similarities between milled and brown rice regarding time, temperature, and maximum predicted values. Germinated rice requires more time and less temperature to obtain higher values of GABA than the other matrices; however, despite the maximum α-amylase activity occurring at the same time and temperature, the values are very low compared to the other matrices. Despite the low α-amylase activity, germinated rice was selected to produce the *Amazake* beverage, as pilot studies showed that germinated rice α-amylase could produce satisfactory results on the enzymatic hydrolysis.

Table 3. Predicted maximum GABA and α-amylase activity according to the simultaneous linear optimisation of both adjusted quadratic models (Table 2).

	Temperature (°C)	Time (h)	GABA (mg/100 g)	α-Amylase Activity (CU/g)
Brown rice	37.2	40.3	251.6	70.4
Milled Rice	37.5	41.8	221.6	79.0
Germinated rice	32.0 *	60.0 *	290.5	4.0

* Within the experimental range.

The predicted results were confirmed by a GABA and α-amylase activity analysis, showing results near the expected: regarding GABA, a concentration of 251.52 ± 6.3 mg/100 g, and regarding α-amylase activity, 4.04 ± 0.07 CU/g. The starch content was also assessed in the selected *Koji*, presenting a predicted value of 55.89 g/100g, and a response of 59.11 ± 0.16 g/100 g.

3.4. *Amazake* Preparation

Amazake, a beverage naturally sweetened by the action of enzymes, can be produced by adding *Koji* and water and exposing it to a controlled temperature over a specific period of time. While the water addition influences texture, it will also change the concentration of α-amylase in the system, manipulating the sweetness of *Amazake*. Time is also a key factor, as sweetness can be controlled by the time the enzymes take to hydrolyse the starch to glucose [48]. The starch enzymatic hydrolysis is carried out in two stages: dextrinisation (or liquefaction) and saccharification. During dextrinisation, there is a breakdown of starch in oligosaccharides, polysaccharides, or maltodextrin with a loss in viscosity; in the saccharification phase, the maltodextrins are mainly converted into glucose [48,49]. Figure 4 shows the °Brix and viscosity values progression over time in the same *Koji*:water samples subjected to 50 °C in RVA, where the rapid decrease of viscosity represents the liquefaction phase, and the slower increase of °Brix represents the saccharification stage.

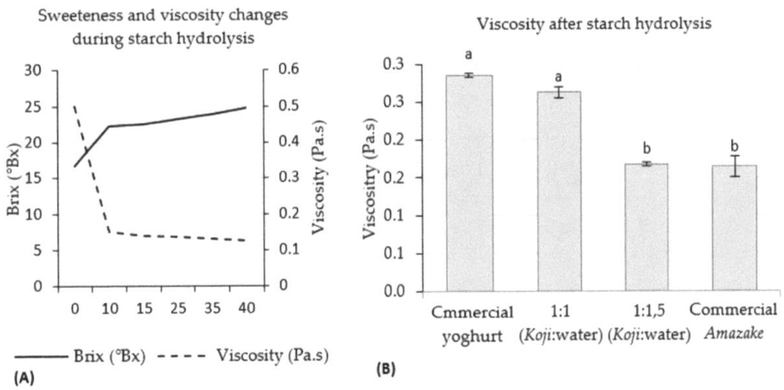

Figure 4. (**A**) Sweetness and viscosity over starch hydrolysis time. (**B**) Viscosity after hydrolysis in the selected *Koji*:water concentration and its comparison with drinking yoghurt and *Amazake*. a, b—homogeneous groups according to the Tukey's post hoc test at a 95% confidence level.

The texture and sweetness of *Amazake* were first tested on a small scale using an RVA at a constant temperature and speed. The optimum temperature for *A. oryzae* α-amylase to act on the substrate is reported to be 50–60 °C [24]; due to the reported low α-amylase activity, the system temperature was maintained at 50 °C. The texture was first improved to reach the yoghurt viscosity (0.285 Pa.s). As expected, the viscosity decreased with the water addition (Figure 4A). As the 1:1 *Koji*:water concentration showed a viscosity similar to the commercial yoghurt (Figure 4B), it was selected for the *Amazake* production at a larger scale for sensory evaluation.

The selected *Koji*:water concentration was then subjected to starch hydrolysis over time, also using the RVA, to obtain the °Brix value. Although °Brix is not a very accurate measure, in this case it was very useful, as a rapid measurement was preferred to a more accurate one due to the rapid action of α-amylase, as it could modify the results in a short amount of time. *Amazake* °Brix was set considering the reference of the commercial *Koji* °Brix, as the other compounds influence the sweet flavour in the *Amazake*, namely volatile components that are largely dependent on lipid oxidation in the *Koji* fermentation stage [50]. Therefore, instead of 14.3 °Brix of yoghurt, a 21 °Brix of commercial *Koji* was considered.

The *Amazake*, produced in the large batches (2 L) for consumers' sensory evaluation, had a relatively small change in the texture (from 0.255 ± 0.007 to 0.345 ± 0.035 Pa.s); in the large batches, the starch hydrolysis was carried out over five hours, with hourly °Brix measurements.

3.5. Effect of Germination and Rice Koji Fermentation on GO and GABA of Rice after Cooking

Figure 5 compares the effects of *Amazake*'s processing stages (cooking, germination, fermentation and starch hydrolysis) on GO and GABA contents with the traditional milled rice *Amazake*. As expected, steam cooking did not affect GO and GABA content, since both are resistant to boiling temperature [51,52]. With germination, both compounds increased, as was already reported in Figure 2; however, while for GO the value was 78.3% higher, for GABA the increase was not significant ($p > 0.05$), and represented only 17.2%.

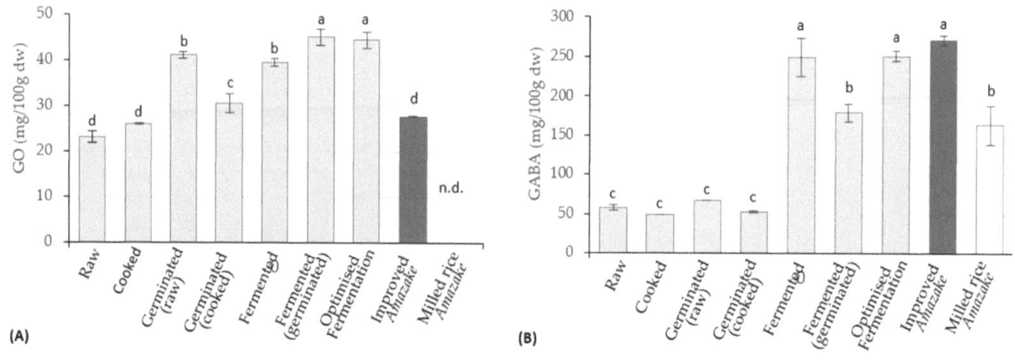

Figure 5. Effect of the various procedures on brown rice on (**A**) GO and (**B**) GABA contents. (n.d. stands for non-detectable). a, b, c, d—homogeneous groups according to the Tukey's post hoc test at a 95% confidence level. The improved *Amazake* is shown in dark grey and the traditional milled rice *Anazake* in shown in white.

Despite the higher GO content in germinated rice when raw, the amount decreased after cooking, which can be related to any leaching of the compounds; the differences between the effect of boiling in raw and germinated rice may be related to the different matrix that germination generates. GO values increased with rice *Koji* fermentation and even more when germination and fermentation were used, as well as in the optimised time × temperature samples (37 °C/20 h).

Regarding GABA, the results show that fermentation greatly impacted the concentration of the compound: non-germinated brown rice fermented at 37 °C for 40 h presents a GABA value of 249.95 ± 24.31 mg/100g; when fermented after germination, the content of GABA decreases; however, after optimisation, (32 °C for 60 h), the GABA content increased to the values of the non-germinated brown rice. The reported values may be explained by starch availability, reported in Table 2 and Figure 3, as *A. oryzae* fermentation occurs based on carbohydrate metabolism [17,18], the production of compounds may be compromised or, in this case, takes longer, derived from the changes occurred during the germination process. Wang, et al. [53] studied the effect of germination and fermentation on GO and GABA contents in rice inoculated with *Bacillus subtilis Natto*. The authors reported a significant decrease of GO and GABA after fermentation in both germinated and non-germinated rice, which is understandable as the starter cultures are different, thus having different metabolisms.

Figure 5 also shows that the GO content decreases sharply after starch hydrolysis, which may be related to the action of other enzymes (e.g., lipases that are also formed during fermentation [19,20]). Regarding GABA, there is a slight content increase after hydrolysis. When comparing the optimised *Amazake* with the one made with milled rice *Koji* under traditional conditions (37 °C/40 h), the optimisation process improved GO and GABA contents.

3.6. Consumer Evaluations

The final *Amazake* product was presented to the consumers at an open science fair. At first, no information was available about the product, and the consumer tried and evaluated the product hedonically. Secondly, the product was shown bottled under the brand DIS!Amasake with a full label and according to the EU labelling legislation for food products [54].

Although the Portuguese gastronomic tradition is rich in rice meals and even deserts [55], eating germinated or fermented rice is not a habit, and to our knowledge, there are no industries in the country producing those products. Despite this, it is possible to find *Koji* products in the market, although only in niche shops or online. Thus, the *Koji* flavour is unknown to most Portuguese consumers.

Overall liking, before elicitation, had a relatively low score, as 42.3% of the participants rated the beverage in the lower range of the scale, and only 37.8% rated the beverage in the higher range (Figure 6). The median and mode score was five, meaning that the product was neither liked nor disliked. The observed values may be related to various factors, including the novelty of the product, and/or cultural differences, as *Amazake* is a traditional drink in Japan. Various authors [56–60] have reported how cultural differences impact food preferences in a variety of cross-cultural studies. Prescott et al. [60] found that Japanese subjects had a higher preference for acidic and umami flavours at higher concentrations than Australian subjects. The Japanese beverage *Amazake* presents an acidic flavour due to the compounds formed naturally in the fermentation process.

Forty consumers filled out the optional open comments section, resulting in a total of 47 different entries. The data were grouped into four categories: taste (n = 23), texture (n = 11), aroma (n = 4), and general attributes (n = 9). The prominent taste characteristics were "bitter" (n = 5), "strong" (n = 4) and "yeast" (n = 4); however, those aspects are not correlated to the liking parameters, but rather to the individual preferences and habits. Regarding texture, the most reported characteristic was "thick" (n = 7), which may be related to the expectation of the testers given their knowledge of rice beverages in the market, which are texturally similar to milk. The aroma was described as "unpleasant" (n = 4), a preference parameter. The participants who described the beverage regarding general attributes reported it as "different/unknown/unexpected" (n = 6), and a participant commented that it "requires habituation". This last comment is interesting, as the literature reports that food preferences are acquired over time [61]; liking a substance tends to increase with familiarity and exposure.

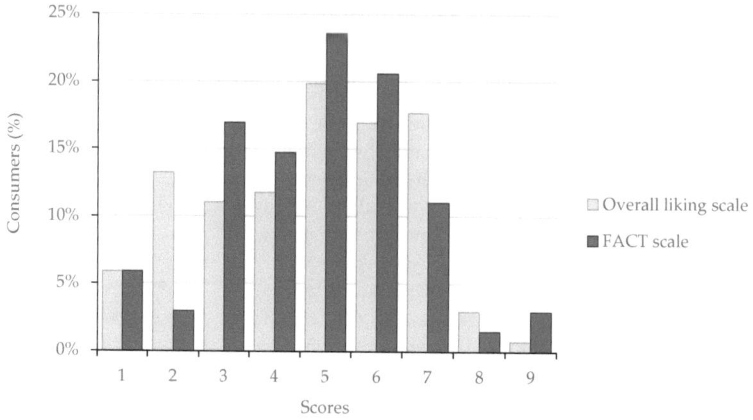

Figure 6. Distribution of consumers' responses regarding overall liking and acceptance (FACT).

Beverage acceptance after exposure to the prototype and related nutritional benefits shows a slight increase when compared with the overall liking scores (Figure 6). Thus, when comparing the differences between liking and acceptance, the participants who most disliked the product were positively impacted by the nutritional elicitation, and the participants with higher liking scores showed a more dispersed attitude; however, the overall liking scores were positively correlated with the FACT scores (R = 0.743).

As for overall liking scores, the impact of nutritional elicitation can be influenced by cultural aspects [62]. When comparing cross-cultural applications of the Food Choice Questionnaire, Cunha, et al. [63] showed that there is a visible difference between the food choice determinants of Japanese and Portuguese consumers: for the Japanese, the health and natural content are more important than sensorial traits, while for the Portuguese the sensorial traits are the most important. Therefore, Portuguese consumers may be less prone to accept unknown health-based foods with a perceived lower sensory appeal.

4. Conclusions and Future Work

The results show that all the measured compounds are affected by the rice raw material (germinated, brown or milled), particularly regarding germinated rice, and also by further fermentation. The germinated rice *Koji* was selected for the *Amazake* preparation due to its higher γ-oryzanol (GO) and γ-aminobutyric acid (GABA) contents. In the selected sample, the GO concentration increased with the germination process in raw samples (from 23.09 ± 1.19 to 41.17 ± 0.84 mg/100 g), and after fermentation (to 44.17 ± 1.8 mg/100 g); however, after the starch hydrolysis process, the values of GO decreased to 27.65 ± 0.23 mg/100 g. The use of the traditional milled rice as raw material for *Amazake* presented a non-detectable GO, while significant GO amounts can be obtained using germinated rice. The GABA content in the selected sample increased with fermentation alone from 57.783 ± 3.60 to 249.95 ± 24.31 mg/100 g, and for the predicted optimum conditions (271.53 ± 5.7 mg/100 g), maintaining the value after starch hydrolysis. Compared to the GABA content of the milled rice *Amazake* (163.95 ± 24.7 mg/100 g), it also presented a great increment. Despite the nutritional improvement associated with the bioactive compounds obtained in the *Amazake*, when presented to the consumers, the overall liking scores were low, which can be attributed to the product novelty and low familiarity with the concept of fermented rice beverages for Portuguese consumers. Nevertheless, as there is room for improvement and even adaptation of the formulation to the Portuguese consumers' habits, further sensory improvement may be sought in the future with different variants of the fermented beverage, as well as the evaluation of acceptance within different consumption contexts.

Supplementary Materials: The following supporting information can be downloaded at: https://www.mdpi.com/article/10.3390/foods12071476/s1, Table S1: Raw data used to fit the response surface design model according to Table 1, RSD (%), and Grubbs distance for the three repetitions. Numbers in bold represent the excluded outliers and criteria of exclusion. When the RSD criteria (>20%) was used, the selected outlier was identified through the highest Grubbs' distance (underlined values).

Author Contributions: Conceptualisation: A.C., C.B. and L.M.C.; methodology, A.C., C.P., M.L., C.B. and L.M.C.; validation, L.M.C., C.B., M.L. and J.C.O.; formal analysis, A.C. and L.M.C.; investigation, A.C., C.P. and M.L.; resources, C.B.; writing—original draft preparation, A.C.; writing—review and editing, C.P., M.L., C.B., J.C.O. and L.M.C.; supervision, C.B. and L.M.C.; project administration, C.B.; funding acquisition, C.B. and L.M.C. All authors have read and agreed to the published version of the manuscript.

Funding: The study was funded by the Foundation for Science and Technology, Portugal, through PhD grant number SFRH/BD/120929/2016 to Ana Castanho, PhD grant number 2020.09555BD to Cristiana Pereira, and through projects UIDB/04551/2020 (GREEN-IT, Bioresources for Sustainability), UIDB/05748/2020 and UIDP/05748/2020 (GreenUPorto). The study was also funded by the project TRACE-RICE: Tracing rice and valorising side streams along Mediterranean blockchain, grant no. 1934, of the PRIMA Programme, supported under Horizon 2020, the European Union's Framework Programme for Research and Innovation.

Institutional Review Board Statement: Ethical review and approval were waived for this study due to the nature of the product, the existance of adequate conditions at the INIAV food science laboratory to prepare the samples under hygienic conditions and to evaluate food safety requirement. The research team ensured the protection and confidentiality of data following EU Regulation 2016/679. All recruited participants were adults or adolescents 16 years of age or above who were willing to participate after being informed that they would be tasting an innovative rice-based drink and that all the results were kept anonymous. All of the adolescent minors participating in this study were accompanied by their consenting parents or educators.

Informed Consent Statement: Informed consent was obtained from all subjects involved in the study.

Data Availability Statement: Data is contained within the article or supplementary material.

Acknowledgments: The authors acknowledge all the consumers that kindly agreed to participate in this study.

Conflicts of Interest: The authors declare no conflict of interest.

References

1. Castanho, A.; Lageiro, M.; Calhelha, R.C.; Ferreira, I.C.F.R.; Sokovic, M.; Cunha, L.M.; Brites, C. Exploiting the bioactive properties of γ-oryzanol from bran of different exotic rice varieties. *Food Funct.* **2019**, *10*, 2382–2389. [CrossRef]
2. Goufo, P.; Trindade, H. Rice antioxidants: Phenolic acids, flavonoids, anthocyanins, proanthocyanidins, tocopherols, tocotrienols, γ-oryzanol, and phytic acid. *Food Sci. Nutr.* **2014**, *2*, 75–104. [CrossRef]
3. Yu, S.G.; Nehus, Z.T.; Badger, T.M.; Fang, N.B. Quantification of vitamin E and gamma-oryzanol components in rice germ and bran. *J. Agric. Food Chem.* **2007**, *55*, 7308–7313. [CrossRef]
4. Shinomiya, M.; Morisaki, N.; Matsuoka, N.; Izumi, S.; Saito, Y.; Kumagai, A.; Mitani, K.; Morita, S. Effects of γ-Oryzanol on Lipid Metabolism in Rats Fed High-Cholesterol Diet. *Tohoku J. Exp. Med.* **1983**, *141*, 191–197. [CrossRef]
5. Sakamoto, K.; Tabata, T.; Shirasaki, K.; Inagaki, T.; Nakayama, S. Effects of gamma-oryzanol and cycloartenol ferulic acid ester on cholesterol diet induced hyperlipidemia in rats. *Jpn. J. Pharmacol.* **1987**, *45*, 559–565. [CrossRef]
6. Jolfaie, N.R.; Rouhani, M.H.; Surkan, P.J.; Siassi, F.; Azadbakht, L. Rice Bran Oil Decreases Total and LDL Cholesterol in Humans: A Systematic Review and Meta-Analysis of Randomized Controlled Clinical Trials. *Horm. Metab. Res.* **2016**, *48*, 417–426. [CrossRef] [PubMed]
7. Bumrungpert, A.; Chongsuwat, R.; Phosat, C.; Butacnum, A. Rice Bran Oil Containing Gamma-Oryzanol Improves Lipid Profiles and Antioxidant Status in Hyperlipidemic Subjects: A Randomized Double-Blind Controlled Trial. *J. Altern. Complement. Med.* **2019**, *25*, 353–358. [CrossRef] [PubMed]
8. Lageiro, M.M.; Castanho, A.; Pereira, C.; Calhelha, R.C.; Ferreira, I.C.F.R.; Brites, C. Assessment of gamma oryzanol variability, an attractive rice bran bioactive compound. *Emir. J. Food Agric.* **2020**, *32*, 38–46. [CrossRef]
9. Kumar, S.S.; Iruthayarajan, M.W.; Bakrutheen, M. Investigations on the suitability of rice bran oil and corn oil as alternative insulating liquids for transformers. *IEEJ Trans. Electr. Electron. Eng.* **2016**, *11*, 10–14. [CrossRef]

10. Valitova, J.N.; Sulkarnayeva, A.G.; Minibayeva, F.V. Plant sterols: Diversity, biosynthesis, and physiological functions. *Biochemistry* **2016**, *81*, 819–834. [CrossRef] [PubMed]
11. Nishimura, M.; Yoshida, S.; Haramoto, M.; Mizuno, H.; Fukuda, T.; Kagami-Katsuyama, H.; Tanaka, A.; Ohkawara, T.; Sato, Y.; Nishihira, J. Effects of white rice containing enriched gamma-aminobutyric acid on blood pressure. *J. Tradit. Complement. Med.* **2016**, *6*, 66–71. [CrossRef] [PubMed]
12. Zhang, H.; Yao, H.-y.; Chen, F.; Wang, X. Purification and characterization of glutamate decarboxylase from rice germ. *Food Chem.* **2007**, *101*, 1670–1676. [CrossRef]
13. Tsuchiya, K.; Nishimura, K.; Iwahara, M. Purification and Characterization of Glutamate Decarboxylase from *Aspergillus oryzae*. *Food Sci. Technol. Res.* **2003**, *9*, 283–287. [CrossRef]
14. Abdou, A.M.; Higashiguchi, S.; Horie, K.; Kim, M.; Hatta, H.; Yokogoshi, H. Relaxation and immunity enhancement effects of gamma-aminobutyric acid (GABA) administration in humans. *Biofactors* **2006**, *26*, 201–208. [CrossRef] [PubMed]
15. Treiman, D.M. GABAergic mechanisms in epilepsy. *Epilepsia* **2001**, *42* (Suppl. S3), 8–12. [CrossRef] [PubMed]
16. Oh, S.-J.; Kim, H.S.; Lim, S.-T.; Reddy, C.K. Enhanced accumulation of gamma-aminobutyric acid in rice bran using anaerobic incubation with various additives. *Food Chem.* **2019**, *271*, 187–192. [CrossRef] [PubMed]
17. Gomi, K. Aspergillus oryzae. In *Encyclopedia of Food Microbiology*; Elsevier Inc.: Amsterdam, The Netherlands, 2014; pp. 92–96. [CrossRef]
18. Chang, P.K.; Horn, B.W.; Abe, K.; Gomi, K. Introduction. In *Encyclopedia of Food Microbiology*; Elsevier Inc.: Amsterdam, The Netherlands, 2014; pp. 77–82. [CrossRef]
19. Soccol, C.R.; Costa, E.S.F.d.; Letti, L.A.J.; Karp, S.G.; Woiciechowski, A.L.; Vandenberghe, L.P.d.S. Recent developments and innovations in solid state fermentation. *Biotechnol. Res. Innov.* **2017**, *1*, 52–71. [CrossRef]
20. Allwood, J.G.; Wakeling, L.T.; Bean, D.C. Fermentation and the microbial community of Japanese *Koji* and miso: A review. *J. Food Sci.* **2021**, *86*, 2194–2207. [CrossRef] [PubMed]
21. Okuda, M.; Iizuka, S.; Xu, Y.; Wang, D. Rice in brewing. In *Rice*, 4th ed.; Bao, J., Ed.; AACC International Press: Washington, DC, USA, 2019; pp. 589–626.
22. Kusumoto, K.-I.; Yamagata, Y.; Tazawa, R.; Kitagawa, M.; Kato, T.; Isobe, K.; Kashiwagi, Y. Japanese Traditional Miso and *Koji* Making. *J. Fungi* **2021**, *7*, 579. [CrossRef]
23. Kurahashi, A.; Yonei, Y. Effects and safety of *Koji Amazake*: An excessive intake test. *J. Brew. Soc. Jpn.* **2019**, *114*, 654–662.
24. Kurahashi, A.; Enomoto, T.; Oguro, Y.; Kojima-Nakamura, A.; Kodaira, K.; Watanabe, K.; Ozaki, N.; Goto, H.; Hirayama, M. Intake of *Koji Amazake* improves defecation frequency in healthy adults. *J. Fungi* **2021**, *7*, 782. [CrossRef]
25. Nagao, Y.; Takahashi, H.; Kawaguchi, A.; Kitagaki, H. Effect of Fermented Rice Drink "*Amazake*" on Patients with Nonalcoholic Fatty Liver Disease and Periodontal Disease: A Pilot Study. *Reports* **2021**, *4*, 36. [CrossRef]
26. WHO. Toxicological evaluation of certain food additives. In Proceedings of the 31st Meeting of the joint FAO/WHO Expert Committee on Food Additives, Geneva, Switzerland, 16–25 February 1987.
27. Lee, I.H.; Chou, C.C. Distribution profiles of isoflavone isomers in black bean *Kojis* prepared with various filamentous fungi. *J. Agric. Food Chem.* **2006**, *54*, 1309–1314. [CrossRef] [PubMed]
28. Kikushima, K.; Nakagawa, T.; Shimizu, K.; Ohnuki, K. Regular Intake of Japanese Traditional Rice Fermented Beverage, *Koji Amazake* for 4 weeks Decreases Systolic Blood Pressure A Randomized, Double-blind, Placebo-controlled, Parallel-group Comparative Study. *Jpn. Pharmacol. Ther.* **2020**, *48*, 305–312.
29. Kurahashi, A.; Nakamura, A.; Oguro, Y.; Yonei, Y. Safety evaluation of a long-term intake of *Koji Amazake*. *J. Brew. Soc. Jpn.* **2020**, *115*, 159–172.
30. Jannoey, P.; Niamsup, H.; Lumyong, S.; Suzuki, T.; Katayama, T.; Chairote, G. Comparison of gamma-aminobutyric acid production in Thai rice grains. *World J. Microbiol. Biotechnol.* **2010**, *26*, 257–263. [CrossRef]
31. Saigusa, N.; Ohba, R. Effects of *Koji* production and saccharification time on the antioxidant activity of *Amazake*. *Food Sci. Technol. Res.* **2007**, *13*, 162–165. [CrossRef]
32. Box, G.E.P.; Hunter, W.H.; Hunter, S. *Statistics for Experimenters*; John Wiley and Sons: New York, NY, USA, 1978; Volume 664.
33. Narahara, H. Growth and Enzyme Production in a Solid-State Culture of *Aspergillus oryzae*. *J. Ferment. Technol.* **1982**, *60*, 311–319.
34. Hong, H.H.; Kim, M.K. Physiochemical Quality and Sensory Characteristics of *Koji* Made with Soybean, Rice, and Wheat for Commercial doenjang Production. *Foods* **2020**, *9*, 975. [CrossRef] [PubMed]
35. Sampaio, P.S.; Soares, A.; Castanho, A.; Almeida, A.S.; Oliveira, J.; Brites, C. Optimization of rice amylose determination by NIR-spectroscopy using PLS chemometrics algorithms. *Food Chem.* **2018**, *242*, 196–204. [CrossRef] [PubMed]
36. Peryam, D.R.; Pilgrim, F.J. Hedonic scale method of measuring food preferences. *Food Technol.* **1957**, *11* (Suppl. S1), 9–14.
37. Schutz, H.G. Food action rating scale for measuring food acceptance. *J. Food Sci.* **1965**, *30*, 365–374. [CrossRef]
38. Ribeiro, J.C.; Lima, R.C.; Maia, M.R.; Almeida, A.A.; Fonseca, A.J.; Cabrita, A.R.J.; Cunha, L.M. Impact of defatting freeze-dried edible crickets (*Acheta domesticus* and *Gryllodes sigillatus*) on the nutritive value, overall liking and sensory profile of cereal bars. *LWT* **2019**, *113*, 108335. [CrossRef]
39. Rojas-Rivas, E.; Rendón-Domínguez, A.; Felipe-Salinas, J.A.; Cuffia, F. What is gastronomy? An exploratory study of social representation of gastronomy and Mexican cuisine among experts and consumers using a qualitative approach. *Food Qual. Prefer.* **2020**, *83*, 103930. [CrossRef]

40. Carter, N.; Bryant-Lukosius, D.; DiCenso, A.; Blythe, J.; Neville, A.J. The Use of Triangulation in Qualitative Research. *Oncol. Nurs. Forum* **2014**, *41*, 545–547. [CrossRef] [PubMed]
41. Akaike, H. Information Theory and an Extension of the Maximum Likelihood Principle. In *Selected Papers of Hirotugu Akaike*; Parzen, E., Tanabe, K., Kitagawa, G., Eds.; Springer: New York, NY, USA, 1998; pp. 199–213. [CrossRef]
42. Wu, N.N.; Li, R.; Li, Z.J.; Tan, B. Effect of germination in the form of paddy rice and brown rice on their phytic acid, GABA, gamma-oryzanol, phenolics, flavonoids and antioxidant capacity. *Food Res. Int.* **2022**, *159*, 111603. [CrossRef] [PubMed]
43. Munarko, H.; Sitanggang, A.B.; Kusnandar, F.; Budijanto, S. Germination of five Indonesian brown rice: Evaluation of antioxidant, bioactive compounds, fatty acids and pasting properties. *Food Sci. Technol.* **2022**, *42*. [CrossRef]
44. Summpunn, P.; Panpipat, W.; Manurakchinakorn, S.; Bhoopong, P.; Cheong, L.Z.; Chaijan, M. Comparative Analysis of Antioxidant Compounds and Antioxidative Properties of Thai Indigenous Rice: Effects of Rice Variety and Processing Condition. *Molecules* **2022**, *27*, 5180. [CrossRef]
45. Nonogaki, H.; Bassel, G.W.; Bewley, J.D. Germination—Still a mystery. *Plant Sci.* **2010**, *179*, 574–581. [CrossRef]
46. Xu, J.; Zhang, H.; Guo, X.; Qian, H. The impact of germination on the characteristics of brown rice flour and starch. *J. Sci. Food Agric.* **2012**, *92*, 380–387. [CrossRef]
47. Elkhalifa, A.E.O.; Bernhardt, R. Influence of grain germination on functional properties of sorghum flour. *Food Chem.* **2010**, *121*, 387–392. [CrossRef]
48. Tomasik, P.; Horton, D. Enzymatic conversions of starch. In *Advances in Carbohydrate Chemistry and Biochemistry*; Horton, D., Ed.; Academic Press: Cambridge, MA, USA, 2012; Volume 68, pp. 59–436.
49. Azmi, A.S.; Malek, M.I.A.; Puad, N.I.M. A review on acid and enzymatic hydrolyses of sago starch. *Int. Food Res. J.* **2017**, *24*, 265–273.
50. Feng, Y.; Cai, Y.; Su, G.; Zhao, H.; Wang, C.; Zhao, M. Evaluation of aroma differences between high-salt liquid-state fermentation and low-salt solid-state fermentation soy sauces from China. *Food Chem.* **2014**, *145*, 126–134. [CrossRef]
51. Srisang, N.; Chungcharoen, T. Quality attributes of parboiled rice prepared with a parboiling process using a rotating sieve system. *J. Cereal Sci.* **2019**, *85*, 286–294. [CrossRef]
52. Srisaipet, A.; Nuddagul, M. Influence of Temperature on Gamma-Oryzanol Stability of Edible Rice Bran Oil during Heating. *Int. J. Chem. Eng. Appl.* **2014**, *5*, 303–306. [CrossRef]
53. Wang, H.J.; Chang, L.; Lin, Y.S. Changes in Functionality of Germinated and Non-Germinated Brown Rice Fermented by Bacillus natto. *Foods* **2021**, *10*, 2779. [CrossRef] [PubMed]
54. Montanari, F.; Moura, A.P.D.; Cunha, L.M. *Manual de Rotulagem Alimentar, Abordagem Integrada Desde os Operadores Económicos ao Consumidor*; Edições Afrontamento: Porto, Portugal, 2022.
55. Cabral, D.; Fonseca, S.C.; Moura, A.P.; Oliveira, J.C.; Cunha, L.M. Conceptualization of Rice with Low Glycaemic Index: Perspectives from the Major European Consumers. *Foods* **2022**, *11*, 2172. [CrossRef] [PubMed]
56. Hong, J.H.; Yoon, E.K.; Chung, S.J.; Chung, L.; Cha, S.M.; O'Mahony, M.; Vickers, Z.; Kim, K.O. Sensory characteristics and cross-cultural consumer acceptability of Bulgogi (Korean traditional barbecued beef). *J. Food Sci.* **2011**, *76*, S306–S313. [CrossRef] [PubMed]
57. Laureati, M.; Pagliarini, E.; Bassoli, A.; Borgonovo, G. Sensory and hedonic perceptions of italian and korean subjects: A Cross-Cultural study of Perilla Frutescens. *Food Sci. Biotechnol.* **2014**, *23*, 1111–1120. [CrossRef]
58. Moskowitz, H.W.; Kumaraiah, V.; Sharma, K.N.; Jacobs, H.L.; Sharma, S.D. Cross-cultural differences in simple taste preferences. *Science* **1975**, *190*, 1217–1218. [CrossRef]
59. Zellner, D.A.; Garriga-Trillo, A.; Rohm, E.; Centeno, S.; Parker, S. Food Liking and Craving: A Cross-cultural Approach. *Appetite* **1999**, *33*, 61–70. [CrossRef] [PubMed]
60. Prescott, J.; Laing, D.; Bell, G.; Yoshida, M.; Gillmore, R.; Allen, S.; Yamazaki, K.; Ishii, R. Hedonic responses to taste solutions: A cross-cultural study of Japanese and Australians. *Chem. Senses* **1992**, *17*, 801–809. [CrossRef]
61. Pliner, P. The Effects of Mere Exposure on Liking for Edible Substances. *Appetite* **1982**, *3*, 283–290. [CrossRef] [PubMed]
62. Miraballes, M.; Fiszman, S.; Gámbaro, A.; Varela, P. Consumer perceptions of satiating and meal replacement bars, built up from cues in packaging information, health claims and nutritional claims. *Food Res. Int.* **2014**, *64*, 456–464. [CrossRef]
63. Cunha, L.M.; Cabral, D.; Moura, A.P.; de Almeida, M.D.V. Application of the Food Choice Questionnaire across cultures: Systematic review of cross-cultural and single country studies. *Food Qual. Prefer.* **2018**, *64*, 21–36. [CrossRef]

foods

MDPI

Article

Effects of Far-Infrared Radiation Drying on Starch Digestibility and the Content of Bioactive Compounds in Differently Pigmented Rice Varieties

Jiranan Ratseewo [1], Frederick Jame Warren [2], Naret Meeso [3] and Sirithon Siriamornpun [4,*]

[1] Division of Food Science and Technology, Faculty of Liberal Arts and Science, Sisaket Rajabhat University, Sisaket 33000, Thailand
[2] Quadram Institute Biosciences, Food and Health, Norwich Research Park, Norwich NR47 UA, UK
[3] Research Unit of Drying Technology for Agricultural Products, Faculty of Engineering, Mahasarakham University, Kantarawichai, Maha Sarakham 44150, Thailand
[4] Research Unit of Thai Food Innovation, Department of Food Technology and Nutrition, Mahasarakham University, Kantarawichai, Maha Sarakham 44150, Thailand
* Correspondence: sirithon.s@msu.ac.th; Tel.: +66-43-754085

Citation: Ratseewo, J.; Warren, F.J.; Meeso, N.; Siriamornpun, S. Effects of Far-Infrared Radiation Drying on Starch Digestibility and the Content of Bioactive Compounds in Differently Pigmented Rice Varieties. *Foods* **2022**, *11*, 4079. https://doi.org/10.3390/foods11244079

Academic Editors: John C Beaulieu, Ming Hsuan Chen, Ida Wenefrida and Piebiep Goufo

Received: 22 November 2022
Accepted: 14 December 2022
Published: 16 December 2022

Abstract: Far infrared radiation (FIR) was applied to six rice varieties with different coloring of the pericarp (purple, red or non-pigment). Changes were determined in amylose content, in gelatinization parameters, in the content of bioactive compounds, in antioxidant activity and in the in vitro digestibility of pigmented rice as affected by FIR. The highest contents of amylose, total phenolic (TPC), total flavonoid (TFC) and total anthocyanins (TAC) were found in the purple and red varieties. Overall, FIR increased TPC, TFC and TAC, including antioxidant capacity. Quercetin and apigenin contents were increased while rutin and myricetin decreased significantly ($p < 0.05$) in all FIR dried samples. Dephinidin, cyanidin-3-glucosides and pelargonidin increased after FIR treatment. Mostly, FIR-treated samples were found to have greater gelatinization enthalpy, compared with unheated rice samples. FIR-dried rice showed lower starch digestibility (25–40%) than unheated rice. This research suggested that the specific genotype of rice had the greatest influence on amylose content in pigmented rice, while FIR drying had no further effect. Our results suggest that FIR could enhance the content of the bioactive compounds capable of inhibiting α-amylase, thereby lowering starch digestibility. Hence, FIR may be considered as an appropriate drying method for pigmented rice regarding health benefits.

Keywords: pigmented rice; starch digestion; phenolics; anthocyanins; far-infrared radiation

1. Introduction

Rice (*Oryza sativa* L.) is an important cereal crop and is widely consumed in many countries of the world. Mostly, white rice (non-pigmented or normal rice) is widely consumed as a staple food in Asia. However, over the past decade, pigmented rice (purple, black, red coloring, etc.) has been gaining more attention from consumers because of its perceived health benefits. The pigments on the seed coat have been implicated for enhancing the function of the immune system. Furthermore, pigments of the rice-bran fractions may inhibit allergic reactions in vitro [1]. The various types of bioactive substances in rice are generally found in the germ, bran and outermost aleurone layers of the grain, and they include colored pericarp layers, depending on the varieties of rice [2–4].

The postprandial blood-glucose-raising potential [5] of rice has been established as an indicator of starch digestibility, thus enabling the selection of foods containing abundant amounts of carbohydrates, therefore avoiding the accrued risk of non-communicable diseases, such as type-2 diabetes, cardiovascular disease and obesity, which may be related to the continuous consumption of high-starch digestion or high glycemic index (GI) foods [5]. The pigmented pericarp of rice comprises various phytochemicals, mainly phenolic compounds and anthocyanins [6,7]. These compounds have been reported to possess anti-oxidative effects and are capable of reducing the intensity of oxidative free radicals [1,8]. Many studies have indicated the significant biological value of compounds such as antioxidants and compounds with anti-cancer, anti-viral, or anti-inflammatory activities, thus offering health-promoting properties [1,9,10]. Furthermore, there has been a number of studies reporting that the starch digestibility of pigmented rice is less than that of normal rice. Such pigmented varieties include pigmented Indian rice [11] and pigmented Thai rice varieties [12,13]. Moreover, researchers reported that phenolic compounds, including anthocyanins, are potent inhibitors of α-amylase and α-glucosidase, these being important enzymes associated with carbohydrate digestion and sugar absorption, thus related to diabetes mellitus. They can be supportive for hyperglycemia control and type-2 diabetes-mellitus treatment through lowering blood-glucose levels [14–17].

Nevertheless, alterations in the quantity and composition of these compounds may be affected by drying methods currently used in the cereal-crops industry. The drying process plays a major role in decreasing moisture content for safe storage. Far-infrared radiation (FIR) has been applied for drying biomaterials, including agricultural products used for food [18,19]. FIR drying involves heat from rays transferred smoothly to the centre of the material being dried, thus causing vibration of the bonds linking molecules or components within the material, without (or only slightly) decomposing molecules on the material's surface. Recently, several studies have reported changes in physical properties, in bioactive compounds and in the antioxidant activity of agricultural products as a result of FIR treatment. Examples include water extraction from peanut hulls [20], Kaprow leaves [19], rice bran [18] and pigmented rice in Thailand [4]. In addition, Tangkhawanit et al. [21] have reported that the effects of FIR-HA drying increase the content of bioactive components, such as total phenolic, flavonoid and isoflavone in whole soybean and soy residue, thereby providing improved inhibition of the enzyme in soymilk residue extracts and lowering starch digestion. From our previous studies, we had reported that the starch digestibly of six varieties of pigmented rice was lower than that for normal rice. These changes were associated with higher amounts of polyphenols and anthocyanins present in those pigmented rice varieties [4]. Later, we reported on how hot air drying and FIR drying affected the bioactive compounds, as well as antioxidant activity in three varieties of pigmented rice. We found that FIR had shown remarkably positive effects on those parameters that were tested [13]. According to our data, it is doubtful whether these results would be consistent during different years of production (climate changes). Although changes due to FIR drying on bioactive compounds and antioxidant activities were reported in some varieties, there have been some pigmented rice varieties not yet studied. Furthermore, the effects of FIR drying methods on starch digestibility have not been reported elsewhere on these varieties of Thai pigmented rice.

This present study aimed to explore the effects of FIR drying methods on starch digestibility and bioactive compounds, namely, phenolics, flavonoids and anthocyanins, along with the antioxidant activities of Thai rice varieties with red and purple pericarp coloring as well as non-pigmented rice. Furthermore, the comparison contrasted unheated and heated grain, thus to providing useful data for food and agricultural processing, engineering, industrial practice and the incorporation of functional food materials into human diets with health-promoting benefits.

2. Materials and Methods

2.1. Chemicals and Reagents

Standards of phenolic acids were used, namely ferulic, *p*-coumaric, sinapic, gallic, caffeic, syringic, protocatechuic, vanillic and chlorogenic acids. Standards of flavonoids were used, namely apigenin, quercetin, myricetin and rutin. Standards of anthocyanins were used, namely malvidin, pelargonidin and cyanidin 3-glucoside. All of the standards were purchased from the Sigma–Aldrich Co. (St. Louis, MO, USA). All other solvents and chemicals used were analytical reagent grade.

2.2. Sample Preparation

Six rice varieties, Hom Mali or KDML 105 (non-pigmented rice); Sung Yod, Mun Poo, Mali Dang (red rice); and Hom Nil and Riceberry (purple rice), were collected from fields located in Thailand, during the 2018–2019 growing season. Each paddy of pigmented rice was de-husked to provide brown rice. Once obtained, the FIR-dried rice was prepared by FIR and combined with hot air (HA), according to our previous study [4]. The dryer consists of a stainless-steel drying chamber (30 cm × 51 cm × 50 cm), a sample tray (25.4 cm × 37 cm), a centrifugal fan and a far-infrared heater (122 mm × 60 mm). Two sets of three-FIR heaters were placed, one at the bottom and another one at the top of the drying chamber. The sample tray was set midway between, and parallel to, the top and bottom heaters. The hot air was circulated through the drying chamber with a fan. The temperature of inlet air flowing through a hot-air heater was controller with a PID controller (accuracy of ±1 °C). The de-husked rice (10 g) was dried in a drying chamber and then treated with FIR at an air velocity of 1.5 m/s, 40 °C and an intensity of 2 kW/m^2, 250 W for 2 h. The moisture content (wet basis) of obtained samples were in a range of 12–14%. Samples were ground into a fine powder and then sieved (using an 80 mesh sieve). The FIR treatment was conducted in triplicate. Subsequently, both unheated and FIR-treated rice powder were kept at 4 °C in three different bags before further analysis.

2.3. Determination of Total Phenolic Content

The total phenolic content (TPC) was determined according to the method described by Ratseewo et al. [22], using the Folin–Ciocalteu method with gallic acid as a standard. The absorbance of samples was read at 725 nm using a spectrophotometer (Beckman Coulter, Fullerton, CA, USA). The quantity of total phenolics was expressed as mg gallic acid equivalents per 100 g of dry weight (mg GAE/100 g DW).

2.4. Determination of Total Flavonoid Content

The total flavonoid content (TFC) was determined using the method of Ratseewo et al. [13], by a colorimetry method. The absorbance of samples was read at 510 nm using a spectrophotometer. The content of flavonoid was expressed as mg rutin equivalents per 100 g dry weight (mg RE/100 g DW).

2.5. Determination of Total Anthocyanin Content

The total anthocyanin content (TAC) was determined spectrophotometrically according to Ratseewo et al. [4]. The absorbance was read at 534 nm using a spectrophotometer. The value of total anthocyanin was determined as mg cyanidin equivalents per 100 g dry weight (mg CyGE/100 g DW).

2.6. Determination of Antioxidant Activity Using DPPH Radical-Scavenging Activity

The determination of antioxidant activity was involved the 2,2-diphenyl-1-picrylhydrazyl (DPPH) radical scavenging activity assay [13]. The absorbance was determined at 517 nm using a spectrophotometer. All analyses were done in triplicate.

The calculation of DPPH radical scavenging activity as percent inhibition was performed using this following formula:

$$(\%) = [1 - (A(sample) - A(control))] \times 100. \quad (1)$$

2.7. Amylose Content

The determination of amylose content involved the iodine dye-binding method according to Knutson [23] with some modification. The samples were weighed (each 5 g) and were mixed with 6 mM iodine and 90% (*v/v*) dimethyl sulfoxide (DMSO) in distilled water (10 mL). Rice samples were incubated at room temperature overnight and shaken at a 180°

angle and 20 rpm. After 30 min, the samples were determined using a spectrophotometer at 600 nm (Libra 550, Biochrom, Holliston, MA, USA). The percentages of apparent amylose and amylopectin contents were calculated using an amylose standard from potato by the equation:

$$\% \text{ Amylose} = \frac{\% \text{ Apparent amylose } - 6.2}{93.8} \quad (2)$$

2.8. Identification of Phenolic Compounds

The authentic phenolic acids were identified using an HPLC (high performance liquid chromatography) system (Shimadzu LC-20 AC pumps). Phenolic acids were separated on a C18 column (4.6 mm × 250 mm) with a particle size of 5 μm (Hichrom Limited, Berks, UK) and were detected at 280 nm. The HPLC analysis used a diode array detection (SPD-M20 A, DAD), as described by Ratseewo et al. [4]. The quantity of individual phenolic acids was calculated as external standards curves.

2.9. Identification of Flavonoids

The authentic individual flavonoids were identified using an HPLC (high performance liquid chromatography) system (Shimadzu LC-20 AC pumps). The individual flavonoids were separated on a C18 column (4.6 mm × 250 mm) with a particle size of 5 μm (Hichrom Limited, Berks, UK) and were detected at 370 nm. The HPLC analysis used diode array detection (SPD-M20 A, DAD), using the same equipment as for phenolic acids. The gradient elution of HPLC system was described by Ratseewo et al. [4].

2.10. Identification of Anthocyanins

The analysis of anthocyanins was performed by an RP-HPLC method, using the same equipment as for phenolic acids and flavonoids. The extraction and a gradient elution procedure were described by Ratseewo et al. [4]. The UV-array detector was set at 520 nm. Anthocyanins were quantified by comparing them with their authentic standards curves. The average peak areas were used in calculations.

2.11. Differential Scanning Calorimetry (DSC)

Differential scanning calorimetry (DSC) analysis was used to determine the thermal behavior of the rice-starch samples using a Multi-Cell DSC (TA Instruments, Elstree, UK), as described by Ratseewo et al. [13]. Pigmented rice flour materials were extracted to obtain rice starch [13] prior to DSC analysis. The parameters, namely onset temperature (T_o), peak temperature (T_p), and conclusion temperature (T_c) including enthalpy of gelatinization $\Delta_{gel}H$ (J_{g-1}), were recorded from each thermogram of the samples.

2.12. Starch Digestion

The kinetics of in vitro starch digestion were studied according to the method of Ratseewo et al. [13] and Warren et al. [24]. To start the starch hydrolysis, α-amylase (80 units in PBS, pH 7.4, Sigma-Aldrich) in 400 μL was added to each rice powder. The starch hydrolysis was monitored by taking sub-samples at pre-defined time points from 1 to 60 min (at 0, 1, 2, 3, 4, 5, 7, 10, 13, 15, 20, 45 and 60 min). The starch hydrolysis products were boiled at 100 °C in a water bath for 5 min and were determined using reducing sugar analysis the 4-hydroxybenzoic acid hydrazide (PAHBAH) reducing sugar assay. The analysis of maltose, the amylolysis product, was performed at an absorbance of 405 nm.

The estimate of the kinetics of in vitro starch digestion was calculated by the first-order equation [25]:

$$C_t = C_\infty \left(1 - e^{-kt}\right) \quad (3)$$

In this equation, the content of maltose (starch digest product) at time t was C_t; the equivalent content of maltose (starch digest product) at the end point of the reaction was

C_∞; the constant of digestibility rate was k. These values were obtained using logarithm of slope (LOS) analysis, as described previously [24,26].

$$\ln\left(\frac{dC}{dt}\right) = \ln(C_\infty k) - kt \tag{4}$$

In addition, a single reaction which is the situation where multiple first-order rate constants are observed can be described by this equation [27]:

$$C_t = \begin{cases} C_{1\infty}\left(1 - e^{-k_1 t}\right) if\ t \le t_{int} \\ C_{int} + C_{2\infty}\left(1 - e^{-k_2(t-t_{int})}\right) if\ t \ge t_{int} \end{cases} \tag{5}$$

This formula exhibited the parameter, namely, the time of intersection of the two plots was t_{int} and the content of product at t_{int} was C_{int}. $C_{1\infty}$ was the end point of the reaction of single phase or first phase. $C_{2\infty}$ was the end point of the reaction of second phase. k_1 and k_2 were the first- and second-rate constants of each phase. Total C_∞ is the sum of $C_{1\infty}$ and $C_{2\infty}$ and represents the total extent of starch amylolysis.

2.13. Statistical Analysis

The data are expressed as the mean ± standard deviation (SD). Analysis of variance (ANOVA) was analyzed using the statistical analysis system (SPSS 11.5 for Windows; SPSS Inc., Chicago, IL, USA). All of the experiments were tested in triplicate. Duncan's multiple range tests were analyzed. The confidence limits used in this study were based on 95% ($p < 0.05$).

3. Results and Discussion

3.1. Total Phenolic Content (TPC)

The total phenolic content of unheated pigmented rice ranged from 112 mg GAE/100 g dw in KDML105 (non-pigmented rice) to 680 mg GAE/100 g dw in Mali dang (Figure 1A). The total phenolic content was significantly different among the rice samples studied. All pigmented rice samples had higher TPC than white rice. Similarly, a range of results for TPC in pigmented rice was observed in rice from different growth sites, i.e., China, Sri Lanka, Thailand [13,28], and Australia [2]. FIR-dried samples had higher TPC than unheated rice, ranging from 10% (KDML 105) to 16% (Hom Nil). Similar findings were reported in many research reports. The amount of TPC increased after drying by FIR in all six rice varieties, with increases from 7.5% to 18% in pigmented rice [4] and 25% in soy-milk residue extracts [21], 8% in marigold flowers [29], 5% in Kaprow leaves [19] and a dramatic increase (10 times) in infrared-dried strawberry [30].

The explanation of these phenomena could be that the heat generated by FIR via molecular vibrations is absorbed rapidly and steadily to the centre of the rice grains evenly during drying. This treatment may result in the breaking down of the covalent bonds. Hence, FIR heating could release and activate the small-molecule phenolic compounds in biomaterials [18,20] which would increase the TPC of dried pigmented rice in our present study. In addition, phenolic groups could increase after thermal processing. Possibly these phenolic compounds could be liberated from wall structures and detached from other macro molecules such as proteins, fibers and pectins via hydrolysis, due to the breaking of covalent bonds [31]. However, in some cases, thermal instability may cause an increase or decrease of total phenolics [9]. We speculate that changes in the internal structures and compounds in pigmented rice were significantly affected by FIR treatment, due to increases in extractable matter, resulting in increased extraction yields of phenolics and other components.

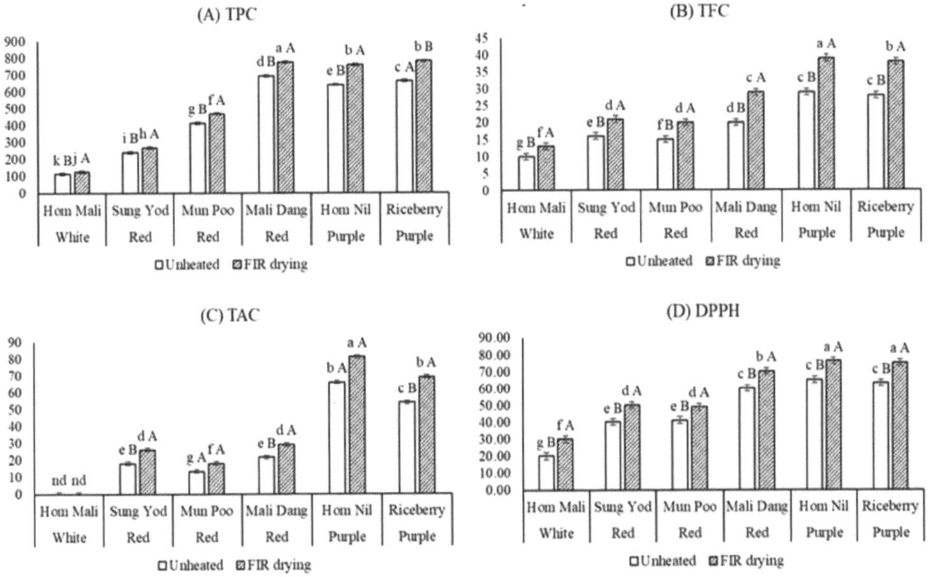

Figure 1. Effect of FIR drying on (**A**) total phenolic content; TPC (mg GAE/100 g DW), (**B**) total flavonoid content; TFC (mg RE/100 g DW), (**C**) total anthocyanin content (TAC) (mg CyGE/100 g) and (**D**) antioxidant capacity (DPPH) (%) of pigmented rice. Values are expressed as mean ± standard deviation (n = 3). Means with different capital letters of unheated and FIR drying rice in the same rice variety as affected by different drying methods were significantly different at the level ($p < 0.05$). Lowercase superscripts are used for comparison of all samples.

3.2. Total Flavonoid Content (TFC)

The total flavonoid contents of the six rice varieties ranged from 10.01 mg RE/g DW in KDML 105 to 39.12 mg RE/g DW in Hom Nil (Figure 1B). TFC of FIR-treated rice to different thermal processing was significantly ($p < 0.05$) higher than that of unheated pigmented rice. The content of total flavonoids increased by approximately 1.3–1.4-fold in FIR-dried rice, compared with unheated rice. The TFC of rice grains seemed to increase when irradiation of far infrared, combined with hot air, was applied. Similarly, to trends in the results of TPC, TFC was enhanced by FIR drying and the proposed reasons could be explained as the same with that of phenolic compounds. Generally, flavonoids are a group of polyphenol secondary metabolites which have many functions including being natural antioxidants in plant materials [18]. The irradiation by far infrared may vibrate covalent bonds, causing the release of lower molecular weight phenolics and flavonoid compounds from the matrix and structure of rice [13]. Many studies have also reported an increase in the TFC of heated plants, such as Thai rice [13], grape seeds [32] and peanut hulls [20], due to the liberation of flavonoid compounds and some other phenolics.

3.3. The Total Anthocyanin Content (TAC)

The TAC of pigmented rice varieties at different thermal treatments are presented in Figure 1C. The TAC of all unheated samples ranged from non-detected in normal rice (KDML 105) to 69 mg/100 g dw in purple rice (Hom Nil). Similar results were shown for TAC values of those varieties, but higher amounts were found in FIR-treated rice ranging from 0–81 mg/100 g. The highest concentration of total anthocyanin was observed in purple rice, followed by red-rice samples. A previous study reported that untreated black/purple rice from Thailand and China showed higher TAC values than red-rice varieties [28]. Our results in this study were similar to the previous study [28]. After FIR treatment, all six rice varieties showed significant increases in the TAC ($p < 0.05$). FIR-dried rice may increase

TAC that are similar to TPC and TFC. Similar results were observed by the increases in anthocyanins after the FIR treatment of pigmented rice [4], including cyanidin-3-glucoside, one of the major anthocyanins of pigmented rice, which may account for increases in the total content of anthocyanins in this research. Our findings were consistent with a previous report [33], that thermal treatment commonly induced an increase in the main phenolic compounds such as the anthocyanins and total cinnamates in orange juice.

3.4. Antioxidant Activity

The DPPH radical scavenging activity of the unheated and FIR-treated pigmented and non-pigmented rice varieties is presented in Figure 1D. Both red and purple samples had higher values of inhibition (%) than white rice. Similar trends were observed in TPC, TFC and TAC, depending on the pericarp color. A similar result was reported in previous studies [2,4,34]. The inhibition (%) of DPPH radical pigmented rice depends on the pericarp of the grains, especially the pigmentation of the pericarp. After dried rice, the highest values of DPPH radical scavenging activity were observed in all FIR-dried samples when compared to unheated samples. The results indicated FIR radiation increased TPC, TFC, TAC and antioxidant activity. Similar results were detected for the antioxidant activity which increased after FIR treatments of rice bran (increased by 20%) and rice husk (increased by 35%) [18], tomato (21%) and papaya (21%) [35]. In addition, Adak et al. [30] reported a massive increase of antioxidant activity in strawberry as affected by infrared drying (1.5 $m{\cdot}s^{-1}$, 200 W, 100 °C). Generally, most natural antioxidants are bound in covalent form to insoluble polymers. When FIR drying is applied to the biomaterial containing those compounds, the radiation of far-infrared is able to activate and release low molecular-weight natural antioxidants, such as phenolic acids, flavonoids, etc., in samples if this bonding is attacked [20].

3.5. Phenolic Compounds

The individual phenolic compounds of unheated and FIR-treated rice samples are shown in Table 1. The predominant phenolic acids found in non-pigmented rice included FA, p-CA and PCCA whereas those compounds, along with VA, were found in all pigmented varieties studied, like the results reported for Australia rice [2].

Table 1. Identifications of phenolic acids in pigmented rice varieties as affected by FIR.

Rice Varieties	Treatments	Phenolic Acids (µg/100 g)									
		GA	PCCA	VA	ChA	CFA	SyA	*p*-CA	FA	SNA	Total
KDML105	Unheated	2.42 ± 0.07 f	14.48 ± 0.56 d	2.24 ± 0.08 g	2.66 ± 0.12 c	1.45 ± 0.04 c	1.53 ± 0.09 f	3.32 ± 0.26 h	16.50 ± 0.63 e	1.14 ± 0.06 j	45.75 ± 2.42 e
(White)	FIR	2.98 ± 0.11 e	21.12 ± 0.11 a	2.49 ± 0.09 g	2.99 ± 0.19 ab	1.46 ± 0.01 c	1.66 ± 0.04 ef	3.55 ± 0.09 h	22.12 ± 0.87 c	1.21 ± 0.08 i	59.58 ± 2.52 c
Sung Yod	Unheated	2.35 ± 0.07 f	13.58 ± 0.19 e	3.49 ± 0.07 f	1.41 ± 0.09 e	1.26 ± 0.07 d	1.51 ± 0.05 f	3.72 ± 0.17 g	16.85 ± 0.21 e	1.49 ± 0.04 h	45.66 ± 3.45
(Red)	FIR	2.97 ± 0.13 e	20.32 ± 0.13 b	5.81 ± 0.21 c	1.45 ± 0.12 de	1.29 ± 0.03 d	1.74 ± 0.06 de	3.98 ± 0.07 f	22.33 ± 1.43 c	1.62 ± 0.05 g	61.51 ± 4.58 c
Mun Poo	Unheated	2.29 ± 0.25 f	12.64 ± 0.67 f	3.74 ± 0.12 e	2.52 ± 0.05 c	1.46 ± 0.05 bc	1.55 ± 0.07 e	5.51 ± 0.22 e	16.83 ± 0.15 e	1.50 ± 0.09 h	48.04 ± 3.71 e
(Red)	FIR	2.97 ± 0.14 e	18.38 ± 0.12 d	4.24 ± 0.22 d	2.61 ± 0.11 c	1.44 ± 0.02 c	1.75 ± 0.05 d	6.12 ± 0.12 d	23.57 ± 1.22 c	1.65 ± 0.04 g	62.35 ± 4.79 c
Mali Dang	Unheated	3.31 ± 0.11 d	13.21 ± 0.26 ef	3.76 ± 0.09 e	1.50 ± 0.07 de	1.54 ± 0.05 b	1.63 ± 0.05 e	5.43 ± 0.23 e	17.71 ± 0.26 d	2.32 ± 0.05 f	50.42 ± 4.91 d
(Red)	FIR	4.98 ± 0.11 c	18.87 ± 0.13 c	4.31 ± 0.18 d	1.68 ± 0.11 d	1.58 ± 0.05 b	1.89 ± 0.01 c	5.77 ± 0.11 e	22.65 ± 1.27 c	2.61 ± 0.18 e	64.34 ± 4.31 c
Hom Nil	Unheated	6.29 ± 0.15 b	14.82 ± 0.72 d	6.71 ± 0.06 b	2.75 ± 0.08 bc	1.71 ± 0.06 a	1.74 ± 0.08 d	6.58 ± 0.23 c	33.11 ± 0.42 b	5.01 ± 0.17 d	78.72 ± 4.28 b
(Purple)	FIR	6.96 ± 0.21 a	20.98 ± 0.15 a	7.01 ± 0.31 a	3.21 ± 0.15 a	1.72 ± 0.09 a	1.99 ± 0.01 b	6.81 ± 0.21 c	40.22 ± 1.35 a	5.32 ± 0.11 c	94.22 ± 4.71 a
Riceberry	Unheated	6.24 ± 0.09 b	14.76 ± 0.34 d	6.70 ± 0.08 b	2.76 ± 0.04 bc	1.76 ± 0.04 a	1.97 ± 0.05 a	7.21 ± 0.13 b	32.32 ± 0.37 b	5.14 ± 0.26 b	78.76 ± 3.98 b
(Purple)	FIR	6.87 ± 0.16 a	21.07 ± 0.15 a	7.12 ± 0.24 a	3.11 ± 0.16 a	1.76 ± 0.07 a	2.03 ± 0.06 a	7.66 ± 0.21 a	40.28 ± 2.39 a	5.43 ± 0.11 a	95.33 ± 4.64 a

GA (Gallic), PCCA (Protocatechuic), VA (Vanillic), ChA (chlorogenic), CFA (Caffeic), SyA (Syringic), p-CA (p-Coumaric), FA (Ferulic) and SNA (Sinapic) acids. Mean values within a column superscripted by the small letter are significantly different at $p < 0.05$.

These results have been consistent with the data from previous research in different years of production that the major phenolic acid in non-pigmented and pigmented rice types (Hom Nil and Riceberry), is FA. It is normally present in the rice bran layers, which may be derived from arabinoxylans, pectin and/or cross-linked to polysaccharides in cell walls [36]. All rice samples treated with FIR had greater phenolic acid contents than untreated rice. GA, FA and SNA were significantly ($p < 0.05$) increased by FIR in all treated samples. Similarly, previous research reported that ChA and p-CA acids of tomato and papaya were significantly increased by FIR treatment [35].

3.6. Composition of Flavonoids

When compared between FIR and untreated rice, the highest values of quercetin were found in FIR-dried samples of the six rice varieties, ranging from 5.43 to 23.94 µg/100 g (Table 2). Rutin and myricetin were significantly reduced ($p < 0.05$) by FIR treatment in all pigmented rice varieties, while apigenin and quercetin were significantly increased ($p < 0.05$) when compared to unheated samples. Apigenin and quercetin contents were increased significantly in all purple rice samples, by 18% and 41%, as affected by FIR, respectively.

Table 2. Identifications of flavonoids and anthocyanin in pigmented rice varieties as affected by FIR.

Rice Varieties	Treatments	Flavonoids (µg/100 g)				Anthocyanin (µg/100 g)		
		Rutin	Myricetin	Quercetin	Apigenin	Cyanidin-3-glucoside	Pelargonidin	Malvidin
KDML105	Unheated	2.73 ± 0.04^{g}	9.43 ± 0.11^{e}	5.43 ± 0.21^{e}	2.42 ± 0.06^{f}	ND	ND	ND
(White)	FIR	2.02 ± 0.07^{h}	7.87 ± 0.15^{f}	7.52 ± 0.22^{d}	2.76 ± 0.04^{d}	ND	ND	ND
Sung Yod	Unheated	3.72 ± 0.14^{c}	15.32 ± 0.45^{c}	9.12 ± 1.08^{c}	2.44 ± 0.01^{f}	9.56 ± 1.43^{g}	0.43 ± 0.04^{f}	1.36 ± 0.07^{g}
(Red)	FIR	2.96 ± 0.11^{f}	10.32 ± 1.17^{e}	15.37 ± 1.27^{b}	3.44 ± 0.07^{c}	20.44 ± 0.75^{e}	3.43 ± 0.09^{d}	1.47 ± 0.02^{e}
Mun Poo	Unheated	3.32 ± 0.09^{de}	15.32 ± 1.26^{b}	10.48 ± 1.68^{c}	2.54 ± 0.03^{e}	9.55 ± 1.48^{g}	0.24 ± 0.02^{g}	1.43 ± 0.08^{f}
(Red)	FIR	2.76 ± 0.08^{g}	9.91 ± 1.18^{e}	15.21 ± 1.82^{b}	3.64 ± 0.11^{c}	18.31 ± 0.33^{f}	3.21 ± 0.01^{e}	1.58 ± 0.05^{d}
Mali Dang	Unheated	4.38 ± 0.21^{b}	17.55 ± 1.76^{b}	14.38 ± 1.96^{b}	2.49 ± 0.09^{ef}	10.45 ± 1.32^{g}	0.45 ± 0.02^{f}	1.52 ± 0.04^{d}
(Red)	FIR	3.21 ± 0.02^{e}	12.43 ± 1.64^{de}	23.43 ± 1.39^{a}	3.54 ± 0.08^{c}	25.76 ± 0.04^{d}	4.11 ± 0.07^{c}	1.73 ± 0.09^{c}
Hom Nil	Unheated	5.51 ± 0.02^{a}	22.32 ± 2.11^{a}	15.34 ± 1.36^{b}	4.54 ± 0.10^{b}	41.56 ± 3.87^{c}	2.36 ± 0.13^{b}	3.62 ± 0.12^{b}
(Purple)	FIR	3.41 ± 0.04^{d}	15.94 ± 1.15^{b}	23.96 ± 1.32^{a}	5.42 ± 0.11 a	127.70 ± 0.13^{a}	6.32 ± 0.27^{a}	4.87 ± 0.14^{a}
Riceberry	Unheated	5.43 ± 0.13^{a}	20.31 ± 2.09^{a}	14.02 ± 2.33^{b}	4.43 ± 0.05^{b}	37.29 ± 2.94^{c}	2.46 ± 0.09^{b}	3.78 ± 0.29^{b}
(Purple)	FIR	3.32 ± 0.17^{de}	14.66 ± 1.07^{b}	23.94 ± 2.72^{a}	5.46 ± 0.14^{a}	61.71 ± 0.14^{b}	5.87 ± 0.39^{a}	4.85 ± 0.11^{a}

Mean values within a column superscripted by the small letter are significantly different at $p < 0.05$. ND is not detected.

This result agrees with our previous study of three varieties of Thai pigmented rice harvested in the previous year [13]; however, a slightly smaller increase was observed. In addition, FIR treatment has been reported to enhance some flavonoids in other food materials: for example, increased myricetin in papaya (1.2 times); quercetin (20.2 times), apigenin in tomato (3.18 times) [35] and quercetin (4.7 times) in buckwheat sprouts [37]. According to the results obtained from our present study, the explanation of increases in quercetin and apigenin accumulation may be caused by the rupture of the glycoside bonds of rutin by FIR [35]. Rutin has a chemical structure similar to that of quercetin, however, it has an extra glycone flavonoid, whereas quercetin does not. A further consideration is that rutin (glycoside bonds) was thought to be the hydrolysis pathway to quercitin (aglycoside bonds) [38]. Previous research [4,20,35] has also explained that the covalent bonds of polymerized polyphenols are broken, thus causing the alteration of HMW (high-molecular weight) to LMW (low-molecular weight) compounds.

3.7. Composition of Anthocyanin

Table 2 classified the individual anthocyanin in different rice varieties studied, compared with the authentic standards. The results showed that anthocyanins were absent in non-pigmented rice (KDML 105), while all three anthocyanin derivatives were found in pigmented rice varieties studied. Cyanidin 3-glucoside was higher in purple rice varieties than others. A previous study by Abdel-Aal et al. [6] reported similar results for Indian rice varieties, namely that anthocyanins (particularly cyanidin-3-O-glucoside) were found to the greatest extent in purple and black rice cultivars. The different anthocyanin contents of the various rice varieties may be caused by genetic or environmental effects. They also found that malvidin was present at a high content in all red varieties [6,7]. We previously reported on how FIR treatment affected the anthocyanins in three pigmented rice varieties: Mali dang, Hom Nil and Riceberry [13] and found that malvidin, cyanidin-3-glucosides and pelargonidin increased after FIR treatment. When compared to our findings in this present study with the same varieties but in a different production year and for two more varieties (Mun Poo and Sung Yod), the results were comparable. The explanation of this occurrence affected by FIR treatment may be similar to that in phenolic acid and flavonoids. Scalzo et al. stated that thermal treatment generally induced an increase in the main phenolic substances of orange juice, such as the anthocyanins and total cinnamates [33].

3.8. *Amylose Content*

Amylose content is one of the most significant parameters indicating the quality of rice, because it affects the cooking quality [39]. All red and purple rice varieties had the highest amylose contents; while KDML105, a non-pigmented variety, had the lowest content (Table 3). The amylose content of all pigmented rice varieties was not significantly different when compared to unheated rice. Several studies have determined the content of amylose in different Thai rice varieties and the values ranged from 2% in white to 19% in purple rice [12,13,40].

Table 3. Gelatinization parameters of pigmented rice varieties as affected by FIR.

Samples	Treatments	% Amylose Content	To (°C)	Tp (°C)	Tc (°C)	Δ_{gel}H (J g^{-1} Starch)
KDML105	Unheated	12.36 ± 0.12 ^{d}	61.19 ± 0.12 ^{e}	67.88 ± 0.11 ^{e}	74.11 ± 0.10 ^{d}	9.63 ± 0.36 ^{g}
(White)	FIR	12.45 ± 0.11 ^{d}	61.12 ± 0.05 ^{e}	67.64 ± 0.07 ^{e}	74.10 ± 0.07 ^{d}	10.33 ± 0.21 ^{ef}
Sung Yod	Unheated	14.42 ± 0.13 ^{bc}	74.18 ± 0.08 ^{a}	79.18 ± 0.09 ^{a}	82.90 ± 0.23 ^{a}	12.55 ± 0.42 ^{b}
(Red)	FIR	14.55 ± 0.12 ^{b}	73.73 ± 0.05 ^{b}	78.81 ± 0.05 ^{b}	82.72 ± 0.11 ^{a}	13.54 ± 0.02 ^{a}
Mun Poo	Unheated	15.77 ± 0.18 ^{a}	62.45 ± 0.15 ^{c}	69.09 ± 0.04 ^{d}	74.96 ± 0.16 ^{c}	9.55 ± 0.39 ^{g}
(Red)	FIR	15.75 ± 0.12 ^{a}	62.39 ± 0.19 ^{cd}	69.10 ± 0.54 ^{cd}	74.90 ± 0.12 ^{c}	10.51 ± 0.32 ^{e}
Mali Dang	Unheated	15.65 ± 0.14 ^{a}	62.21 ± 0.11 ^{cd}	68.77 ± 0.11 ^{d}	74.95 ± 0.29 ^{c}	11.53 ± 0.43 ^{d}
(Red)	FIR	15.70 ± 0.14 ^{a}	62.20 ± 0.09 ^{d}	68.87 ± 0.53 ^{d}	74.86 ± 0.24 ^{c}	12.44 ± 0.33 ^{b}
Hom Nil	Unheated	14.20 ± 0.18 ^{c}	60.59 ± 0.11 ^{f}	69.54 ± 0.12 ^{c}	75.74 ± 0.12 ^{b}	9.78 ± 0.76 ^{fg}
(Purple)	FIR	14.31 ± 0.15 ^{c}	60.43 ± 0.11 ^{f}	69.32 ± 0.54 ^{cd}	75.70 ± 0.21 ^{b}	11.02 ± 0.31 ^{d}
Riceberry	Unheated	14.08 ± 0.12 ^{c}	58.04 ± 0.08 ^{g}	65.75 ± 0.08 ^{f}	73.82 ± 0.25 ^{d}	10.81 ± 0.39 ^{de}
(Purple)	FIR	14.18 ± 0.12 ^{c}	58.08 ± 0.11 ^{g}	65.64 ± 0.05 ^{f}	73.38 ± 0.06 ^{e}	11.74 ± 0.28 ^{c}

Onset (T_o), peak (T_p) and concluding (T_c) temperatures of gelatinization are shown. Δ_{gel}H is the enthalpy change associated with the gelatinization of 1 g of purified starch. Mean values within a column superscripted by the small letter are significantly different at $p < 0.05$.

The contents of amylose that were in the range between 10 and 20% are classified as a low-amylose-content rice [41]. According to the data obtained from our present study, all varieties are classified as low-amylose types (below 20%). In general, typical Thai rice contains low-amylose content, thus providing a relatively soft and adhesive texture after cooking [42]. Therefore, this research suggested that different reactions to FIR drying do not relate to amylose content in pigmented rice varieties.

3.9. *Differential Scanning Calorimetry (DSC)*

Differential scanning calorimetry (DSC) has demonstrated different values in onset (T_o), peak (T_p) and concluding temperatures (T_c) and gelatinization enthalpy (Δ_{gel}H) parameters for extracted samples of rice starch (Table 3). The onset, peak and conclusion temperatures (T_o, T_p and T_c) of unheated rice starch ranged from 58.04–74.18 °C to 65.75–79.18 °C and 73.82–82.90 °C. For FIR-treated samples, those parameters were in ranges of 58.08–73.73 °C, 65.64–78.81 °C and 73.38–82.72 °C, respectively. The Δ_{gel}H of unheated rice ranged from 9.55 in Mun Poo (red) to 12.55 in Sung yod (red); among those, the Δ_{gel}H of KDML105 was not significantly different with Mun Poo. After FIR treatment, there was a significant increase in Δ_{gel}H, with Sung yod (13.54 J g^{-1}), Mali Dang

(12.44 J g^{-1}) and Riceberry (11. 74 J g^{-1}) having greater values of $\Delta_{gel}H$ than any other rice variety. However, the gelatinization temperatures, T_o and T_p of Sung Yod as treated by FIR were slightly decreased when compared to unheated samples (74.18 to 73.73 °C) while, the reducing of T_c was observed in Ricberry from 73.82 to 73.38 °C ($p < 0.05$). Sung Yod was also observed with the highest temperatures for T_o, T_p and T_c.

The relationships between the gelatinization parameters including T_o, T_p, and T_c and the internal crystalline structure of starch and its heat stability, as well as the ΔH reflects the energy needed for double helical dissociation and hence the crystallinity of starch [43,44]. Gelatinization behavior may be influenced by starch characteristics; however, the starches chosen for our present study were comparatively similar in many aspects. In this study, the rice starch was extracted from rice flour for investigation only of starch behavior as affected by FIR. The starches of pigmented rice had similar $\Delta_{gel}H$ (Jg_{-1}) though they had higher gelatinization temperatures. As can be observed, FIR-treated pigmented-rice varieties were found to have greater gelatinization enthalpy values, compared with unheated samples. Therefore, this research suggested that FIR drying gave greater effects on gelatinization behavior in pigmented rice varieties than normal rice.

3.10. *In Vitro Digestibility*

The starch digestibility of pigmented rice cultivars, as affected by FIR treatment, are presented in Table 4. The values of digested starch (60 min) in unheated rice ranged from 9.57–16.17% while digested starch content in FIR-treated rice ranged from 7.33–12.17%. The values of digested starch (%) in unheated rice were greater in the following order: KDML 105 > Mun Poo > Mali Dang and Riceberry > Hom Nil > Sung Yod ($p < 0.05$). In the case of FIR-treated rice, those values were significantly lower in the following order: KDML 105 (non-pigmented rice) > Mun Poo > Sung Yod, Mali Dang, Riceberry and Hom Nil. We characterized the digestion of the rice samples as low rates during the digestion process at small intestine (approximately after 60 min), as shown in Table 4. The rice digestibility was determined using single and two phases by plotting LOS analysis of each sample. For the single phase of samples, the value of k_1 (the digestibility rate constant) for pigmented rice (Mun Poo, red rice) and non-pigmented rice (KDML105) being treated by FIR were substantially lower compared to unheated rice. This reflects their slower digestion rates by FIR treatment, with values ranging from 0.028 to 0.041 min^{-1} in Mun Poo and from 0.028 to 0.071 min^{-1} in KDML 105. In the present study, the normal rice exhibited the highest of C_∞ value (17.11%). When considering the digestion kinetics, Hom Nil, Mali Dang, Riceberry, and Sung Yod could be the most suitable as modelled by performing first order rate constants twice [27]. As a result, the estimated values of k (rate of starch digestion) and C_∞ (extent of amylolysis) for variables in Equation (3) for those samples are shown in Table 4. According to observations, when the starch digestion occurred in two phases, this indicates a slower phase in digestion. The k_2 values of unheated rice samples varieties were found highest in Mali Dang, followed by Hom Nil, Riceberry and Sung Yod, respectively. These values can be compared to k_1 values, where amylolysis occurs as a single-phase process in a certain sample. In the case of FIR-dried samples, the second rate (k_2) values were found to be the highest in Mali Dang (0.052), followed by Riceberry (0.041), Hom Nil (0.039) and Sung Yod (0.030).

The C_∞ (extent of amylolysis) decreased similarly in FIR-treated rice when compared to unheated rice. The first-order kinetic data explains the amylolysis of starch in rice from white, red and purple rice varieties, as are explained in this paper. The values (C_∞ and k) of all parameters related to starch digestion (in vitro) were significantly lower for all FIR-treated pigmented samples, when compared to unheated rice. Although the results of enthalpy in both unheated and FIR-treated samples showed the highest values in Sung Yod varieties, the percentages of starch digested were lower for the purple-rice varieties (Hom Nil and Riceberry). In addition, the results of DSC of unheated and FIR-dried rice samples (Table 3) also showed the lower values of gelatinization enthalpy for normal rice. The gelatinization enthalpy increased in all samples after FIR treatment. DSC analysis was used for rice starch determination, hence the influence of starch structure for any variety of rice was a significant factor of starch digestibility. In this present study, the normal or non-pigmented rice sample (KDML105) had the highest contents of digestible starch in both unheated and FIR-treated form. When compared with the other rice varieties studied, the KDML105, as a non-pigmented rice, contains other phenolic compounds, although it has no anthocyanins. This may cause the higher digestibility of the KDML105 relative to the pigmented varieties. These results indicated that starch digestibility in pigmented rice could be associated with phenolic compounds and anthocyanins. Anthocyanins have been exhibited to possess in vitro inhibition of α-glucosidase activity. Previous studies reported significant activity of α-glucosidase inhibition of polyphenols and anthocyanins from pigmented potato extract and these compounds were able to reduce starch digestibility [16,17]. Moreover, many studies have shown that the main anthocyanins in pigmented rice were peonidin 3-glucoside and cyanidin 3-glucoside, while these compounds were absent in non-pigmented or white rice [12,28,45]. Among anthocyanin derivatives, the most effective inhibitor of α-glucosidase activity is cyanidin 3-glucoside [46], a phenolic compound in millet extracts that could inhibit both enzymes α-amylase and α-glucosidase. Additionally, Tadera et al. [47] have reported flavonoid compounds (namely, genistein, daidzein, kaempferol, naringenin and apigenin) as having the capacity of α-glucosidase inhibition. Recently, Tangkhawanit et al. [21] have observed that TPC, TFC and flavonoids increased in soymilk residue after FIR treatment and that the extracts were more efficient in inhibiting α-amylase when compared with hot air-dried extract; thus, drawing the conclusion of them being involved in lowering starch digestion. In our present study, we found that ferulic acid (FA) is an abundant phenolic acid in rice. FA has been reported to be beneficial in the treatment of type-2 diabetes [48], since it controls blood-glucose levels (BGLs) by increasing glucokinase functions and producing glycogen in the liver. Furthermore, FA significantly lowered BGLs and raised levels of plasma insulin in mice [48]. Our research has suggested that phenolic compounds, flavonoids and anthocyanins may play important roles in decreasing starch amylolysis and the absorption of glucose in the small intestine; therefore, they may play a possible role in preventing diabetes by controlling post-prandial glycaemia. According to our data, it could be stated that phenolic compounds have the capacity to combine with an amino group in protein or glycoside bonds in conjugated polysaccharides by covalent and non-covalent bonds, thus inactivating a digestive enzyme [21].

Table 4. The estimate values from LOS plots for all pigmented rice varieties as affected by FIR.

Sample	Treatments	%Starch Digestedat 60 min.	Single or First Phase		Second Phase		Total
			$C_{1\infty}$ (%)	k_1 (min^{-1})	$C_{2\infty}$ (%)	k_2 (min^{-1})	C_∞ (%)
KDML105	Unheated	16.21 ± 0.19 ^{a}	17.21 ± 0.59 ^{a}	0.071 ± 0.001 ^{f}	N/A	N/A	17.44 ± 0.59 ^{a}
(White)	FIR	12.17 ± 0.19 ^{c}	12.81 ± 0.59 ^{b}	0.059 ± 0.001 ^{g}	N/A	N/A	12.81 ± 0.10 ^{e}
Sung Yod	Unheated	9.57 ± 0.02 ^{d}	3.46 ± 0.11 ^{f}	0.198 ± 0.001 ^{d}	8.99 ± 0.12 ^{b}	0.041 ± 0.002 ^{c}	12.45 ± 0.75 ^{e}
(Red)	FIR	7.43 ± 0.02 ^{f}	2.32 ± 0.11 ^{g}	0.185 ± 0.002 ^{e}	5.64 ± 0.11 c	0.030 ± 0.001 ^{d}	7.96 ± 0.18 ^{g}
Mun Poo	Unheated	13.31 ± 0.40 ^{b}	13.54 ± 2.23 ^{b}	0.057 ± 0.003 ^{h}	N/A	N/A	13.54 ± 0.11 ^{d}
(Red)	FIR	8.55 ± 0.32 ^{e}	8.68 ± 1.11 ^{c}	0.044 ± 0.002 ^{i}	N/A	N/A	8.68 ± 0.06 ^{f}
Mali Dang	Unheated	12.13 ± 1.35 ^{bc}	6.00 ± 0.71 ^{d}	0.364 ± 0.003 ^{a}	9.76 ± 0.15 ^{a}	0.065 ± 0.001 ^{a}	15.76 ± 0.23 ^{b}
(Red)	FIR	7. 54 ± 0.04 ^{f}	3.32 ± 0.21 ^{f}	0.221 ± 0.003 ^{c}	4.57 ± 0.11 ^{d}	0.052 ± 0.001 ^{b}	7.89 ± 0.11 ^{g}
Hom Nil	Unheated	11.69 ± 1.11 ^{c}	3.89 ± 0.49 ^{f}	0.362 ± 0.002 ^{a}	9.57 ± 0.14 ^{a}	0.058 ± 0.002 ^{b}	13.46 ± 0.13 ^{d}
(Purple)	FIR	7.33 ± 0.30 ^{f}	3.28 ± 0.29 ^{f}	0.211 ± 0.001 ^{c}	4.61 ± 0.31 ^{d}	0.039 ± 0.001 ^{c}	7.79 ± 0.17 ^{g}
Riceberry	Unheated	12.50 ± 1.32 ^{bc}	4.96 ± 0.07 ^{e}	0.313 ± 0.003 ^{b}	9.71 ± 0.17 ^{a}	0.059 ± 0.001 ^{b}	14.67 ± 0.14 ^{c}
(Purple)	FIR	7.65 ± 0.81 ^{f}	3.34 ± 0.07 ^{f}	0.219 ± 0.004 ^{c}	4.59 ± 0.07 ^{d}	0.041 ± 0.004 ^{c}	7.93 ± 0.13 ^{g}

The amylolysis occurred by a single-phase process, and that no second phase was observed. Value for the second phase is therefore not applicable (N/A). Mean values within a column superscripted by the same letter are significantly different at $p < 0.05$.

The results of our present study may support the literature reports that bioactive compounds in rice, such as ferulic acid, protocatechuic acid, quercetin, cyanidin 3-glucoside, apigenin and malvidin, may play an important role in inhibiting α-amylase activity in pigmented rice. Thus, our present study has established that starch structure, along with bioactive compounds of pigmented rice, may have an impact on lower digestibility. Furthermore, heat treatment using FIR could be a new factor to assist in aiding the extractable yield of those compound which may be associated with greater capacity for retarding the starch-digestion rate of pigmented rice. Consumption of slowly digested rice starches is related to a reduced risk of developing type-2 diabetes and cardiovascular disease. Accordingly, our discoveries have offered valuable data for the development of slowly digested starch of biomaterials for application in the practice of providing nutritional options for individuals with metabolic disorders such as type-2 diabetes and obesity, via the use of drying processing such as FIR.

4. Conclusions

Our results have indicated that FIR provided both negative and positive impacts of FIR treatment on starch digestion and bioactive compounds of pigmented rice and normal rice. For a negative impact, FIR reduced the content of certain heat-sensitive substances such as sinapic acid and rutin. On the other hand, the application of FIR increased the levels of total phenolic, flavonoid and anthocyanin contents in all rice samples. However, the changes of individual compounds such as phenolic acids and flavonoids were varied. FIR increased the functional properties of antioxidants by scavenging DPPH-radical reducing power, capable of inhibiting α-amylase, thereby lowering the starch digestibility of rice. These data indicate that the reduction of the starch digestion rate may result from the association of starch structure and the inhibitory effects of polyphenols in pigmented rice.

This study suggests that FIR may have a favorable effect on extraction yield for bioactive compounds of the grain by assisting the release of small or free molecules from the complex compounds. Hence, the selection of an appropriate drying method could be a smart way for enhancing the functional properties and nutritional quality of rice, thus contributing to consumer health benefits. In addition, our results have indicated that the lower gelatinization enthalpy and higher starch digestibility of white or non-pigmented rice varieties, when compared with those of pigmented rice, could be affected by the absence of anthocyanins, even though it contains a comparable number of polyphenols and flavonoids. However, white or non-pigmented rice may have more desirable sensory quality (e.g., being stickier). Our results suggest that Hom Nil (purple rice) may be the best variety for drying with FIR treatment for future works because Hom Nil was enhanced the most of cyanidin 3-glucoside (67%) by FIR, which led to a decreasing digestibility rate constant (32%) and less starch digested (37%). This research delivers important indications that FIR could be considered as an appropriate drying method for pigmented rice regarding health benefits. Our findings are expected to be supportive for applications to the food, nutraceutical and pharmaceutical industries.

Author Contributions: Conceptualization, J.R. and S.S.; methodology, F.J.W., N.M. and S.S.; software, F.J.W., N.M. and S.S.; validation, F.J.W. and S.S.; formal analysis, J.R.; investigation, J.R.; resources, J.R., F.J.W., N.M. and S.S.; data curation, J.R. and S.S.; writing—original draft preparation, J.R.; writing—review and editing, S.S.; visualization, J.R.; supervision, S.S.; project administration, S.S.; funding acquisition, F.J.W. and S.S. All authors have read and agreed to the published version of the manuscript.

Funding: This research was funded by Royal Golden Jubilee (RGJ) Ph.D. Program under Thailand Research Fund (TRF) grant number PHD/0136/2556, Thailand and the Newton Fund via Industry Academia Partnership Program grant number IAPPT2\100012 and BBSRC Institute Strategic Programme Grant Food Innovation and Health grant number BB/R012512/1 and its constituent projects BBS/E/F/000 PR10345 (Theme 2, Digestion in the Upper GI Tract). The APC was funded by Foods.

Data Availability Statement: Data is contained within the article.

Acknowledgments: We would like to thank Colin Wrigley, University of Queensland, Australia, for his valuable comments and English proofreading.

Conflicts of Interest: No conflict of interest is declared for any of the authors.

References

1. Choi, Y.; Jeong, H.S.; Lee, J. Antioxidant activity of methanolic extracts from some grains consumed in Korea. *Food Chem.* **2007**, *103*, 130–138. [CrossRef]
2. Zhou, Z.; Robards, K.; Helliwell, S.; Blanchard, C. The distribution of phenolic acids in rice. *Food Chem.* **2004**, *87*, 401–406. [CrossRef]
3. Wolever, T.M.; Jenkins, D.J.; Vuksan, V.; Josse, R.G.; Wong, G.S.; Jenkins, A.L. Glycemic index of foods in individual subjects. *Diabetes Care* **1990**, *13*, 126–132. [CrossRef]
4. Ratseewo, J.; Meeso, N.; Siriamornpun, S. Changes in amino acids and bioactive compounds of pigmented rice as affected by far-infrared radiation and hot air drying. *Food Chem.* **2020**, *306*, 125644. [CrossRef]
5. Jenkins, D.J.; Kendall, C.W.; Augustin, L.S.; Mitchell, S.; Sahye-Pudaruth, S.; Mejia, S.B.; Vidgen, E. Effect of legumes as part of a low glycemic index diet on glycemic control and cardiovascular risk factors in type 2 diabetes mellitus: A randomized controlled trial. *Arch. Intern. Med.* **2012**, *172*, 1653–1660. [CrossRef]
6. Abdel-Aal, E.S.M.; Young, J.C.; Rabalski, I. Anthocyanin composition in black, blue, pink, purple, and red cereal grains. *J. Agric. Food Chem.* **2006**, *54*, 4696–4704. [CrossRef] [PubMed]
7. Chen, X.Q.; Nagao, N.; Itani, T.; Irifune, K. Anti-oxidative analysis, and identification and quantification of anthocyanin pigments in different coloured rice. *Food Chem.* **2012**, *135*, 2783–2788. [CrossRef] [PubMed]
8. Adom, K.K.; Liu, R.H. Antioxidant activity of grains. *J. Agric. Food Chem.* **2002**, *50*, 6182–6187. [CrossRef]
9. Ferreira, C.D.; Ziegler, V.; Bubolz, V.K.; Da Silva, J.; Cardozo, M.M.C.; Elias, M.C.; De Oliveira, M. Effects of the roasting process over the content of secondary metabolites from peanut grains (*Arachishypogaea* L.) with different colorations of testa. *J. Food Qual.* **2016**, *39*, 685–694. [CrossRef]
10. Daiponmak, W.; Senakun, C.; Siriamornpun, S. Antiglycation capacity and antioxidant activities of different pigmented Thai rice. *Int. J. Food Sci. Technol.* **2014**, *49*, 1805–1810. [CrossRef]
11. Deepa, G.; Singh, V.; Naidu, K.A. A comparative study on starch digestibility, glycemic index and resistant starch of pigmented ('Njavara'and 'Jyothi') and a non-pigmented ('IR 64') rice varieties. *J. Food Sci. Technol.* **2010**, *47*, 644–649. [CrossRef]
12. Ponjanta, J.; Chomsri, N.O.; Meechoui, S. Correlation of pasting behaviors with total phenolic compounds and starch digestibility of indigenous pigmented rice grown in upper Northern Thailand. *Funct. Foods Health Dis.* **2016**, *6*, 133–143. [CrossRef]
13. Ratseewo, J.; Warren, F.J.; Siriamornpun, S. The influence of starch structure and anthocyanin content on the digestibility of Thai pigmented rice. *Food Chem.* **2019**, *298*, 124949. [CrossRef] [PubMed]
14. Ademiluyi, A.O.; Oboh, G. Soybean phenolic-rich extracts inhibit key-enzymes linked to type 2 diabetes (α-amylase and α-glucosidase) and hypertension (angiotensin I converting enzyme) in vitro. *Exp. Toxicol. Pathol.* **2013**, *65*, 305–309. [CrossRef]
15. Shobana, S.; Sreerama, Y.N.; Malleshi, N.G. Composition and enzyme inhibitory properties of finger millet (*Eleusinecoracana* L.) seed coat phenolics: Mode of inhibition of α-glucosidase and pancreatic amylase. *Food Chem.* **2009**, *115*, 1268–1273. [CrossRef]
16. McDougall, G.J.; Shpiro, F.; Dobson, P.; Smith, P.; Blake, A.; Stewart, D. Different polyphenolic components of soft fruits inhibit α-amylase and α-glucosidase. *J. Agric. Food Chem.* **2005**, *53*, 2760–2766. [CrossRef] [PubMed]
17. Ramdath, D.D.; Padhi, E.; Hawke, A.; Sivaramalingam, T.; Tsao, R. The glycemic index of pigmented potatoes is related to their polyphenol content. *Food Funct.* **2014**, *5*, 909–915. [CrossRef]
18. Wanyo, P.; Meeso, N.; Siriamornpun, S. Effects of different treatments on the antioxidant properties and phenolic compounds of rice bran and rice husk. *Food Chem.* **2014**, *157*, 457–463. [CrossRef]
19. Raksakantong, P.; Siriamornpun, S.; Meeso, N. Effect of drying methods on volatile compounds, fatty acids and antioxidant property of Thai kaffir lime (*Citrus hystrix* DC). *Int. J. Food Sci. Technol.* **2012**, *47*, 603–612. [CrossRef]
20. Lee, S.C.; Jeong, S.M.; Kim, S.Y.; Park, H.R.; Nam, K.C.; Ahn, D.U. Effect of far-infrared radiation and heat treatment on the antioxidant activity of water extracts from peanut hulls. *Food Chem.* **2006**, *94*, 489–493. [CrossRef]
21. Tangkhawanit, E.; Meeso, N.; Siriamornpun, S. Changes in bioactive components, biological activities and starch digestibility of soymilk residues as affected by far-infrared radiation combined with hot-air and hot-air drying. *Dry. Technol.* **2022**, 1–14. [CrossRef]
22. Ratseewo, J.; Tangkhawanit, E.; Meeso, N.; Kaewseejan, N.; Siriamornpun, S. Changes in antioxidant properties and volatile compounds of kaffir lime leaf as affected by cooking processes. *Int. Food Res. J.* **2016**, *23*, 188.
23. Knutson, C.A.; Grove, M.J. Rapid method for estimation of amylose in maize starches. *Cereal Chem.* **1994**, *71*, 469–471.
24. Warren, F.J.; Zhang, B.; Waltzer, G.; Gidley, M.J.; Dhital, S. The interplay of α-amylase and amyloglucosidase activities on the digestion of starch in in vitro enzymic systems. *Carbohydr. Polym.* **2015**, *117*, 192–200. [CrossRef] [PubMed]
25. Goñi, I.; García-Alonso, A.; Saura-Calixto, F. A starch hydrolysis procedure to estimate glycemic index. *Nutr. Res.* **1997**, *17*, 427–437. [CrossRef]
26. Butterworth, P.J.; Warren, F.J.; Grassby, T.; Patel, H.; Ellis, P.R. Analysis of starch amylolysis using plots for first-order kinetics. *Carbohydr. Polym.* **2012**, *87*, 2189–2197. [CrossRef]
27. Edwards, C.H.; Warren, F.J.; Campbell, G.M.; Gaisford, S.; Royall, P.G.; Butterworth, P.J.; Ellis, P.R. A study of starch gelatinization behaviour in hydrothermally processed plant food tissues and implications for in vitro digestibility. *Food Funct.* **2015**, *6*, 3634–3641. [CrossRef]
28. Sompong, R.; Siebenhandl-Ehn, S.; Linsberger-Martin, G.; Berghofer, E. Physicochemical and antioxidative properties of red and black rice varieties from Thailand, China and Sri Lanka. *Food Chem.* **2011**, *124*, 132–140. [CrossRef]

29. Kaisoon, O.; Siriamornpun, S.; Weerapreeyakul, N.; Meeso, N. Phenolic compounds and antioxidant activities of edible flowers from Thailand. *J. Funct. Foods* **2011**, *3*, 88–99. [CrossRef]
30. Adak, N.; Heybeli, N.; Ertekin, C. Infrared drying of strawberry. *Food Chem.* **2017**, *219*, 109–116. [CrossRef]
31. Zhang, Z.; Lv, G.; Pan, H.; Fan, L. Optimisation of the microwave-assisted extraction process for six phenolic compounds in *Agaricus blazei* murrill. *Int. J. Food Sci. Technol.* **2012**, *47*, 24–31. [CrossRef]
32. Kim, S.Y.; Jeong, S.M.; Park, W.P.; Nam, K.C.; Ahn, D.U.; Lee, S.C. Effect of heating conditions of grape seeds on the antioxidant activity of grape seed extracts. *Food Chem.* **2006**, *97*, 472–479. [CrossRef]
33. Scalzo, R.L.; Iannoccari, T.; Summa, C.; Morelli, R.; Rapisarda, P. Effect of thermal treatments on antioxidant and antiradical activity of blood orange juice. *Food Chem.* **2004**, *85*, 41–47. [CrossRef]
34. Pengkumsri, N.; Chaiyasut, C.; Saenjum, C.; Sirilun, S.; Peerajan, S.; Suwannalert, P.; Sivamaruthi, B.S. Physicochemical and antioxidative properties of black, brown and red rice varieties of northern Thailand. *Food Sci. Technol.* **2015**, *35*, 331–338. [CrossRef]
35. Siriamornpun, S.; Ratseewo, J.; Kaewseejan, N.; Meeso, N. Effect of osmotic treatments and drying methods on bioactive compounds in papaya and tomato. *RSC Adv.* **2015**, *5*, 18579–18587. [CrossRef]
36. Clifford, M.N. Chlorogenic acids and other cinnamates–nature, occurrence and dietary burden. *J. Sci. Food Agric.* **1999**, *79*, 362–372. [CrossRef]
37. Ghimeray, A.K.; Sharma, P.; Phoutaxay, P.; Salitxay, T.; Woo, S.H.; Park, S.U.; Park, C.H. Far infrared irradiation alters total polyphenol, total flavonoid, antioxidant property and quercetin production in tartary buckwheat sprout powder. *J. Cereal Sci.* **2014**, *59*, 167–172. [CrossRef]
38. Vetrova, E.V.; Maksimenko, E.V.; Khizrieva, S.S.; Bugaeva, A.F.; Borisenko, N.I.; Minkin, V.I. A simple way for the preparation of natural antioxidant quercetin from rutin by subcritical water. *J. Nat. Sci. Biol. Med.* **2017**, *8*, 213. [CrossRef]
39. Adu-Kwarteng, E.; Ellis, W.O.; Oduro, I.; Manful, J.T. Rice grain quality: A comparison of local varieties with new varieties under study in Ghana. *Food Control* **2003**, *14*, 507–514. [CrossRef]
40. Saikia, S.; Dutta, H.; Saikia, D.; Mahanta, C.L. Quality characterisation and estimation of phytochemicals content and antioxidant capacity of aromatic pigmented and non-pigmented rice varieties. *Food Res. Int.* **2012**, *46*, 334–340. [CrossRef]
41. IRRI. 2018. Available online: http://www.knowledgebank.irri.org/millingprocess/index.php/ricequality-mainmenu-281/quality-characteristics-of-milled-rice-mainmenu-283 (accessed on 5 October 2022).
42. Chatthongpisut, R.; Schwartz, S.J.; Yongsawatdigul, J. Antioxidant activities and antiproliferative activity of Thai purple rice cooked by various methods on human colon cancer cells. *Food Chem.* **2015**, *188*, 99–105. [CrossRef] [PubMed]
43. Ding, Y.; Cheng, J.; Lin, Q.; Wang, Q.; Wang, J.; Yu, G. Effects of endogenous proteins and lipids on structural, thermal, rheological, and pasting properties and digestibility of adlay seed (*Coix lacryma-jobi* L.) starch. *Food Hydrocoll* **2021**, *111*, 106254. [CrossRef]
44. Tarahi, M.; Shahidi, F.; Hedayati, S. Physicochemical, Pasting, and Thermal Properties of Native Corn Starch–Mung Bean Protein Isolate Composites. *Gels* **2022**, *8*, 693. [CrossRef] [PubMed]
45. Melini, V.; Acquistucci, R. Health-Promoting Compounds in Pigmented Thai and Wild Rice. *Foods* **2017**, *6*, 9. [CrossRef] [PubMed]
46. Adisakwattana, S.; Charoenlertkul, P.; Yibchok-anun, S. α-Glucosidase inhibitory activity of cyanidin-3-galactoside and synergistic effect with acarbose. *J. Enzym. Inhib. Med. Chem.* **2009**, *24*, 65–69. [CrossRef]
47. Tadera, K.; Minami, Y.; Takamatsu, K.; Matsuoka, T. Inhibition of α-glucosidase and α-amylase by flavonoids. *J. Nutr. Sci. Vitaminol.* **2006**, *52*, 149–153. [CrossRef]
48. Jung, E.H.; Ran Kim, S.; Hwang, I.K.; Youl Ha, T. Hypoglycemic effects of a phenolic acid fraction of rice bran and ferulic acid in C57BL/KsJ-db/db mice. *J. Agric. Food Chem.* **2007**, *55*, 9800–9804. [CrossRef]

foods

MDPI

Article

Lipid Profiles in Preliminary Germinated Brown Rice Beverages Compared to Non-Germinated Brown and White Rice Beverages

John C. Beaulieu [1,*], Robert A. Moreau [2], Michael J. Powell [2] and Javier M. Obando-Ulloa [3]

1 Food Processing & Sensory Quality Research Unit, United States Department of Agriculture, Agricultural Research Service, Southern Regional Research Center, 1100 Robert E. Lee Blvd., New Orleans, LA 70124, USA
2 Sustainable Biofuels and CoProducts Research Unit, Eastern Regional Research Center, USDA, ARS, 600 East Mermaid, Lane, Wyndmoor, PA 19038, USA; bobmoreauzz@gmail.com (R.A.M.); michael.powell@usda.gov (M.J.P.)
3 Doctorate Program in Natural Science for Development (DOCINADE) and Agronomy Engineering School, Costa Rica Institute of Technology (ITCR), San Carlos Technology Local Campus, P.O. Box 223-21001, Ciudad Quesada, San Carlos 30101, Alajuela, Costa Rica; jaobando@itcr.ac.cr
* Correspondence: John.Beaulieu@usda.gov; Tel.: +1-504-286-4471

Abstract: Brown rice is nutritionally superior to white rice, yet oil rancidity can be problematic during processing and storage regarding sensory attributes. Germinating brown rice is known to generally increase some health-promoting compounds. In response to increasing the consumption of plant-based beverages, we sprouted unstabilized brown rice, using green technologies and saccharification enzymes for value-added beverages. 'Rondo' paddy rice was dehulled, sorted and germinated, and beverages were produced and compared against non-germinated brown and white brewers rice beverages. The preliminary germinated brown rice beverage contained significantly higher concentrations of total lipids, diacylglycerols, triacylglycerols, free sterols, phytosterol esters and oryzanols than both non-germinated brown and white rice beverages. White rice beverages had significantly higher free fatty acids. Significant lipid losses occurred during sieving, yet novel germinated brown rice beverages contained appreciable levels of valuable health-beneficial lipids, which appeared to form natural emulsions. Further pilot plant investigations should be scaled-up for pasteurization and adjusted through emulsification to ameliorate sieving losses.

Keywords: enzymatic saccharification; functional beverages; germination; lipids; sprouting; value-added

Citation: Beaulieu, J.C.; Moreau, R.A.; Powell, M.J.; Obando-Ulloa, J.M. Lipid Profiles in Preliminary Germinated Brown Rice Beverages Compared to Non-Germinated Brown and White Rice Beverages. *Foods* **2022**, *11*, 220. https://doi.org/10.3390/foods11020220

Academic Editor: Joan M. King

Received: 15 September 2021
Accepted: 7 January 2022
Published: 14 January 2022

Publisher's Note: MDPI stays neutral with regard to jurisdictional claims in published maps and institutional affiliations.

1. Introduction

Rice feeds approximately half the world's population and is the main food crop in developing nations [1]. However, the majority of rice consumed is white rice, which is not nutritionally dense and considered a starchy food source. Whole grain brown rice (BR) is superior to white rice (WR) since most nutrients, such as the oils, fatty acids, proteins, vitamins, fiber, micronutrients and antioxidants are retained in the bran [2]. BR containing bran, embryo and aleurone delivers substantial proteins and lipids that convey health-promoting nutritional constituents for consumers [3–6]. Rice bran also contains high amounts of fiber and bioactive phytochemicals, such as tocopherols, tocotrienols, oryzanols, vitamin B complex, phytosterols (β-sitosterol, campesterol and stigmasterol), carotenoids and phenolic compounds [7]. Several such bioactive compounds have long been recognized to improve human health through antioxidant activities, including scavenging free radicals, enhancing the immune system and reducing the development of cancer and heart disease [8–11]. Rice flour and rice bran, and certain other grains, are known to contain high levels of lipolytic enzymes that require thermal and non-thermal methods to stabilize these materials [2]. Unfortunately, storage, milling and further food processing affects the lipids, starch and protein, often resulting in undesirable sensory, textural and nutritional changes to the final rice products in the marketplace.

In the past, BR and bran, although nutritious, were usually not consumed because of their high fiber content and possible hull contamination [12], notorious oxidation, off-flavor issues [13] and lengthy cooking time [14]. For marketability and consumer acceptance, stabilization or inactivation of lipase and the inhibition of the formation of free fatty acids (FFA) is considered necessary, immediately after milling. As most older BR protein extraction methods negatively affect the functional and nutritional properties of the proteins, less harsh procedures involving enzymes have been used to extract oil and/or protein from rice flour and bran [2] (pp. 143–162). In full-fat or only partially defatted rice bran, it was noted that the liquid phase could be further processed and stabilized using amylase and amyloglucosidase into rice beverage products [15]. Ironically, the vast majority of plant-based beverages have exogenous oils added toward the end of processing to assist emulsification. Globally, the dairy-alternatives plant beverage market is forecast to surpass USD 34 B by 2024 [16] and, according to the Information Resources Institute, US plant-based beverage sales attained USD 1.7 B, representing a 6.2% increase in one year, through 2019 [17].

Germination (sprouting) is a low-cost technology that initiates with seed water uptake, ultimately followed by the protrusion of the radicle from the seed. Along with strengthened health trends, the advent of "sprouted" whole grain products has markedly increased in the food and beverage marketplace [18]. The content of γ-aminobutyric acid (GABA), free sterols and phytosterol esters, free fatty acids, soluble fiber, γ-oryzanol and antioxidants, such as vitamin E, phenolic compounds and other bioactive compounds, usually increase during BR germination [3–6,19]. However, there are often different results concerning increased or decreased levels of key nutrients based on the grain type and/or variable soaking and germinating conditions and times [4,19–21]. Specifically, the reports addressing the changes in free fatty acids during BR germination also have contradictions [20,21].

Rice bran oil (RBO) offers several health-benefits due to the presence of ferulics, sterols, tocopherols, γ-oryzanol and tocotrienols, which convey antioxidant characteristics and stability along with several health-promoting prebiotic and probiotic benefits [9]. Very few food constituents have been granted the European Commission and the US Food and Drugs Administration approval to use health claims. Plant sterols have attained such approval status due to their proven cholesterol-lowering properties and, recently, attention has refocused on plant sterols regarding the anticarcinogenic and anti-inflammatory effects [7,22]. Subsequently, BR was germinated and assessed for beverage formulations with respect to lipid characteristics. Germinated (sprouted) brown 'Rondo' rice (GBR) was softened, wet milled, gelatinized and enzymatically converted into beverages. The methodology has no added oils or salts, additives or fortification. We previously characterized the germination conditions, processing itself, phytic acid and quality parameters in the preliminary beverages [23], along with some key health-beneficial compounds and arsenic levels [24]. GBR beverages should contain significantly higher concentrations of health-beneficial lipids compared to non-germinated brown and white rice beverages. Herein, we investigated this assumption and report several classes of lipids in preliminary beverages from germinated brown rice (GBR), compared to non-germinated brown 'Rondo' (BRR) and white rice (WR) beverages.

2. Materials and Methods

2.1. Rice Material, Germination, Softening, Wet-Milling and Enzyme Processing

Sourcing the 'Rondo' rice and methods used to de-hull, mill, sprout, and the novel protocol used to deliver a free-flowing soluble matrix through thermal softening, wet-milling, gelatinization and saccharification, were previously reported [24]. Briefly, 'Rondo' was grown at the Dale Bumpers National Rice Research Center (Stuttgart, AR, USA), harvested and dried to 12% moisture, and the paddy rice was cleaned using a screen cleaner (Model MICRO-224-LH, Crippen Northland Superior Supply Co., Winnipeg, MB, Canada) and stored at 60% RH at 4 °C. The same seed lot was dehulled using a Yamamoto

Impeller Type Husker (Model FC2K, Calibration Plus, Woodland, CA, USA) and milled into white rice (WR), including broken, Yamamoto Miller Rice Pal (Model VP-32T, Calibration Plus, Woodland, CA, USA) in Stuttgart, AR, then shipped to the Southern Regional Research Center (SRRC) in New Orleans, LA, USA.

Beverage processing methods were optimized for GBR (n = 6 per treatment) then compared against non-germinated BRR and WR beverages (n = 3 per treatment) using the same methods. For WR, the milled and polished rice was not sorted, which deliberately delivered brewers rice or "seconds" that is often used in commercial beverages. Freshly de-hulled (Satake Husk Aspirator, HA 60B, Higashi-Hiroshima, Japan) sorted and graded (Clipper 400 Office Tester Cleaner; A. T. Ferrell Company, Bluffton, IN, USA) BRR was treated with a peracetic acid food-safety rinse [23], then soaked in a rice:water ratio of 1:1 at 35 °C, followed by germination with rinsing every 4 h, for a total of 48 h, attaining the GBR. Then, WR, BRR and GBR were thermally softened (1:2 rice:water ratio) at temperatures just below the 'Rondo' WR gelatinization temperature (<70 °C). After thermal softening, wet-milling in a 4 L blender (Waring Commercial, CB15V, Torrington, CN, USA) with additional water dilution established a free-flowing liquid that additionally avoided gelatinization. Post wet-milling (PWM) samples passed a 30-mesh sieve (0.595 mm or 595 μm, Gilson Co. Inc., Lewis Center, OH, USA). Then, the free flowing beverages were heated to 80 °C for starch gelatinization, followed by liquefaction at ~55 °C using glucoamylase (EC 3.2.1.3) and α-amylase (EC 3.2.1.1) at 300 μL per 100 g starch, followed by sieving through a 140-mesh sieve (0.105 mm or 105 μm, Gilson Co., Inc.). The preliminary beverage (prior to homogenization or pasteurization) was termed as post enzyme (PNZ) beverages. The PWM sieving loss (PWM-SL) and PNZ sieving loss (PNZ-SL) were also analyzed.

Control, pre-processed rice crude fat content were measured [23] and compared, herein, to the total lipids assessed per HPLC (below). Thereafter, the controls were ground into flours using a cyclone sample mill (UDY model 3010-080P, UDY Corporation, Ft. Collins, CO, USA) and all experimental samples were freeze dried into powders and stored at −80 °C for later analyses. No commercial rice beverages were compared since virtually all the products found in local stores had added oils (e.g., safflower and/or canola and/or sunflower) and many also contain additives (fortification). However, two commercial white flours (CRF) were used as comparisons to the in-house developed WR, BRR and GBR flours that were produced after freeze drying, using a cyclone sample mill (UDY model 3010-080P, UDY Corporation, Ft. Collins, CO, USA). A Rivland RL-100 long grain rice flour (Riviana Foods Inc., Houston, TX, USA) and Remyflo R7-150T high amylose rice flour (Remy/Beneo, Morris Plains, Fairfield, NJ, USA) were chosen due to similar characteristics compared to 'Rondo' white rice [25].

2.1.1. Lipid Characterization

After all the experiments were completed, stored freeze-dried samples were evaluated for total lipids, free palmitic, free stearic, free linoleic and free linolenic acids, unknown free fatty acids, free sterols, phytosterol esters (which include some very nonpolar lipids), diacylglycerols (DAG), triacylglycerols (TAG) and oryzanol, as described below. Moisture content of each sample was used to calculate the compounds and compound classes on a dry weight basis.

2.1.2. Accelerated Solvent Extraction

The initial samples were present in a ground state as either flours or freeze-dried powders and, therefore, no further grinding was needed. Duplicate 1.0 g samples were prepared. The samples were mixed with uniform 20–30 mesh Ottawa sand (Thermo Fisher Scientific, Waltham, MA, USA) in a beaker and transferred to 11 mL cells. The cells were topped off with Ottawa sand and bottomed off prior to filling; containing cellulose filters at both the top and bottom. Ottawa sand is a very clean, inert, uniformly shaped material that helps to prevent the sample from clumping, and ensures a good flow of solvents in the extraction vessel. The ASE (Accelerated Solvent Extractor, Model 200, Dionex, Sunnyvale,

CA, USA) was operated, as previously described [26]. The parameters were as follows: pressure, 1000 psi; temperature, 100 °C; preheat time, 0; heat time; 5 min; static time; 10 min; static cycles, 3; flush volume, 100% (11 mL cell); purge time, 60 sec and the solvent was hexane. All the extracts were dried under a stream of N_2 in a heated water bath to obtain the total extract weight. For storage, each sample extract was then dissolved in about 10 mg/mL in 85:15 chloroform–methanol with 0.01% butylated hydroxytoluene.

2.1.3. Nonpolar Lipid Analysis

Injections (100 µL) were made of each solvent extracted sample on a Thermo Ultimate 3000 HPLC (Thermo-Fisher, Sunnyvale, CA, USA) using an updated method, as previously reported [27]. The column was a 100 × 3 mm LiChrosorb 5 DIOL (Chrompack, Raritan, NJ, USA), using a 0.5 mL/min flow rate and the following binary gradient: A: 1000:1 hexane:acetic acid; B: 100:1 hexane:isopropyl alcohol with the following ramp: 0 min, 100/0 A/B; 8 min, 100/0; 10 min, 75/25; 40 min, 75/25; 41 min, 100/0 and 60 min, 100/0. Detection was accomplished by diode array detectors (DAD) at 205 nm and 320 nm, and also with the Thermo Charged Aerosol Detector (CAD). Oleic acid, linoleic acid, linolenic acid and the unknown potential free fatty acids (FFA), using linoleic calibration, were appraised at 205 nm and oryzanol at 320 nm. The 254 nm and 280 nm responses were also recorded by the DAD for monitoring, but were not used for any peak analyses. Detection was accomplished by CAD for the steryl esters (StE), triacylglycerols (TAG), stearic/palmitic acid, 1,3-diacylglycerols and 1,2-diacylglycerols (DAG), both using sterol calibration. There were multiple iterations of the various possible DAG compounds, and peaks were therefore broad sets that separated out as 1,2 versus 1,3. The instrument numbers generated were reliable for showing how the classes delivered trends through processing.

2.2. Data Analysis and Statistics

Data were initially analyzed to assure a normal distribution using the Shapiro–Wilk W hypothesis test, and outliers were removed to avoid their effect on the results using JMP® 13 PRO for Windows (SAS Institute Inc., Cary, NC, USA). Thereafter, data were analyzed using the analysis of variance (ANOVA) in JMP® 13 PRO for Windows. Comparisons were made across the whole experiment, per treatment, for which there is technically not an overall control (e.g., WR cannot be germinated, there was one non-germinated BRR beverage and another BR was germinated into GBR; Tables 1–3). Since the same lot of rice was used for all beverages, data were also presented per rice beverage type (WR, BRR, GBR), based on each unique initial starting material unique control (Supplemental Tables S1 and S2). Throughout, two sets of symbol and letter designations were purposely used in the tabulated data. When statistically significant differences were found, the means were compared against the control by the Dunnett's test at $p < 0.05$ (illustrated by asterisks). In the cases in which a control was not available, treatment differences were evaluated by the Tukey–Kramer HSD (Honestly Significant Difference) test at $p < 0.05$.

Table 1. Total lipids (weight %, dwb) and free fatty acids (mg/100 g) in three beverages prepared from white, brown and germinated brown 'Rondo' rice, with commercial flour comparisons.

Treatments	Total Lipid/Oil Weight % (dwb)	Palmitic and Stearic	Oleic	Linoleic	Linolenic
BRR (→ GBR) [1]	*3.95 ± 0.72 z* [2]	*29.98 ± 5.64 z*	*27.67 ± 10.45 z*	*26.64 ± 4.46 z*	*1.61 ± 0.56 z*
GBR (control)	2.46 ± 0.18 yb	24.69 ± 1.58 yb	11.83 ± 5.69 yb	9.44 ± 1.49 yb	0.50 ± 0.14 yb
BRR (control)	2.78 ± 0.15 a	21.59 ± 3.62 b	15.74 ± 5.59 b	13.18 ± 1.88 b	0.71 ± 0.20 b
WR (control)	0.77 ± 0.09 c	138.98 ± 29.41 a	199.17 ± 36.27 a	178.59 ± 35.77 a	6.09 ± 1.42 a
GBR, PWM	1.98 ± 0.04 *B	14.45 ± 1.72 *A	5.37 ± 1.31 *A	8.81 ± 1.30 A	0.52 ± 0.05 A
BRR, PWM	2.17 ± 0.16 *A	6.85 ± 1.07 *B	1.18 ± 0.57 *B	1.51 ± 0.34 *C	0.16 ± 0.09 *B
WR, PWM	0.14 ± 0.02 *C	5.93 ± 0.45 *B	7.09 ± 0.37 *A	5.57 ± 0.79 *B	0.18 ± 0.05 *B
GBR, PNZ	0.62 ± 0.12 *t	20.85 ± 2.02 *s	6.72 ± 1.56 *s	26.03 ± 3.69 *t	1.40 ± 0.20 *t
BRR, PNZ	0.23 ± 0.02 *s	5.61 ± 1.01 *r	1.91 ± 0.42 *r	7.40 ± 1.03 *s	0.43 ± 0.08 s

Table 1. *Cont.*

Treatments	Total Lipid/Oil Weight % (dwb)	Palmitic and Stearic	Oleic	Linoleic	Linolenic
WR, PNZ	0.14 ± 0.00 *s	31.46 ± 0.51 *t	13.87 ± 4.06 *t	29.07 ± 4.56 *t	1.33 ± 0.24 *t
GBR, PWM-SL	4.77 ± 0.41 *Z	40.32 ± 3.36 *Z	16.14 ± 2.02 *Z	33.66 ± 3.02 *Z	1.98 ± 0.10 *Z
BRR, PWM-SL	3.93 ± 0.16 *Y	18.96 ± 1.39 Y	3.92 ± 0.72 *Y	5.90 ± 0.21 *Y	0.42 ± 0.01 Y
WR, PWM-SL	i.s.[3]	i.s.	i.s.	i.s.	i.s.
GBR, PNZ-SL	n.s.[3]	n.s.	n.s.	n.s.	n.s.
BRR, PNZ-SL	3.26 ± 0.09 *T	66.05 ± 3.01 *S	24.05 ± 3.04 * S	116.84 ± 1.71 *S	5.83 ± 0.09 *S
WR, PNZ-SL	0.94 ± 0.10 S	123.49 ± 28.11 T	82.50 ± 19.97 * T	170.52 ± 23.43 T	7.35 ± 0.61 T
CRF R7-150T	0.33 ± 0.01	85.13 ± 7.78	72.91 ± 6.52	71.72 ± 7.83	2.05 ± 0.19
CRF RL-100	0.89 ± 0.01	189.2 ± 19.81	226.08 ± 24.10	223.24 ± 24.58	6.58 ± 0.32

[1] Treatment acronyms: BRR, brown 'Rondo' rice; GBR, germinated brown rice; PWM, post wet-milling; PNZ, post enzymes; PWM-SL, post wet-milling sieving loss; PNZ-SL, post enzymes sieving loss; WR; white rice and CRF, commercial rice flour. The *BRR (→ GBR)* is italicized since it was the original starting material to generate GBR but, it is technically not the GBR beverage control. [2] Means highlighted with an asterisk (*) are significantly different from the rice type control (GBR, BRR or WR) according to a Dunnett's test at $p < 0.05$. Control and treatment means not connected by the same letter are significantly different among them, according to a Tukey–Kramer HSD test at $p < 0.05$. z, y indicates the significant differences between the germinated GBR control versus the initial BRR used for germination; a, b, c indicates the differences among the GBR, BRR and WR controls; A,B,C indicates the differences among GBR, BRR and WR for the PWM treatments; X,Y,Z indicates the differences among GBR, BRR and WR for the PWM-SL treatments; r, s, t indicates the differences among GBR, BRR and WR for the PNZ treatments; and R,S,T indicates the differences among GBR, BRR and WR for the PNZ-SL treatment. The data represent the means from independent comparisons, where $n = 3$ or $n = 6$ ± standard deviation. [3] i.s. indicates insufficient sample to collect, whereas n.s. indicates not sampled.

Table 2. Free fatty acids and acylglycerols (mg/100 g) in the three beverages prepared from white, brown and germinated brown 'Rondo' rice, grouped by treatment, with commercial flour comparisons.

Treatments	Unknown FFA	Total FFAs	TAG (Triacylglycerols)	1,3-DAG (Diacylglycerols)	1,2-DAG (Diacylglycerols)
BRR (→ GBR) [1]	2.04 ± 1.00 z [2]	87.93 ± 19.59 z	3163.52 ± 31.50 z	24.04 ± 7.98 z	116.62 ± 24.92 z
GBR (control)	1.33 ± 0.46 yb	47.80 ± 6.86 yb	1919.15 ± 6.86 yb	9.97 ± 4.78 yb	52.24 ± 9.66 yb
BRR (control)	1.94 ± 0.24 a	53.15 ± 10.86 b	2356.24 ± 205.96 a	23.49 ± 6.48 a	80.16 ± 15.93 a
WR (control)	0.23 ± 0.04 c	523.06 ± 102.37 a	167.89 ± 27.23 c	13.39 ± 4.15 b	3.40 ± 0.48 c
GBR, PWM	0.70 ± 0.12 *B	29.85 ± 3.02 *A	1583.44 ± 67.07 *B	15.05 ± 3.00 *B	64.48 ± 10.33 A
BRR, PWM	1.64 ± 0.17 A	11.33 ± 2.00 *B	1844.73 ± 104.49 *A	32.03 ± 3.06 A	43.49 ± 6.32 *B
WR, PWM	0.08 ± 0.01 *C	18.85 ± 0.96 *B	94.31 ± 15.93 C	8.73 ± 2.74 C	3.13 ± 0.81 C
GBR, PNZ	0.33 ± 0.06 *t	55.33 ± 5.51 s	438.85 ± 82.06 *t	9.37 ± 1.33 t	7.84 ± 1.60 *t
BRR, PNZ	0.20 ± 0.02 *s	15.55 ± 2.44 *r	171.39 ± 26.26 *s	5.39 ± 1.19 *s	2.24 ± 0.55 *s
WR, PNZ	0.22 ± 0.02 t,s	75.95 ± 8.37 *t	39.59 ± 12.53 *s	3.64 ± 1.59 s	1.71 ± 1.07 s
GBR, PWM-SL	3.66 ± 0.41 *Y	101.53 ± 4.84 *Z	3697.75 ± 403.22 *Z	19.61 ± 3.58 *Y	177.72 ± 11.82 *Z
BRR, PWM-SL	4.64 ± 0.28 *Z	33.841.15 *Y	2999.40 ± 4.47 *Y	36.56 ± 5.95 *Z	99.90 ± 16.28 Y
WR, PWM-SL	i.s. [3]	i.s.	i.s.	i.s.	i.s.
GBR, PNZ-SL	n.s. [3]	n.s.	n.s.	n.s.	n.s.
BRR, PNZ-SL	2.37 ± 0.18 *T	215.15 ± 3.25 *S	2578.79 ± 241.27 T	107.36 ± 13.16 *T	36.42 ± 8.67 *T
WR, PNZ-SL	0.82 ± 0.08 *S	384.68 ± 70.68 T	410.12 ± 83.03 *S	46.63 ± 11.04 *S	16.21 ± 2.43 *S
CRF R7-150T	0.89 ± 0.03	232.71 ± 22.06	18.32 ± 2.28	0.51 ± 0.13	0.56 ± 0.06
CRF RL-100	0.68 ± 0.14	645.80 ± 68.64	159.14 ± 16.45	6.38 ± 2.54	3.38 ± 0.43

[1] Treatment acronyms: BRR, brown 'Rondo' rice; GBR, germinated brown rice; PWM, post wet-milling; PNZ, post enzymes; PWM-SL, post wet-milling sieving loss; PNZ-SL, post enzymes sieving loss; WR; white rice and CRF, commercial rice flour. Measured factor acronyms: FFA, free fatty acid(s). The *BRR → GBR* is italicized since it was the original starting material to generate GBR but, it is technically not the GBR beverage control. [2] Means highlighted with an asterisk (*) are significantly different from the rice type control (GBR, BRR or WR) according to a Dunnett's test at $p < 0.05$. Control and treatment means not connected by same letter are significantly different among them, according to a Tukey–Kramer HSD test at $p < 0.05$. z, y indicates the significant differences between the germinated GBR control versus the initial BRR used for germination; a, b, c indicates the differences among the GBR, BRR and WR controls; A,B,C indicates the differences among GBR, BRR and WR for the PWM treatments; X,Y,Z indicates the differences among GBR, BRR and WR for the PWM-SL treatments; r, s, t indicates the differences among GBR, BRR and WR for the PNZ treatments; and R,S,T indicates the differences among GBR, BRR and WR for the PNZ-SL treatment. The data represent the means from independent comparisons, where $n = 3$ or $n = 6$ ± standard deviation. [3] i.s. indicates insufficient sample to collect, whereas n.s. indicates not sampled.

Table 3. Phytosterol esters (nonpolar lipids), free sterols, oryzanols and summed lipids/oils (mg/100 g) in the three beverages prepared from white, brown and germinated brown 'Rondo' rice, grouped by treatment, with commercial flour comparisons.

Treatments	StE (Phytosterol Esters)	Free Sterol	Oryzanol	Sum Classes (Recovered)
BRR (→ GBR) [1]	*138.50 ± 14.12 z* [2]	*35.30 ± 4.84 z*	*51.62 ± 11.84 z*	*3617.52 ± 756.10 z*
GBR (control)	97.29 ± 10.18 ya	26.92 ± 2.52 ya	25.10 ± 3.79 yb	2178.45 ± 152.70 yb
BRR (control)	102.05 ± 14.97 a	22.28 ± 1.75 b	35.67 ± 4.35 a	2673.07 ± 244.00 a
WR (control)	34.42 ± 1.81 b	3.35 ± 0.21 c	5.66 ± 0.96 c	751.17 ± 135.59 c
GBR, PWM	81.72 ± 5.02 *A	24.59 ± 0.42 A	18.11 ± 1.42 *B	1817.23 ± 77.78 *B
BRR, PWM	92.62 ± 9.91 A	22.66 ± 1.83 A	27.51 ± 1.39 A	2074.40 ± 116.70 *A
WR, PWM	17.25 ± 3.05 *B	2.40 ± 0.42 B	1.01 ± 0.25 *C	145.67 ± 21.34 *C
GBR, PNZ	29.02 ± 1.72 *t	8.74 ± 1.54 *t	5.84 ± 0.93 *t	554.98 ± 92.22 *t
BRR, PNZ	20.36 ± 5.89 *s	3.43 ± 0.53 *s	1.15 ± 0.06 *s	219.50 ± 36.39 *s
WR, PNZ	19.30 ± 0.82 *s	2.03 ± 0.27 s	0.36 ± 0.15 *s	142.55 ± 6.43 *s
GBR, PWM-SL	238.55 ± 20.89 *Z	50.84 ± 2.71 *Z	165.86 ± 11.43 *Z	4695.43 ± 448.29 *Z
BRR, PWM-SL	170.97 ± 16.21 *Y	43.62 ± 4.88 *Y	132.32 ± 12.56 *Y	3516.60 ± 52.56 *Y
WR, PWM-SL	i.s. [3]	i.s.	i.s.	i.s.
GBR, PNZ-SL	n.s. [3]	n.s.	n.s.	n.s.
BRR, PNZ-SL	137.21 ± 10.44 *T	41.91 ± 3.47 *T	69.20 ± 6.45 *T	3186.03 ± 254.20 *T
WR, PNZ-SL	33.42 ± 3.77 S	17.23 ± 2.53 *S	11.42 ± 1.07 *S	919.70 ± 33.94 S
CRF R7-150T	17.97 ± 2.98	7.03 ± 0.13	1.22 ± 0.06	278.31 ± 2.98
CRF RL-100	36.40 ± 16.45	8.57 ± 2.54	3.28 ± 0.43	862.96 ± 2.08

[1] Treatment acronyms: BRR, brown 'Rondo' rice; GBR, germinated brown rice; PWM, post wet-milling; PNZ, post enzymes; PWM-SL, post wet-milling sieving loss; PNZ-SL, post enzymes sieving loss; WR; white rice and CRF, commercial rice flour. The *BRR (→ GBR)* is italicized since it was the original starting material to generate GBR but, it is technically not the GBR beverage control. [2] Means highlighted with an asterisk (*) are significantly different from the rice type control (GBR, BRR or WR) according to a Dunnett's test at $p < 0.05$. Control and treatment means not connected by same letter are significantly different among them, according to a Tukey–Kramer HSD test at $p < 0.05$. z, y indicates the significant differences between the germinated GBR control versus the initial BRR used for germination; a, b, c indicates the differences among the GBR, BRR and WR controls; A,B,C indicates the differences among GBR, BRR and WR for the PWM treatments; X,Y,Z indicates the differences among GBR, BRR and WR for the PWM-SL treatments; r, s, t indicates the differences among GBR, BRR and WR for the PNZ treatments; R,S,T indicates the differences among GBR, BRR and WR for the PNZ-SL treatment. The data represent the means from independent comparisons, where $n = 3$ or $n = 6$ ± standard deviation. [3] i.s. indicates insufficient sample to collect, whereas n.s. indicates not sampled.

3. Results and Discussion

3.1. Total Lipid and Proximate Analysis

Total lipids determined by the HPLC from freeze-dried powders indicated that the starting materials contained 0.77%, 2.78%, 3.95% and 2.46% (dry weight basis, dwb), for WR, BRR and BRR sprouted into GBR (often designated as BRR → GBR, for clarity) and the GBR, respectively (Table 1). With the exception of the WR controls, these data corroborated well with the original proximate analyses, per trial. The initial crude fat proximate contents, corresponding to the original raw data per trials compared herein, were 1.19 (WR), 3.04 (BRR) and 3.59% in the BRR used for germination, which resulted in 2.48% for GBR [23]. These were the control, pre-processed rice crude fat contents that were potentially delivered into each rice beverage type. However, as discussed below, the original WR proximate analysis was accomplished with the freshly dehulled rice and shipped WR, whereas the later WR lipid determination was accomplished with the samples from stored rice with "brokens" that were afterwards milled into flour.

There was wide variability in the MC among the initial control samples, post-experimental saved samples and commercial flours, due to sourcing, different freeze-drying runs, storage differences and certainly germinated versus non-germinated effects (data not shown). Therefore, a global correction factor was applied to all the data by calculating each result based on the unique MC of the samples, resulting in the delivery of data as percentage lipid constituents (mg/100 g, dwb). The WR PWM-SL (sieving loss) was negligible (1.05%) and there was an insufficient sample quantity to collect and analyze (Table 1). The GBR

PNZ-SL (sieving loss) samples were unfortunately not collected or lost. Lipids were not measured in any commercial rice beverages because every product found on the grocer's shelf had exogenously added oil (labeled as canola and/or safflower and/or sunflower oil), which would confound data regarding the endogenously present oils and impart obvious lipid, FFA and sterol impurities [22].

3.2. Total HPLC Lipids (and Oils) versus Summed Recovered Compounds

When one compares the total lipids/oil (% dwb) to the summed amount of the compounds recovered (mg/100 g, dwb), the initial HPLC percentages were very congruent with the overall summed lipids recovered and reported (Tables 1 and 3). Due to the utilized units, there is a 1000-fold numerical difference between % dwb total (HPLC lipids) versus the summation of all the recovered compounds and compound classes (mg/100 g). In the 3 PNZ beverages, there was an overall 93.2 ± 9.0% lipid/oil recovery rate. Aside from some FFAs, the levels of total lipids in the GBR PNZ beverages were oftentimes significantly higher than the non-germinated BRR beverage, which was likewise significantly higher than the WR beverage (Tables 1–3). The WR beverage stream contained the least amount of starting lipids and also lost the most significant proportion through processing into the WR PNZ beverage, compared to both the BRR and GBR beverages (Tables 1–3; Supplemental Tables S1 and S2).

An Italian plant-based beverage survey evaluating 72 commercial rice beverages found the average oil content of 1.0 g/100 mL, with a range from 1.0–1.1 g/100 mL with saturates comprising 0.2 g/100 mL [28]. This level corroborates with the level reported in unsweetened rice beverages in the USDA Food Data Central database of 0.97 g (https://ndb.nal.usda.gov/fdc-app.html#/food-details/171942/nutrients) (accessed on 14 September 2021). However, this was a back-calculated value and it is unknown if oil was added in this sample. Oil levels reported elsewhere in commercial samples (generally about 1 g/100 mL) are not on par with the levels found herein (where no exogenous oils were added) in post-enzyme treated (PNZ) WR and BRR beverages, but were closer to the completely endogenous GBR PNZ beverage, containing 0.62 g/100 g (Table 1). Most plant-based beverages are processed with stabilized raw ingredients, which effectively strip away endogenous and natural lipids that cause rancidity issues. Herein, we purposely used native, non-stabilized brown rice that was sprouted and conveyed most endogenous ingredients back into a processed beverage using "green technologies". Subsequently, the 0.62% oils contained in the GBR PNZ beverage is 100% natural, and requires no added ingredients or exogenously added oils to augment emulsion and, hence, has no labeling considerations.

The overall percentage of the material lost during the process through sieving, followed the trend whereby BRR > GBR > WR [23] and PNZ GBR samples that passed through a 140-mesh sieve had a significantly larger mean and D90 cumulative particle size, compared to both the BRR and WR PNZ beverages [24].The GBR PNZ had about half the processing loss compared to the non-germinated BRR starting material, while the WR process resulted in the least losses and better solubilization as almost the entire starting weight was starchy material [23]. Nonetheless, where lipid processing losses were reported (PWM-SL and PNZ-SL), these values were often the most significant percentage of the materials recovered, per beverage type, on a dry weight basis (Supplemental Tables S1 and S2). For example, there was a 5.17% loss of total lipids in the GBR PWM. It is of note that these sieving loss steps inherently have a "concentrating effect", as very low MC residuals (e.g., starchy granules and fiber) have been removed by the sieves during the processing regime, whereas a 5-fold dilution has carried forward the liquid matrix into the resulting beverages.

3.3. Free Fatty Acids (FFA): Saturated (SFA) and Unsaturated (USFA)

The major free fatty acids recovered in the three starting materials (WR, BRR and BRR that was germinated into GBR) were palmitic/stearic, oleic and linoleic acids (Table 1), as generally reported in most rice and BR fractions [2] (pp. 163–190). The method used herein did not fully resolve the two saturated fatty acids, stearic versus palmitic. However, stearic acid generally comprises only 2–4% of the total lipid profile in rice [2] (pp. 163–190). Total FFAs recovered in WR, BRR and GBR were 523.06, 53.15 and 47.80 mg/100 g, respectively (Table 1). In general, there was a fairly even distribution of the 3 main FFA categories across controls and through processing, whereby roughly 20 to 40% was comprised by palmitic (C16:0)/stearic (C18:0), oleic (C18:1) and linoleic (C18:2) acids. Similar ranges were reported in rice bran oil (RBO) [29]. The palmitic/stearic, oleic and linoleic acids composition based on dry weight for the major of the fatty acids recovered in BRR, was 21.59, 15.74 and 13.18 mg/100 g, and in the sprouted method the BRR was 29.98, 27.67 and 26.64 mg/100 g, which gave rise to the GBR containing 24.69, 11.83 and 9.44 mg/100 g, respectively (Table 1).

Germination caused significant decreases in all the free fatty acids evaluated (Table 1). During rice germination, the oleic acid decreased, whereas the palmitic and linoleic acid contents increased [20]. On the other hand, the concentrations of oleic, palmitic and palmitoleic acids increased in the initial stage of germination, but decreased rapidly after 72 h [21]. However, both of these studies measured the fatty acids after transesterification, so they were measuring the esterified fatty acids (in glycerolipids, such as triacylglycerols, glycolipids and phospholipids), whereas the individual free fatty acids were evaluated herein. Furthermore, it appears that the methodology used by [21] to measure FFA would combine the data with other lipids, which also makes comparisons to our data difficult. In another study, the contents of the oil components (palmitic, oleic and linoleic acid), γ-oryzanol, phytosterol, vitamin E and squalene were slightly increased or not changed by germination in the two rice varieties [30]. Nonetheless, in the germination portion of the study, virtually every FFA significantly decreased after the 48 h germination period, and further decreases were generally observed as beverage processing (control → PWM → PNZ) ensued (Table 1).

In general, the 'Rondo' WR controls had significantly higher concentrations of all the FFAs in this study (aside from one reference commercial rice flour, CRF) and the significantly lower total lipid recovered compared to the BRR and GBR controls (Table 1). The WR contained 138.98, 199.17 and 178.59 mg/100 g, palmitic/stearic, oleic and linoleic acids, respectively (Table 1). The two commercial white flours that were included for comparison, likewise, had very high levels of FFAs. Nonetheless, it is not legitimate to make statistical comparisons between unknown commercially processed flour versus a well-characterized experimental variety and process. FFA levels in both CRF's were remarkably higher (232.7–645.8 mg/100 g) than the freshly de-hulled 'Rondo' non-germinated BRR samples (11.3–53.2) and GBR samples, (29.9–87.9), discounting the loss streams (Table 1). Subsequently, the lipids in the endosperm of the WR experienced a fair amount of lipase activity. Although 523 mg FFA/100 g appears to be a high amount, it only translates to 0.523%. Therefore, in the WR, this quantity of FFAs is not really a large proportion of the overall nutrients. As the total lipid in white rice was 774 mg/100 g, then the FFAs appeared to be the most abundant lipid in these WR samples (Table 1).

Prior to converting all the data to a dry weight concentrations basis using the discrete sample MC, the control BRR and GBR samples used herein had free fatty acids (FFAs), free sterols and diacylglycerols (DAGs) levels (data not shown) within the similar ranges previously reported in the control ground "Macia" sorghum [31]. These levels of FFAs indicated higher than normal levels of lipase activity in the sorghum, as was likewise probable in the WR findings. The materials used to produce the WR beverages and stored commercial white flours (CRF) used as a comparison had excessive FFAs. It has long been known, however, that rice (even though it contains a relatively low amount of oil) and RBO, are subject to the rapid accumulation of FFA and lipid oxidation products, due to the exceedingly high lipase levels, even in the mature, dried kernels [2,32] (pp. 143–162).

These data indicate that the WR experienced substantial lipid hydrolysis from the time of milling through shipping/handling and short 4 °C storage (1 month) before use, or due to the sample freeze drying and flour produced thereafter. However, brokens were purposely received and used to make a low cost, value-added beverage. On the other hand, all the BRR and GBR samples were freshly dehulled and immediately utilized, unstabilized, for each experiment, then frozen prior to freeze-drying and flour/powder sample production. Rice has one of the most notoriously active and persistent levels of lipase activity and this apparently resulted in the initial free fatty acid differences during these trials. However, these experiments were optimized for the GBR, and the non-germinated BRR and WR were run as "checks" and helped serve as comparison and for validation purposes.

The individual FFAs, across all three beverages, decreased from controls (WR or BRR or GBR) after wet-milling (PMW), but generally increased after saccharification (PNZ) (Table 1). There was a substantial and significant loss of most FFAs in the WR samples, as controls were heated (softened), wet milled into PWM and enzyme-treated resulting in the PNZ beverage (Table 1). Significant overall losses also occurred oftentimes in both the BRR and GBR beverages. When the values were converted from mg/100 g to relative percentage, the total FFA loss in WR PNZ was 85.5%, without concomitant increase measures in any other analyte to compensate for the mass balance. Total FFA losses in BRR were less (70.7%) and the total FFAs in GBR increased (15.8%), as linoleic acid interestingly increased through processing from PMW to PNZ in all three beverages. Perhaps this is a result of the second heating step (80 °C), in which the starch is purposely hydrated through gelatinization to physically facilitate the saccharification enzyme process, which dissociated TAG and DAG to free more fatty acid moieties. In RBO from the germinated rice, linoleic and linolenic acid composition increased while oleic and palmitic acid decreased [33], which mirrors our trends observed from the GBR controls into the PNZ beverage. Regardless, this is a positive finding since linoleic acid is an essential FA. Overall, there was a general trend in all three beverage types (WR, BRR and GBR), whereby there were marked decreases in almost all the compounds (such as the aforementioned TAG and DAG), except FFAs, through processing (control → PWM → PNZ).

3.4. *Triacylglycerols (TAG) and Diacylglycerols (DAG)*

The BRR used for germination, GBR (BRR → GBR) and BRR controls contained the significantly highest level of TAGs in all the analyzed samples, with 3163.5, 1919.2 and 2356.2 mg/100 g, respectively, and 140.7, 62.2 and 103.6 mg/100 g DAGs (1,3-DAG plus 1,2-DAG), respectively (Table 2). All the samples utilizing BRR and GBR from control through the PNZ beverages (not including the processing losses PMW-SL and PNZ-SL) contained between 81.6–92.6% of the recovered lipids as TAGs and DAGs (Table 4). Before converting data to dwb, approximately 70–80% (wet wt%) of the compounds recovered in the BRR and GBR samples were TAG and DAG, similar to the levels previously reported in RBO [32]. On the other hand, the WR beverage control contained significantly less TAG, 1,3-DAG and 1,2-DAG compounds (167.89, 13.39 and 3.40 mg/100 g, respectively) and markedly reduced the relative percentages (24.6–72.9%) throughout processing (Tables 2 and 4). This indicates that relatively low lipase or oxidation occurred in the BRR and GBR samples compared to the WR treatments. Lipase hydrolysis of esterified fatty acids (FA) from oil triacylglycerols (TAG) produce 1,2,diacylglycerols (DAG) and 1,3,diacylglycerols (DAG), which ultimately leads to the net conversion of oil to sugars during germination [34]. The breakdown in TAGs was the main expected change in FFAs to occur during germination. Indeed, from BRR → GBR, there was a significant decrease in TAGs, 1,3-DAG and 1,2-DAG of 39.3, 58.5 and 55.2%, respectively (Table 2). Except for 1,3-DAG in GBR PWM, the TAGs and DAGs significantly decreased from controls through PWM (presumably due to heating and wet-milling) into each rice type PNZ beverage (Supplemental Table S2).

Table 4. Percentage of compounds and compound classes recovered, based on the total lipids isolated in white, brown and germinated brown 'Rondo' rice beverages.

Treatments	% FFAs	% TAG	% TAGs and DAGs	% StE (Phytosterol Esters)	% Free Sterols	% Oryzanol
BRR (→ GBR) [1]	2.43	87.45	91.34	3.83	0.98	1.43
GBR (control)	2.19	88.10	90.95	4.47	1.24	1.15
PWM	1.64	87.13	91.51	4.50	1.35	1.00
PNZ	9.97	79.07	82.17	5.23	1.58	1.05
PWM-SL	2.20	83.90	88.10	5.08	1.08	3.53
PNZ-SL	n.s. [2]	n.s.	n.s.	n.s.	n.s.	n.s.
BRR (control)	1.99	88.15	92.03	3.82	0.83	1.33
PWM	0.55	88.93	92.57	4.47	1.09	1.33
PNZ	7.08	78.08	81.56	9.27	1.56	0.52
PWM-SL	0.96	85.29	89.17	4.86	1.24	3.76
PNZ-SL	6.75	80.94	85.45	4.31	1.32	2.17
WR (control)	69.63	22.35	24.59	4.58	0.45	0.75
PWM	12.94	64.74	72.88	11.84	1.65	0.70
PNZ	53.27	27.77	31.52	13.54	1.43	0.25
PWM-SL	i.s. [2]	i.s.	i.s.	i.s.	i.s.	i.s.
PNZ-SL	41.83	44.59	51.43	3.63	1.87	1.24

[1] Treatment acronyms: BRR, brown 'Rondo' rice; GBR, germinated brown rice; PWM, post wet-milling; PNZ, post enzymes; PWM-SL, post wet-milling sieving loss; PNZ-SL, post enzymes sieving loss and WR; white rice. Measured factor acronyms: FFA, free fatty acid(s); TAG, triacylglycerols; DAG, diacylglycerols and StE, phytosterol esters, including very nonpolar lipids. [2] n.s. indicates not sampled, whereas i.s. indicates insufficient sample to collect.

The WR samples and CRF, in general, displayed the same trends regarding the classes of FFAs and TAGs/DAGs having high or low concentrations. WR control samples (aside from CRF) had the lowest significant concentration of TAGs and DAGs in the study (Table 2). This appears to be consistent with the FFAs being starch lipids. However, when total TAGs and DAGs were expressed as a relative percentage of the total lipid compounds recovered, the control WR, BRR and GBR contained 24.6, 92.0 and 91.0%, respectively (Table 4). This was due to the fact that the WR had the highest significant quantity (523.06 mg/100 g; Table 1) and relative percentage (69.6%; Table 4) of FFAs, compared to the BRR and GBR. The majority of the total lipids recovered in BRR and GBR were TAGs (92.0 and 91.0%, respectively; Table 4), which concomitantly contained relatively low levels of total FFAs (2.0 and 2.2%, respectively). These results paralleled the above FFA finding, indicating that the WR (as well as stored WR check flours) succumbed to lipid oxidation. Aside from a few exceptions (linoleic acid, linolenic acid and 1,3-DAG in GBR), there were generally significant decreases in most of the parameters measured (FFAs, TAGs, 1,2-DAG, sterols and oryzanols) in all three beverage processes, especially in non-germinated BRR and WR (Supplemental Tables S1 and S2).

3.5. Phytosterol Esters (StE) and Nonpolar Lipids

Similar to the other compound trends, the GBR contained significantly lower levels of phytosterol esters (97.29 mg/100 g), compared to the original BRR starting material (138.50 mg/100 g). Again, there was a general trend whereby the processing caused an initial decrease (in PWM samples) and/or significant decrease in the phytosterol esters in each PNZ beverage (Table 3). In a study looking at the compositional change of policosanols and oils in four varieties of post-germinated brown rice oil, squalene increased 2.4 fold and the phytosterols campesterol, stigmasterol and β-sitosterol increased by 8.3%, 31.6% and 3.3%, respectively, whereas the cycloartenol and 24-methylcycloartanol (probably from the hydrolysis of γ-oryzanol) decreased by 11.0 and 4.5%, respectively [33]. Herein, the phytosterol esters were measured but the peak actually contains other very nonpolar lipids, such as hydrocarbons (including squalene) and wax esters. For example, we reported between 20.4–138.5 mg/100 g of phytosterol esters in the BRR and GBR processing stages, whereas others [??] reported 4.3 mg/100 mL for the rice beverage (principally β-sitosterol, β-sitosterol-β-D-glucoside, campesterol and stigmasterol). The unstabilized 100% natural PNZ beverages delivered 19.3, 20.4 and 29.9 mg/100 g in WR, BRR and GBR, respectively, with the GBR PNZ beverages being significantly higher than the non-germinated beverages (Table 3). Technically, these phytosterol esters can be better classified as "very nonpolar lipids". Although the analytical system herein utilized did not separate the phytosterol esters from the other non-polar lipids, we believe there is still value demonstrating this peak since interesting trends were observed in these health-promoting compounds [7,22]. The beverage processing loss streams (PWM-SL and PNZ-SL) contained the highest significant levels of phytosterol esters per beverage category tested (Table 3).

3.6. Free Sterols

Brown 'Rondo' rice (BRR) contained 35.30 mg/100 g total free sterols, which significantly decreased upon germination to 26.92 mg/100 g in GBR. The other BRR control that was not germinated contained 22.28 mg/100 g and the significantly lowest free sterol level was found in the WR control (3.35 mg/100 g) (Table 3). Through processing, the total free sterols significantly decreased to 8.74, and 3.43 and 2.03 mg/100 g in the post-enzyme treated (PNZ) GBR, BRR and WR, respectively. The concentration in the non-germinated WR and BRR control (PNZ) beverages were on par, compared to the 4.29 mg/100 g reported [22]. However, the exact constituents of those beverages tested (e.g., BR versus sprouted, organic, or what commercial processes were employed; [22]) was not determined. The germinated BRR, used in the free-flowing green process to generate GBR PNZ beverages, contained significantly greater concentrations of total free sterols (8.74 mg/100 g) than the non-germinated beverages and the aforementioned well-characterize rice beverage, with 4.29 mg/100 g [22]. This is a valuable finding regarding the health-related advantages of utilizing this methodology to deliver a fully 100% natural plant-based beverage. Due to the analogous structure of cholesterol, many phytosterols are known to compete and interfere with the absorption and binding of cholesterol in the GI tract, ultimately decreasing low-density lipoprotein (LDL) cholesterol levels, which can decrease the threat of coronary heart failure [7,22]. GBR PNZ delivered the significantly highest level for all the compounds listed in Tables 2 and 3 (TAG, DAG, sterols, oryzanol and the summation of all lipids reported), whereas both the non-germinated BRR and WR beverage quantities were markedly lower. These classes of compounds in the non-germinated BRR PNZ were 1.4- to 5.1-fold lower than the GBR PNZ beverage. Furthermore, the WR PNZ beverage values were even lower, at 1.5- to 16.2-fold lower than GBR PNZ. As with the other assessed lipid categories, the free sterol loss stream (PWM-SL and PNZ-SL) contained the highest significant levels recovered (on a dry weight basis) in each beverage type (Supplemental Tables S1 and S2).

3.7. Oryzanol

Significantly lower oryzanol levels were found in WR (5.66 mg/100 g), compared to BRR (35.67 mg/100 g) and the initial BRR (51.62 mg/100 g) used to sprout the GBR (25.10 mg/100 g) (Table 3). Oryzanol levels followed a trend whereby control BRR > GBR > WR, with 0.8, 1.3, 1.5 and 1.2% of all lipid recovery attributed to oryzanol in WR, BRR and BRR → GBR, respectively (Table 4). Oryzanol levels in BRR and GBR compared well to those in 30 BR varieties grown at different sites and in different seasons that delivered an average 26–63 mg/100 g of γ-oryzanol [35], in 16 Korean rice varieties displaying a range of 26.7–61.6 mg/100 g [36], and in a summary of 59 whole grain BR varieties [37]. The oryzanol concentrations in BRR were 6.3- to 14.5-fold and GBR was 7.6-fold higher than the 'Rondo' WR. Other BR and GBR have been found to contain roughly 5 times more γ-oryzanol than the counterpart polished rice in "Heugkwang" (black rice) and "Keunnun" (giant embryo) [30]. Significantly lower oryzanol concentrations were recovered in WR, and the CRF samples had the lowest oryzanol levels recovered (1.22 and 3.28 mg/100 g) compared to the GBR and BRR (Table 3).

Sprouting conditions used in 'Rondo', resulted in a 2-fold reduction in the oryzanols (Table 3). However, the concentration of γ-oryzanol was previously reported to increase [5], or did not markedly change through the various germination protocols [19,30]. Total oryzanol concentrations conveyed forward into each PNZ beverage also significantly decreased to 5.84, 1.14 and 0.36 mg/100 g in GBR, BRR and WR, respectively. Oryzanol is an ester and it is therefore possible that it can be hydrolyzed by lipases, or degraded as a consequence of the two heating steps during the beverage formation. Researchers have already indicated that the methods to maximize the concentration of γ-oryzanol in germinated rice needs further investigation because initial levels are variety-dependent [35] and germination conditions (the duration and rates of water uptake) are known to affect the metabolic mobilization and concentration changes in this class of compounds during sprouting [5]. In the future, a time-course evaluation of γ-oryzanol and other key phytonutrients is warranted to better optimize the germination stopping point.

All processing loss stages recovered and analyzed had significantly higher levels of oryzanol (Table 3 and Supplemental Table S2). The PWM GBR loss was 165.86 mg/100 g. Unfortunately, the GBR PNZ loss samples were not collected or analyzed to tally the overall oryzanol loss. However, the 2 sieving stages in BRR resulted in 201.5 mg/100 g of lost oryzanol, whereas only 11.42 mg/100 g of oryzanol was lost in the WR PNZ (there was negligible WR PWM-SL and, thus, no samples). Particle size of the materials lost and discarded from the sieves was not measured. Yet, the WR losses (principally gritty, starchy endosperm) had the lowest relative percentage loss, whereas the losses in the BRR were the highest while GBR losses were intermediate. This indicates that the germination and endogenous enzymatic activity must have softened, and solubilized more constituents in the GBR beverage stream compared to the non-germinated BRR beverage [23]. As previously noted, the BRR and GBR materials were slightly gritty and brown, indicating that some fiber-associated bran and aleurone materials and hard starchy endosperm constituents were discarded. This implies that, across the board, the processing regime either failed to sufficiently soften/wet-mill the raw materials and solubilize the majority of oryzanols into the beverages, or sieving needs to be readdressed. For example, 25.5, 3.2 and 6.3% of the original control oryzanol was conveyed into the GBR vs. BRR vs. WR beverages, respectively (calculated from Table 1). Only the germinated, endogenously softened, enzyme-activated GBR materials conveyed a significantly higher relative percentage of the oryzanols into the beverages. This trend was observed throughout all three beverages regarding most recovered lipid categories and compounds.

γ-Oryzanol is technically a mixture of ferulic acid esters of triterpene alcohols and sterols. More specifically, these hydroxycinnamate sterol esters are esters of cycloartenol and 24-methylenecycloartanol in rice and sitostanol in corn, which have also recently been demonstrated to contain coumaric, caffeic and sinapic acids esterified to sterols in rice and corn [7]. The ferulate part of γ-oryzanol is attributed to the antioxidant capacity in this sterol class, according to a linoleic acid model wherein the major compounds of γ-oryzanol (cycloartenyl ferulate, 24-methylenecycloartanyl ferulate and campestanyl ferulate) prevented the ultraviolet-derived oxidation of linoleic acid, although the effects were less pronounced than free ferulic acid and α-tocopherol [9], and make up about 90% of the γ-oryzanol in GBR [30]. Hence, one can make an assertion that several additional health-beneficial low molecular weight compounds are probably made available and remain soluble in the GBR beverage prepared and reported herein. Subsequently, more complete softening, particle size reduction and/or emulsification should be employed in the general processing scheme to ameliorate these losses.

Rice bran contains ferulic acid in an insoluble bound form that is esterified with arabinose or arabinoxylans as feruloylated arabinoxylo-oligosaccharies [38]. Excellent functional and emulsifying oil-in-water properties have been attributed to arabinoxylans in grains [39] due to the cross-linking of their ferulic acids [40], which have a unique capacity to form covalent gels [41]. In this beverage production system, decreasing the starch content through native and exogenous saccharification enzymes, generating lower molecular weight oligosaccharides, would also aid to stabilize an emulsion. We previously speculated how the sugars (~15%), oligosaccharides, fiber, protein and oils remaining in these GBR beverages can lead to a natural emulsion [24]. We believe that the neutral pH, along with endogenously generated catabolically produced polysaccharides and exogenously delivered enzymatic oligosaccharides, interacted positively with the proteins to inhibit oil aggregation, which can lead to undesirable feathering and sedimentation. In short-term GBR PNZ storage, no sedimentation was observed [23,24]. Study of the rheological profiles and particle size will continue in scaled-up pilot plant studies, including emulsification followed by pasteurization.

4. Conclusions

A tenet of the research conducted on the rice beverages presented herein was to use "green technologies" and rice materials that were not stabilized by any chemical or physical treatments, prior to using a "free-flowing" natural (aside from food grade saccharification enzymes) value-added process, using no additives, oils or salt. Older beverage patents and technologies have included steps for the stabilization and/or rice protein, or oil extraction methods relying heavily upon chemical (acid or base) processes that oftentimes negatively affect the functional and nutritional properties of the proteins and remove endogenous oils. Enzymatic methods are also available and likely well suited, yet an enzyme cocktail is needed if the bran/germ has not been removed or germinated. Much of the data presented herein illustrates the significant lipid losses through germination and processing, especially attributed to sieving, which would be desirable to keep in the pipeline through product development. Nonetheless, this report documents how preliminary GBR PNZ beverages contained significantly greater concentrations of total lipids, TAGs, DAGs, free sterols, phytosterol esters and oryzanols, than both non-germinated BRR and WR PNZ beverages. These are valuable findings, considering the possible health-promoting compounds identified and discussed. Free sterols, phytosterols and oryzanol recovered at significantly higher concentrations in the GBR beverage are promising, concerning recently advancing knowledge regarding both the compound characterization and relevance to human health and well-being.

The developed GBR beverage method has low inputs, requires relatively simple and inexpensive equipment and is applicable for both germinated brown and colored rice varieties. Based on the observations and physicochemical data, a significant amount of valuable endogenous lipids are retained in the GBR beverage, which appear to be incorporated into a natural emulsion. In this beverage process, the heat plus oligosaccharides can provide good conditions to form emulsions with the FFA, lipids and bran-associated fiber, protein and reactive antioxidant compounds. Both sieving steps on a 30-mesh sieve (PWM-SL) or 140-mesh sieve (PNZ-SL) resulted in major fiber-related, lipid and protein losses. Subsequently, the process itself needs to be refined to better soften and wet-mill the starting materials. Future work should analyze the arabinoxylans, ferulic acid containing compounds and ferulates, soluble fiber, and characterize better the oryzanols and phytosterols through emulsification and pasteurization in these newly developed all-natural beverages. Modified methods that capture all the previously documented sieving losses and pilot plant scale-up, should position this advantageous green processing methodology to deliver 100% natural, no additives, value-added germinated rice beverages. This is important industrially and economically, considering the burgeoning plant-based beverage market and the desire of industries to capture more non-animal protein and health-related attributes from an agronomic and relatively inexpensive crop, such as rice. Developing plant-based, protein- and lipid-rich functional beverages with rice that has proven health benefits, will have a positive economic impact.

Supplementary Materials: The following supporting information can be downloaded at: https://www.mdpi.com/article/10.3390/foods11020220/s1, Supplemental Table S1: Total lipids (weight %, dwb) and free fatty acids (mg/100 g) in three beverages prepared from white, brown and germinated brown 'Rondo' rice, grouped by rice beverage type (GBR, BRR, WR); Supplemental Table S2:

Acylglycerols, phytosterol esters (nonpolar lipids), free sterols, oryzanols and summed lipids/oils (mg/100 g) in three beverages prepared from white, brown and germinated brown 'Rondo' rice, grouped by rice beverage type (GBR, BRR, WR).

Author Contributions: J.C.B. conceptualized, developed and integrated green technologies with germinating and created the free-flowing process through an empirical trial and error approach, loosely based on patents and the literature. J.C.B. was responsible for the funding, investigation, methodology, project administration, supervision, writing and reviewing the manuscript and all revisions. Original research was conducted entirely at the Southern Regional Research Center, in which data were collected and curated by J.C.B. J.C.B. ran the experiments and, together with M.J.P. and R.A.M., contributed the resources and completed analyses. J.M.O.-U. aided the data curation and performed formal analysis through statistical software and accomplished interpretations, and generated the tabled mean comparisons. All authors have read and agreed to the published version of the manuscript.

Funding: This research received no external funding from funding agencies in the public, commercial, or not-for-profit sectors. The senior author's USDA Federal budget was the sole source of funding promoting the research. This article is a US Government work and is in the public domain in the USA. Mention of a trademark or proprietary product is for identification only and does not imply a guarantee or warranty of the product by the US Department of Agriculture. The US Department of Agriculture prohibits discrimination in all its programs and activities on the basis of race, color, national origin, gender, religion, age, disability, political beliefs, sexual orientation and marital or family status.

Institutional Review Board Statement: Not applicable.

Informed Consent Statement: Not applicable.

Data Availability Statement: The corresponding author will supply the calculated and statistical dataset (in spreadsheet format) upon reasonable and justified request.

Acknowledgments: The authors appreciate the field and laboratory support and help from Anna M. McClung and Laduska Sells (USDA, ARS, Stuttgart, AR, USA), Mila P. Hojilla-Evangelista for performing additional moisture analyses (NCAUR, USDA, ARS, Peoria, IL, USA) and Shawndrika S. Reed (USDA, ARS, New Orleans, LA, USA) for laboratory support.

Conflicts of Interest: The authors declare no conflict of interest.

Abbreviations

BR: brown rice; BRR, brown 'Rondo' rice; CRF, commercial rice flour; DAG, diacylglycerols; FFA, free fatty acids; GABA, γ-aminobutyric acid; GBR, germinated brown rice; MC, moisture content; PWM, post wet-milling; PWM-SL, post wet-milling sieving loss; PNZ, post enzymes; PNZ-SL, post enzymes sieving loss; TAG, triacylglycerols and WR, white rice.

References

1. Hanafiah, N.M.; Mispan, M.S.; Lim, P.E.; Baisakh, N.; Cheng, A. The 21st Century Agriculture: When Rice Research Draws Attention to Climate Variability and How Weedy Rice and Underutilized Grains Come in Handy. *Plants* **2020**, *9*, 365. [CrossRef]
2. Champagne, E.T. *Rice: Chemistry and Technology*, 3rd ed.; American Association of Cereal Chemists, Inc.: St. Paul, MN, USA, 2004; 640p.
3. Cho, D.H.; Lim, S.T. Germinated brown rice and its bio-functional compounds. *Food Chem.* **2016**, *196*, 259–271. [CrossRef] [PubMed]
4. Han, A.; Arijaje, E.O.; Jinn, J.R.; Mauromoustakos, A.; Wang, Y.J. Effects of germination duration on milling, physicochemical, and textural properties of medium- and long-grain rice. *Cereal Chem.* **2016**, *93*, 39–46. [CrossRef]
5. Kim, H.Y.; Hwang, I.G.; Kim, T.M.; Woo, K.S.; Park, D.S.; Kim, J.H.; Kim, D.J.; Lee, J.; Lee, Y.R.; Jeong, H.S. Chemical and functional components in different parts of rough rice (*Oryza sativa* L.) before and after germination. *Food Chem.* **2012**, *134*, 288–293. [CrossRef]
6. Wu, F.; Yang, N.; Toure, A.; Jin, Z.; Xu, X. Germinated brown rice and its role in human health. *Crit. Rev. Food Sci. Nutr.* **2013**, *53*, 451–463. [CrossRef]
7. Moreau, R.A.; Nystrom, L.; Whitaker, B.D.; Winkler-Moser, J.K.; Baer, D.J.; Gebauer, S.K.; Hicks, K.B. Phytosterols and their derivatives: Structural diversity, distribution, metabolism, analysis, and health-promoting uses. *Prog. Lipid Res.* **2018**, *70*, 35–61. [CrossRef]

8. Qureshi, A.A.; Mo, H.; Packer, L.; Peterson, D.M. Isolation and identification of novel tocotrienols from rice bran with hypocholesterolemic, antioxidant, and antitumor properties. *J. Agric. Food Chem.* **2000**, *48*, 3130–3140. [CrossRef]
9. Xu, Z.; Hua, N.; Godber, J.S. Antioxidant activity of tocopherols, tocotrienols, and g-oryzanol components from rice bran against cholesterol oxidation accelerated by 2,2-Azobis(2-methylpropionamidine) dihydrochloride. *J. Agric. Food Chem.* **2001**, *49*, 2077–2081. [CrossRef] [PubMed]
10. Liu, R.H. Whole grain phytochemicals and health. *J. Cereal Sci.* **2007**, *46*, 207–219. [CrossRef]
11. Okarter, N.; Liu, C.S.; Sorrells, M.E.; Liu, R.H. Phytochemical content and antioxidant activity of six diverse varieties of whole wheat. *Food Chem.* **2010**, *119*, 249–257. [CrossRef]
12. Luh, B.S. *Rice Production*; Van Nostrand Reinhold: New York, NY, USA, 1991.
13. Rackis, J.J.; Sessa, D.J.; Honig, D.H. Flavor problems of vegetable food proteins. *J. Am. Oil Chem. Soc.* **1979**, *56*, 262–271. [CrossRef]
14. Guraya, H.S.; Patindol, J.A. Storage stability of flour-blasted brown rice. *Cereal Chem.* **2011**, *88*, 56–63. [CrossRef]
15. Mitchell, C.; Mitchell, P.R.; Nissenbaum, R. Nutritional Rice Milk Production. US Patent 4,744,992, 17 May 1988.
16. Anonymous. Global Dairy Alternatives Market, Consumption Volume (by Source, Region & Application) and 20 Company Profile-Forecast to 2024. *Res. Mark.* **2018**, *172*. Available online: https://www.researchandmarkets.com/reports/4458877/global-dairy-alternatives-market-consumption#pos-1 (accessed on 12 June 2019).
17. Grzebinski, T. *The Surge of Plant-Based Foods*; Information Resources Inc. IRI POS (Point-of-Sale), MURLOC (Multi Outlet plus Convenience Stores): Chicago, IL, USA, 2019.
18. Pagand, J.; Heirbaut, P.; Pierre, A.; Pareyt, B. The magic and challenges of sprouted grains. *Cereal Foods World* **2017**, *62*, 221–226. [CrossRef]
19. Moongngarm, A.; Saetung, N. Comparison of chemical compositions and bioactive compounds of germinated rough rice and brown rice. *Food Chem.* **2010**, *122*, 782–788. [CrossRef]
20. Choi, I.; Suh, S.J.; Kim, J.H.; Kim, S.L. Effects of germination on fatty acid and free amino acid profiles of brown rice 'Keunnun'. *Food Sci. Biotechnol.* **2009**, *18*, 799–802.
21. Shu, X.L.; Frank, T.; Shu, Q.Y.; Engel, K.H. Metabolite profiling of germinating rice seeds. *J. Agric. Food Chem.* **2008**, *56*, 11612–11620. [CrossRef]
22. Decloedt, A.I.; Van Landschoot, A.; Watson, H.; Vanderputten, D.; Vanhaecke, L. Plant-based beverages as good sources of free and glycosidic plant sterols. *Nutrients* **2018**, *10*, 21. [CrossRef]
23. Beaulieu, J.C.; Reed, S.S.; Obando-Ulloa, J.M.; McClung, A.M. Green processing protocol for germinating and wet milling brown rice beverage formulations: Sprouting, milling and gelatinization effects. *Food Sci. Nutr.* **2020**, *2020*, 2445–2457. [CrossRef]
24. Beaulieu, J.C.; Reed, S.S.; Obando-Ulloa, J.M.; Boue, S.M.; Cole, M.R. Green processing, germinating and wet milling brown rice (*Oryza sativa*) for beverages: Physicochemical effects. *Foods* **2020**, *9*, 1016. [CrossRef]
25. Patindol, J.A.; Guraya, H.S.; Champagne, E.T.; McClung, A.M. Nutritionally important starch fractions of rice cultivars grown in southern united states. *J. Food Sci.* **2010**, *75*, H137–H144. [CrossRef]
26. Johnston, D.B.; McAloon, A.J.; Moreau, R.A.; Hicks, K.B.; Singh, V. Composition and economic comparison of germ fractions from modified corn processing technologies. *J. Am. Oil Chem. Soc.* **2005**, *82*, 603–608. [CrossRef]
27. Johnston, D.J.; Moreau, R.A. A comparison between corn and grain sorghum fermentation rates, distillers dried grains with solubles composition, and lipid profiles. *Bioresour. Technol.* **2017**, *226*, 118–124. [CrossRef]
28. Angelino, D.; Rosi, A.; Vici, G.; Dello Russo, M.; Pellegrini, N. Nutritional quality of plant-based drinks sold in Italy: The food labelling of Italian products (FLIP) study. *Foods* **2020**, *9*, 682. [CrossRef]
29. Orsavova, J.; Misurcova, L.; Ambrozova, J.V.; Vicha, R.; Mlcek, J. Fatty acids composition of vegetable oils and its contribution to dietary energy intake and dependence of cardiovascular mortality on dietary intake of fatty acids. *Int. J. Mol. Sci.* **2015**, *16*, 12871–12890. [CrossRef] [PubMed]
30. Jayadeep, A.; Malleshi, N.G. Nutrients, composition of tocotrienols, tocopherols, and g-oryzanol, and antioxidant activity in brown rice before and after biotransformation. *CyTA-J. Food* **2011**, *9*, 82–87. [CrossRef]
31. Moreau, R.A.; Harron, A.F.; Powell, M.J.; Hoyt, J.L. A comparison of the levels of oil, carotenoids, and lipolytic enzyme activities in modern lines and hybrids of grain sorghum. *J. Am. Oil Chem. Soc.* **2016**, *93*, 569–573. [CrossRef]
32. Yoshida, H.; Tanigawa, T.; Kuriyama, I.; Yoshida, N.; Tomiyama, Y.; Mizushina, Y. Variation in fatty acid distribution of different acyl lipids in rice (*Oryza sativa* L.) brans. *Nutrients* **2011**, *3*, 505–514. [CrossRef]
33. Kwak, J.E.; Yoon, S.W.; Kim, D.J.; Yoon, M.R.; Lee, J.H.; Oh, S.K.; Kim, I.H.; Lee, J.S.; Lee, J.S.; Chang, J.K. Changes in nutraceutical lipid constituents of pre- and post-geminated brown rice oil. *Korean J. Food Nutr.* **2013**, *26*, 591–600. [CrossRef]
34. Graham, I.A. Seed storage oil mobilization. *Annu. Rev. Plant Biol.* **2008**, *59*, 115–142. [CrossRef] [PubMed]
35. Miller, A.; Engel, K.H. Content of g-oryzanol and composition of steryl ferulates in brown rice (*Oryza sativa* L.) of European origin. *J. Agric. Food Chem.* **2006**, *54*, 8127–8133. [CrossRef] [PubMed]
36. Kim, H.W.; Kim, J.B.; Cho, S.M.; Cho, I.K.; Li, Q.X.; Jang, H.H.; Lee, S.H.; Lee, Y.M.; Hwang, K.A. Characterization and quantification of -oryzanol in grains of 16 Korean rice varieties. *Int. J. Food Sci. Nutr.* **2014**, *66*, 166–174. [CrossRef] [PubMed]
37. Goufo, P.; Trindade, H. Rice antioxidants: Phenolic acids, flavonoids, anthocyanins, proanthocyanidins, tocopherols, tocotrienols, g-oryzanol, and phytic acid. *Food Sci. Nutr.* **2014**, *2*, 75–104. [CrossRef] [PubMed]
38. Fang, H.Y.; Chen, Y.K.; Chen, H.H.; Lin, S.Y.; Fang, Y.T. Immunomodulatory effects of feruloylated oligosaccharides from rice bran. *Food Chem.* **2012**, *134*, 836–840. [CrossRef] [PubMed]

39. Yadav, M.P.; Hicks, K.B. Isolation of barley hulls and straw constituents and study of emulsifying properties of their arabinoxylans. *Carbohydr. Polym.* **2015**, *132*, 529–536. [CrossRef]
40. Mendez-Encinas, M.A.; Carvajal-Millan, E.; Rascon-Chu, A.; Astiazaran-Garcia, H.F.; Valencia-Rivera, D.E. Ferulated Arabinoxylans and Their Gels: Functional Properties and Potential Application as Antioxidant and Anticancer Agent. *Oxid. Med. Cell. Longev.* **2018**, *2018*, 2314759. [CrossRef]
41. Nino-Medina, G.; Carvajal-Millan, E.; Rascon-Chu, A.; Marquez-Escalante, J.A.; Guerrero, V.; Salas-Munoz, E. Feruloylated arabinoxylans and arabinoxylan gels: Structure, sources and applications. *Phytochem. Rev.* **2010**, *9*, 111–120. [CrossRef]

foods

MDPI

Article

Parboiled Germinated Brown Rice Improves Cardiac Structure and Gene Expression in Hypertensive Rats

Nattira On-Nom [1], Kanoknad Khaengamkham [1], Aikkarach Kettawan [1], Thanaporn Rungruang [2], Uthaiwan Suttisansanee [1], Piya Temviriyanukul [1], Pattaneeya Prangthip [3] and Chaowanee Chupeerach [1,*]

[1] Institute of Nutrition, Mahidol University, Salaya, Phuttamonthon, Nakhon Pathom 73170, Thailand
[2] Department of Anatomy, Faculty of Medicine Siriraj Hospital, Mahidol University, Bangkok 10700, Thailand
[3] Department of Tropical Nutrition and Food Science, Faculty of Tropical Medicine, Mahidol University, Ratchathewi, Bangkok 10400, Thailand
* Correspondence: chaowanee.chu@mahidol.ac.th

Abstract: Hypertension leads to oxidative stress, inflammation, and fibrosis. The suppression of these indicators may be one treatment approach. Parboiled germinated brown rice (PGBR), obtained by steaming germinated Jasmine rice, reduces oxidative stress and inflammation in vivo. PGBR contains more bioactive compounds than brown rice (BR) and white rice (WR). Anti-hypertensive benefits of PGBR have been predicted, but research is lacking. The anti-hypertensive effects of PGBR were investigated in the downstream gene network of hypertension pathogenesis, including the renin–angiotensin system, fibrosis, oxidative stress production, and antioxidant enzymes in *N*-nitro-L-arginine methyl ester (L-NAME)-induced hypertensive rats. To strengthen our findings, the cardiac structure was also studied. PGBR-exposed rats showed significant reductions in systolic blood pressure (SBP) compared to the hypertensive group. WR did not reduce SBP because of the loss of bioactive compounds during intensive milling. PGBR also reduced the expression of the angiotensin type 1 receptor (AT1R), transforming growth factor-β (TGF-β), and nicotinamide adenine dinucleotide phosphate (NADPH) oxidase (NOX4), which contribute to the renin–angiotensin system, fibrosis, and oxidative stress production, respectively. Losartan (Los, an anti-hypertensive drug)-treated rats also exhibited similar gene expression, implying that PGBR may reduce hypertension using the same downstream target as Los. Our data also indicated that PGBR reduced cardiac lesions, such as the cardiomyopathy induced by L-NAME. This is the first report on the anti-hypertensive effects of PGBR in vivo by the suppression of the renin response, fibrosis, and improved cardiac structure.

Keywords: Khao Dawk Mali 105; blood pressure; fibrosis; Sprague-Dawley rat; renin angiotensin system

Citation: On-Nom, N.; Khaengamkham, K.; Kettawan, A.; Rungruang, T.; Suttisansanee, U.; Temviriyanukul, P.; Prangthip, P.; Chupeerach, C. Parboiled Germinated Brown Rice Improves Cardiac Structure and Gene Expression in Hypertensive Rats. *Foods* **2023**, *12*, 9. https://doi.org/10.3390/foods12010009

Academic Editors: John C Beaulieu, Ming Hsuan Chen, Ida Wenefrida, Piebiep Goufo and Gian Carlo Tenore

Received: 30 September 2022
Revised: 30 November 2022
Accepted: 17 December 2022
Published: 20 December 2022

1. Introduction

Hypertension, or high blood pressure, is a serious medical condition that leads to vascular disease, which is the leading cause of death worldwide [1]. The molecular process by which hypertension impacts health is complicated. Simply stated, the sympathetic nervous system (SNS) induces the release of renin, also known as angiotensinogenase, from the kidney, which then hydrolyzes angiotensinogen to produce angiotensin I (Ang I), which is then hydrolyzed by lung angiotensin-converting enzymes (ACEs), yielding angiotensin II (Ang II), a key factor in hypertension and myocardial remodeling. Ang II is a major bioactive peptide contributing to the renin–angiotensin system (RAS) and increases blood pressure by activating the angiotensin II type 1 receptor (AT1R), leading to vascular smooth muscle contraction [2]. Ang II has also been reported to stimulate transforming growth factor-β (TGF-β) into a potent stimulator of collagen-producing cardiac fibroblasts [3,4]. Hypertension provokes oxidative stress by activating a major ROS-producing enzyme, nicotinamide adenine dinucleotide phosphate (NADPH) oxidase, which then activates the redox-sensitive transcription factor nuclear factor kappa B (NF-κB). The activated NF-κB is then translocated to the nucleus and promotes the transcription of several inflammation-associated genes [5]. Finally, prolonged hypertension with tissue damage may result in the accumulation of collagen type I (Col I) and collagen type III (Col III), which are biomarkers of changed cardiac structure and fibrosis [6]. Thus, hypertension has extensive impacts on gene networks involving inflammation, antioxidant enzymes, and fibrosis. The low-grade inflammation resulting from hypertension can chronically damage cells and organs within the body, leading to other health complications such as stroke, renal failure, and dementia [7].

Jasmine rice is Thailand's primary export rice and is renowned for its distinctive flavor and texture. Jasmine rice is typically consumed as white rice; however, white Jasmine rice has a low nutritive value and bioactive compounds with a high glycemic index due to the intensive milling, which removes the rice husk, germ, and bran. By contrast, brown Jasmine rice with the rice germ has a high nutritive value and bioactive compounds with a low glycemic index [8]. Parboiling germinated brown rice increases the bioactive compounds, especially γ-aminobutyric acid (GABA) and total phenolics [9]. Previous studies demonstrated that parboiled germinated brown rice (PGBR), produced from Khao Dawk Mali 105 (*Oryza sativa* L. ssp. *indica* cv. KDML 105) had a higher nutritive value and more bioactive components such as GABA, ferulic acid, γ-oryzanol, γ-tocotrienol, and fiber than brown rice (BR) and white rice (WR) [10]. PGBR also prevented carbon tetrachloride (CCl_4)-induced liver oxidative stress and injury by enhancing the antioxidant capacities in rats [9]. The consumption of PGBR made from KDML 105 also reduced liver inflammation and fibrosis in vivo [11].

The modulation of oxidative stress, inflammation, and fibrosis might be a therapeutic target for hypertension, but the efficacy of PGBR as an anti-hypertensive agent that reduces oxidative stress and inflammation remains questionable. Therefore, this study explored the anti-hypertensive effects of PGBR on a network of genes involved in hypertension complications and cardiac structure. The genes were divided into four groups, including (i) the renin–angiotensin system, such as AT1R; (ii) the TGF-β, collagen type I (Col I), and collagen type III (Col III) genes, which are involved in fibrosis; (iii) the NADPH oxidase (NOX4) gene, which is involved in oxidative stress production; and (iv) antioxidant enzymes and nitric oxide-producing enzymes (glutathione peroxidase (GPx), catalase (CAT), superoxide dismutase (SOD), and endothelial nitric oxide synthase (eNOS)). To strengthen the evidence of PGBR as a promising hypertensive agent, we also investigated the cardiac structure in hypertensive rats. Our findings can be used to enhance the development of functional foods derived from PBGR as a strategy for controlling hypertension.

2. Materials and Methods

2.1. Animals and Diet

Male Sprague-Dawley rats (200–250 g) were obtained from the National Laboratory Animal Center, Mahidol University, Thailand. The rats were maintained in an environmentally controlled room (23 ± 2 °C, with a 12 h light/12 h dark cycle) and given access to food and water ad libitum. All experiments were conducted according to protocols approved by the Siriraj Animal Care and Use Committee (SiACUC) from the Faculty of Medicine of Siriraj Hospital, Mahidol University, and complied with international guidelines for animal research protection, such as the International Guiding Principles for Biochemical Research Involving Animals (SiACUP number: SI-ACAP 008/2557).

Parboiled germinated brown rice (PGBR), brown rice (BR), and white rice (WR) were produced from a rice variety grown in Thailand, Khao Dawk Mali 105 (*Oryza sativa* L. ssp. *indica* cv. KDML 105). PGBR was prepared according to the method developed by Rattanadee et al. (2011) [12]. In brief, all three types of rice were cooked in an electric rice cooker (1.8 L, Sharp KS-19ET) and then freeze-dried (Supplementary Table S1).

The freeze-dried rice powders were stored at −20 °C until use. The proximate compositions of the powders are presented in the Supplementary Materials. The AIN76A diet (basal diet) was composed of corn starch (15%, *w*/*w*), casein (20%, *w*/*w*), DL-methionine (0.3%, *w*/*w*), sucrose (49.9%, *w*/*w*), corn oil (5%, *w*/*w*), cellulose powder (5%, *w*/*w*), mineral mix (3.5%, *w*/*w*), vitamin mix (1%, *w*/*w*), and choline bitartrate (0.2%, *w*/*w*). The rice powders (PGBR, BR, and WR) replaced corn starch in the AIN76A diet (15%) with ad libitum feeding and became the PGBR, BR and WR diets, respectively. The components of each diet were designed based on the macronutrient compositions in the rice (Supplementary Tables S2 and S3).

2.2. Experimental Design

The experimental design and treatment are illustrated in Figure 1. Following a one-week acclimatization period, the rats were randomly divided into six groups with seven animals each. The hypertensive groups were given *N*-nitro-L-arginine methyl ester (L-NAME) in the drinking water for a daily intake of 20 mg/kg throughout the experimental period (12 weeks). The rats started to consume the rice diets after four weeks until the end of the experimental period. Systolic blood pressures were measured weekly by the indirect method of tail-cuff occlusion in conscious animals using a Niprem 645 pressure recorder (Cibertec, Barcelona, Spain) [13].

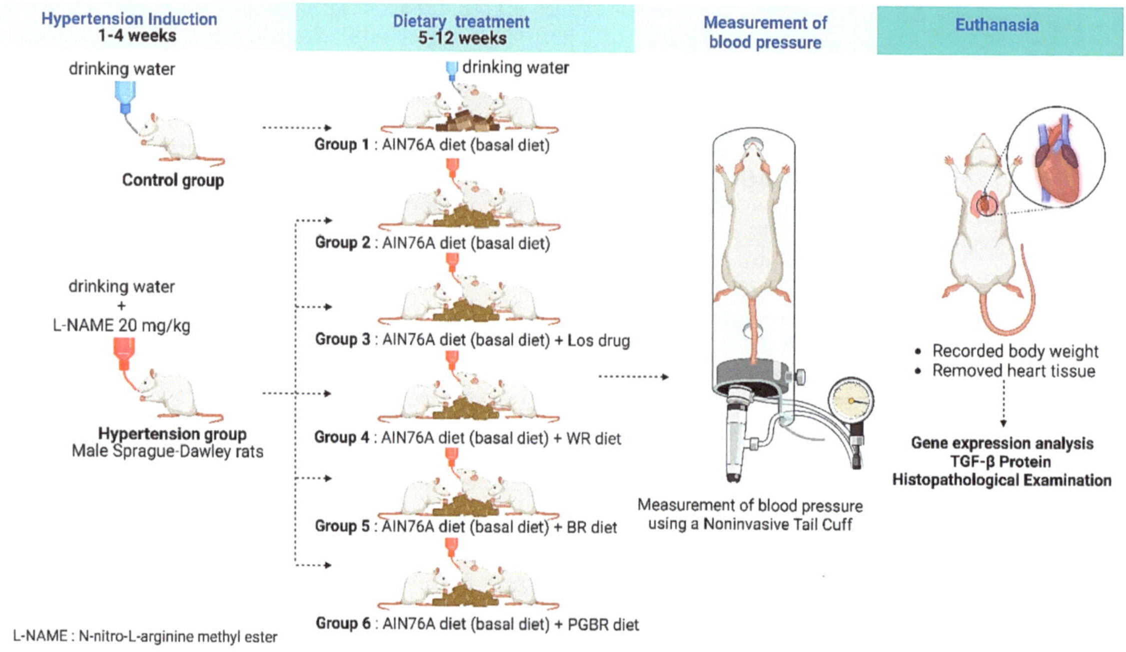

Figure 1. A schematic of the experimental design for inducing hypertension in rats and administering PGBR. The hypertensive groups were given *N*-nitro-L-arginine methyl ester (L-NAME) in the drinking water for a daily intake of 20 mg/kg throughout the experimental period (12 weeks).

The rats were divided into six groups of seven animals as follows:
Group 1: normotensive rats that received the basal diet (control);
Group 2: hypertensive rats that received the basal diet;
Group 3: hypertensive rats that received the basal diet and losartan (an anti-hypertensive drug);
Group 4: hypertensive rats that received the WR diet;
Group 5: hypertensive rats that received the BR diet;
Group 6: hypertensive rats that received the PGBR diet.

At the end of the treatment period of eight weeks, the animals were anesthetized by the inhalation of CO_2. The body weight was recorded, and the heart was removed intact and weighed for each rat. The heart tissues were frozen and stored at −80 °C until analysis.

2.3. Gene Expression Analysis

The tissue for RNA isolation was the left ventricle, and RNA was extracted using an RNA isolation kit (Roche Diagnostics Corp., Indianapolis, IN, USA). The purity and quantity of the RNA were determined using a NanoDrop UV Spectrophotometer (Thermo Scientific, Wilmington, DE, USA). To measure the mRNA level, complementary DNA was amplified by a quantitative real-time polymerase chain reaction (Lightcycler 96) (Roche Diagnostics Corp., Indianapolis, IN, USA) using a SYBR Green kit (Roche Diagnostics Corp., Indianapolis, IN, USA). The forward and reverse primers used in this study [13–16] are shown in Supplementary Table S4. The quantitative fold changes in mRNA expression were determined relative to beta-actin (β-actin) using the $2^{-\Delta\Delta CT}$ method [17].

2.4. Histopathological Examination

The tissue for the histology assay was the left atrium. The tissue was fixed with a 10% formaldehyde buffer and embedded in paraffin. Samples of 5 μm thickness were stained with hematoxylin and eosin, followed by examination under a light microscope (Olympus, Tokyo, Japan). A cross section of the heart was evaluated from photographs of whole tissue sections taken at 40× magnification by a pathologist at the National Laboratory Animal Center, Mahidol University, who was blinded to the treatments.

2.5. Statistical Analysis

Data are presented as means ± standards error of the means (SEM). Statistical comparisons between the groups were performed using the Mann–Whitney U test. All statistical analyses were performed using the Statistical Package for the Social Sciences for Windows (SPSS) version 18.0. The significance level was 0.05 (α = 0.05).

3. Results

3.1. Effects of White Rice, Brown Rice, and Parboiled Germinated Brown Rice Diets on Heart and Body Weight

At the end of the experiment, the heart and body weights of all rats were determined. Table 1 shows that the L-NAME-exposed group (hypertensive rats) acquired considerably more body weight than the control group (388 ± 7.64 vs. 316 ± 53.8 g; p-value < 0.05). This observation was also found in the white rice (WR), brown rice (BR), and parboiled germinated brown rice (PGBR)-treated groups. The L-NAME-treated rats gained body weight, but heart weight and heart/body weight ratio remained unaffected (Table 1). The diet consumption record of all groups is shown in Supplementary Figure S1. The rice consumption amounts in the WR, BR, and PGBR groups were 10.5 ± 1.8, 10.5 ± 2.1, and 11.0 ± 2.3 g/kg body weight, respectively.

Table 1. Heart and body weights of rats after different treatments.

Group	Body Weight (g)	Heart Weight (g) NS	Heart/Body Weight Ratio (g %) NS
Control	315 ± 53.8 *	1.06 ± 0.06	0.34 ± 0.02
L-NAME	388 ± 20.2	1.10 ± 0.07	0.28 ± 0.02
L-NAME + Los	401± 10.0	1.39 ± 0.11	0.35 ± 0.03
L-NAME + WR	408 ± 9.00 *	1.29 ± 0.08	0.32 ± 0.02
L-NAME + BR	413 ± 21.3 *	1.17 ± 0.07	0.28 ± 0.01
L-NAME + PGBR	422 ± 27.3 *	1.23 ± 0.07	0.29 ± 0.02

All data are presented as means ± SEM (n = 7). The * shows significantly different weights of rats under different treatments at $p < 0.05$ compared to those treated with N-nitro-L-arginine methyl ester (L-NAME) using the Mann–Whitney U test (two-tailed). NS shows no significant difference among rat groups. Los: losartan; WR: white rice; BR: brown rice; PGBR: parboiled germinated brown rice.

3.2. Effects of White Rice, Brown Rice, and Parboiled Germinated Brown Rice Diets on Systolic Blood Pressure

To observe the anti-hypertensive properties of the WR, BR, and PGBR treatments, rats were administered L-NAME for four weeks to assure the successful elevation of the systolic blood pressure (SBP). At week 0 in Figure 2, L-NAME significantly induced SBP in rats compared to the control group. After the successful elevation of systolic blood pressure (SBP), rats were exposed to losartan (Los, an anti-hypertensive drug), WR, BR, and PGBR for an additional eight weeks, and the SBP was measured weekly using a noninvasive tail cuff. After treatment with Los, the SBP dropped to normal within two weeks and remained constant until the end of the experiment, confirming its potent anti-hypertensive properties. Interestingly, BR- and PGBR-exposed rats showed a gradual decrease in SBP over the experimental period. PGBR was more potent than BR at week 8 (Figure 2), while the WR diet showed the smallest anti-hypertensive properties when compared with the other rice. Both the BR and PGBR diets showed anti-hypertensive properties, with PGBR being the most potent among the three types of rice.

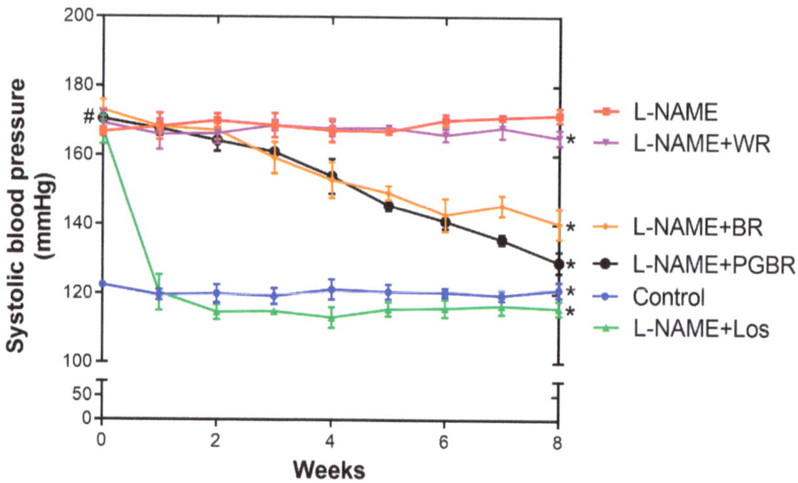

Figure 2. Systolic blood pressure (SBP) of rats during *N*-nitro-L-arginine methyl ester (L-NAME)-induced hypertension after hypertensive inducing for four weeks. Data are presented as means ± SEM ($n = 7$). # shows significantly different SBP in rats under different treatments compared to the control group (no treatment) at $p < 0.05$ using the Mann–Whitney U test (two-tailed), while * shows significantly different SBP in rats under different treatments compared to those treated with L-NAME (L-NAME group) at $p < 0.05$ using the Mann–Whitney U test (two-tailed). Los: losartan; WR: white rice; BR: brown rice; PGBR: parboiled germinated brown rice.

3.3. Effects of White Rice, Brown Rice, and Parboiled Germinated Brown Rice on Gene Expression in Heart Tissue

Figure 2 shows the promising anti-hypertensive effects of BR and PGBR. Therefore, the heart tissues of PGBR-treated rats were further investigated for gene expression, including (i) the renin–angiotensin system, such as AT1R; (ii) the TGF-β, collagen type I (Col I), and collagen type III (Col III) genes, which are involved in fibrosis; (iii) the NADPH oxidase (NOX4) gene, which is involved in oxidative stress production; and (iv) antioxidant enzymes and nitric oxide-producing enzymes (glutathione peroxidase (GPx), catalase (CAT), superoxide dismutase (SOD), and endothelial nitric oxide synthase (eNOS)). Figures 3 and 4 demonstrate that L-NAME-treated rats showed significantly increased gene expression of AT1R, TGF-β, Col I, and GPx. In the Los-exposed group, the expression of AT1R, TGF-β, and NOX4 decreased, even though the expression of most antioxidant enzymes remained unchanged, indicating that in our condition AT1R, TGF-β, and NOX4 might be therapeutic targets for Los in the treatment of hypertension. Interestingly, the PGBR-treated group significantly suppressed AT1R and TGF-β expression at the same level as the control and Los-treated groups (Figure 3). Although NOX4 was not statistically reduced, a clear trend of reduction was obtained. Therefore, PGBR might lower SBP by targeting AT1R, TGF-β, and NOX4, similar to the anti-hypertensive drug (Los), while the antioxidant enzymes may be neglected.

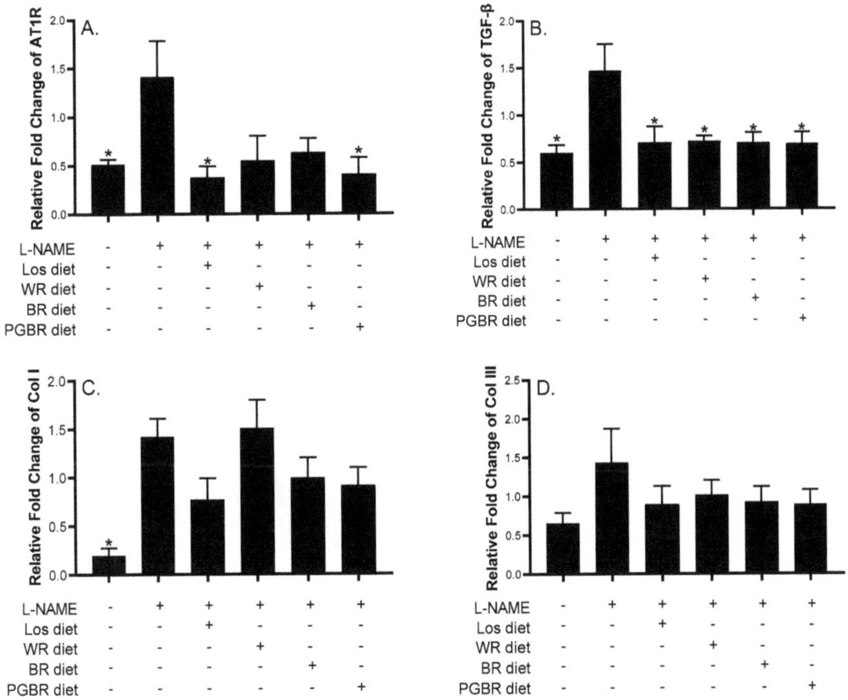

Figure 3. Effects of different diets on gene expression in heart tissue of rats. Data are presented as means ± SEM (n = 7); Gene expression of (**A**) AT1R; (**B**) TGF-β; (**C**) Col I; (**D**) Col III among rat groups. * shows significantly different relative gene expressions of rats treated with different diets compared with those treated with N-nitro-L-arginine methyl ester (L-NAME) at $p < 0.05$ using the Mann-Whitney U test (two-tailed). AT1R: angiotensin II type 1 receptor; TGF-β: transforming growth factor-β; Col I: collagen type I; Col III: collagen type III; Los: losartan; WR: white rice; BR: brown rice; PGBR: parboiled germinated brown rice.

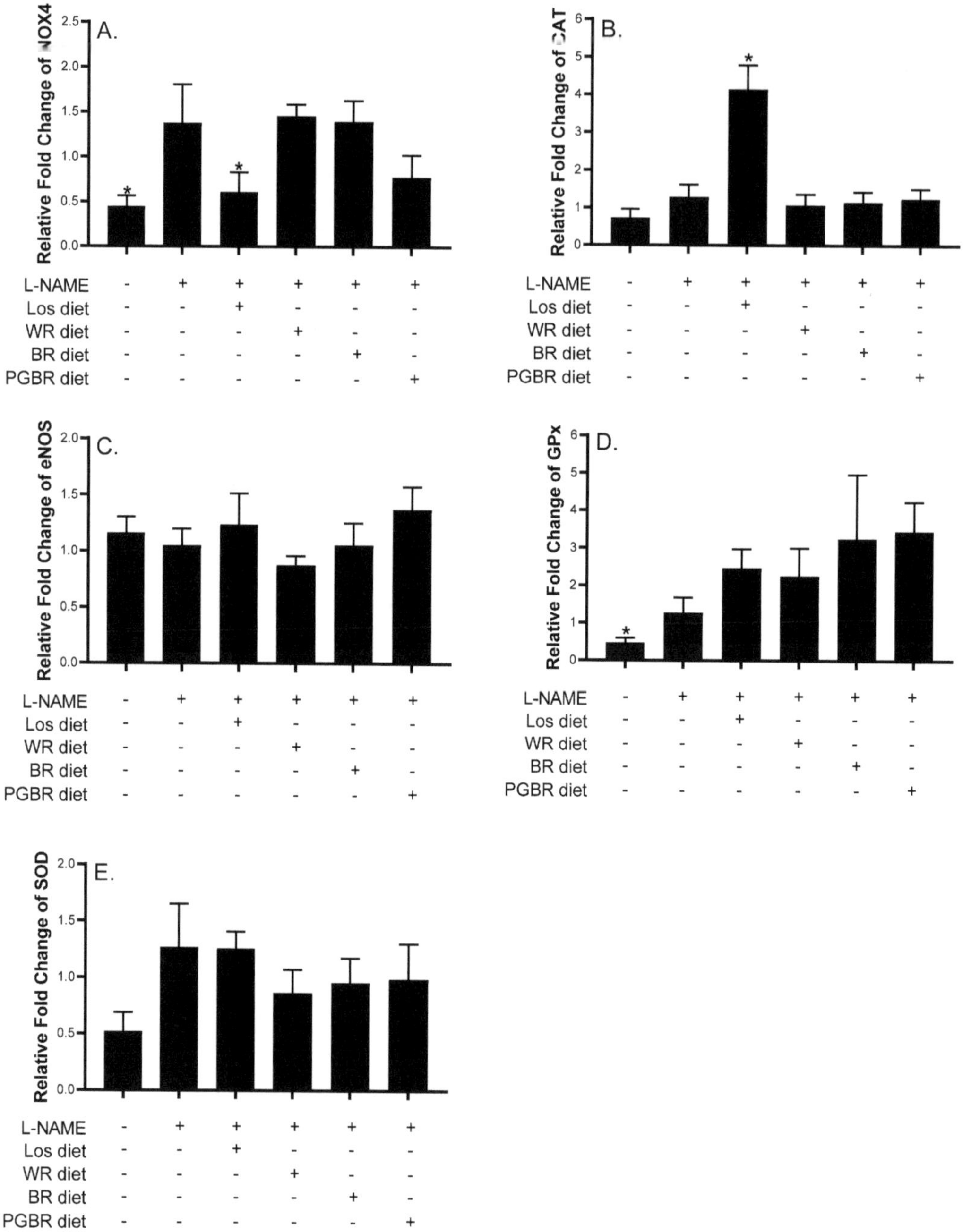

Figure 4. Effects of different diets on gene expression in heart tissue of rats. Data are presented as means ± SEM (n = 7); Gene expression of (**A**) NOX 4; (**B**) CAT; (**C**) eNOS; (**D**) GPx; (**E**) SOD among rat groups. * shows significantly different relative gene expressions of rats treated with different diets compared with those treated with *N*-nitro-L-arginine methyl ester (L-NAME) at $p < 0.05$ using the Mann-Whitney U test (two-tailed). NOX4: NADPH oxidase; CAT: catalase; eNOS: endothelial nitric oxide synthase; GPx: glutathione peroxidase; SOD: superoxide dismutase; Los: losartan; WR: white rice; BR: brown rice; PGBR: parboiled germinated brown rice.

3.4. Effect of PGBR on Heart Histopathological Changes

The heart wall is composed of three layers, from the inside to the outside: the endocardium, myocardium, and epicardium. Several types of lesions are induced during hypertension. Thus, to ensure the anti-hypertensive properties of PGBR, we performed a histopathological analysis using H&E staining for lesion quantification. Representative figures of normal myocardial fibers, myocardial necrosis, inflammatory infiltration, and myocardial fibrosis and scar formation, which were scored in the present study, are shown in Table 2 and Figure 5. In the L-NAME group, lesions were found in all layers, with focally endothelial hyperplasia in the epicardium. In the myocardium, cellular degeneration to necrosis in inflammatory cells was observed. In the sub-endocardium, large focal branching fibrosis, acute to chronic hemorrhages, and focal collagen deposits were frequently found. The outer epicardium showed cellular hyperplasia with mimic mononuclear cell infiltration. A cardiomyopathy analysis revealed degeneration, necrosis, inflammatory cell infiltration, and fibrosis/collagen scar formation. Treatment with PGBR in hypertensive rats showed a lower amount of cardiomyopathy compared with the L-NAME-induced hypertensive rat group, indicating an improvement in cardiac histology from PGBR after hypertension.

Table 2. Microscopic findings of animals in each group, with numbers of heart lesions.

Microscopic Findings	Control	L-NAME	L-NAME + Los	L-NAME + WR	L-NAME + BR	L-NAME + PGBR
Fibrosis, myocardium	0	1	0	0	0	0
Degeneration to necrosis, myocardium	1	1	0	0	1	0
Inflammation to Aschoff body cell infiltration	0	1	2	1	0	0
Collagen deposit, sub-endocardium	0	1	0	0	0	0
* Cardiomyopathy	1	3	2	1	1	0

* Cardiomyopathy combines degeneration, necrosis, inflammatory cell infiltration, and fibrosis/collagen scar formation; a refers to the same animal, number 2/1 ($n = 5$). L-NAME: *N*-nitro-L-arginine methyl ester; Los: losartan; WR: white rice; BR: brown rice; PGBR: parboiled germinated brown rice.

A.

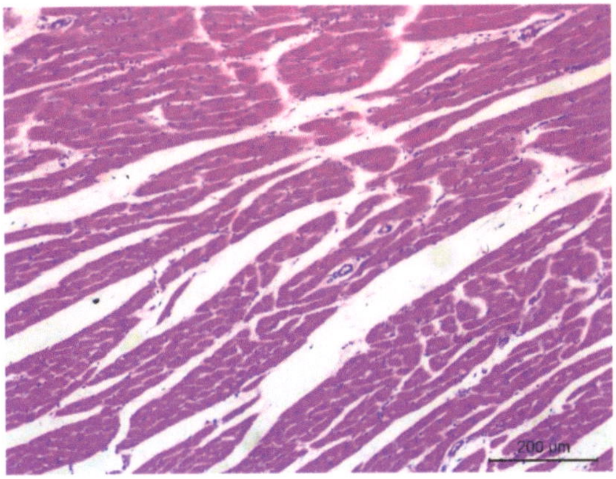

Figure 5. *Cont.*

B.
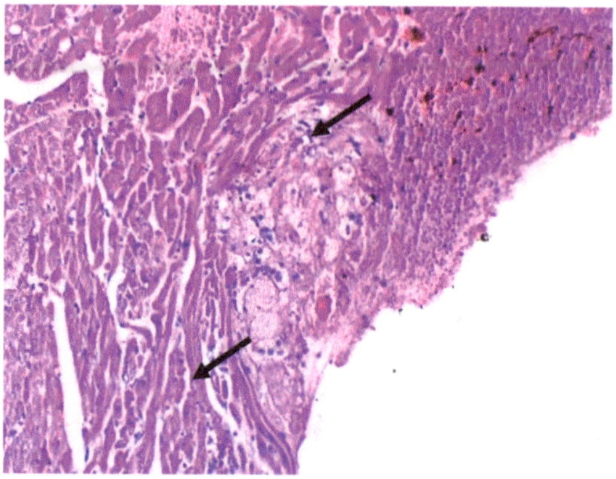

C.
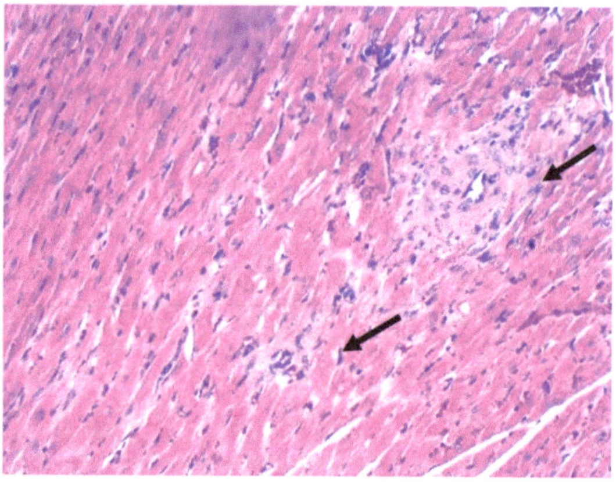

Figure 5. Histopathological changes in the heart tissues. H&E staining under a light microscope: (**A**) a representative figure of normal myocardial fibers: 10×, (**B**) a representative figure of focal myocardial necrosis and inflammatory infiltration: 40×, and (**C**) a representative figure of myocardial fibrosis and scar formation: 40×. The arrows indicate cardiomyopathy in the heart tissue.

4. Discussion

This study reports the anti-hypertensive effect of parboiled germinated brown rice (PGBR) made from the Khao Dawk Mali 105 variety on antioxidant enzymes and gene expression associated with fibrosis in the hearts of hypertensive rats induced by *N*-nitro-L-arginine-methyl ester (L-NAME). Clear beneficial results were obtained when the diet was mixed with PGBR. The treatment with PGBR attenuated blood pressure and decreased myocardial degeneration. In several studies, there were no effects of L-NAME on body weight [18,19]. However, increased body weight was observed in the L-NAME group in this study. It could be that the three rice groups of rats consumed higher amounts of food than the control, L-NAME, and drug groups. Therefore, the weights were higher than in the other groups (Supplementary Figure S1). Moreover, the three different rice groups showed increases in body weight. These might be the effects of the difference in energy from the rice diets. The corn starch in the basal diet was substituted with different rice varieties. Therefore, the difference in energy in the rat diets came from the rice types. The rats in the WR, BR, and PGBR groups received 40.0 ± 6.9, 40.8 ± 8.1, and 42.8 ± 9.1 kcal of energy from rice, respectively (Supplementary Table S2).

Previous studies reported that PGBR contained high amounts of γ-aminobutyric acid (GABA), ferulic acid, γ-oryzanol, γ-tocotrienol, and p-coumaric acid and showed higher antioxidant abilities than white rice and brown rice [9,10]. In agreement with earlier reports, the anti-hypertensive effect of pregerminated brown rice was also evaluated in spontaneously hypertensive rats (SHRs) fed a diet containing 40% pregerminated brown rice for eight weeks. The results showed anti-hypertensive effects compared to the control group [20]. In this study, L-NAME-induced hypertensive rats showed a significantly increased systolic blood pressure (SBP) that significantly decreased after treatment with PGBR. The results agreed with the anti-hypertensive effect of PGBR, which contains high contents of bioactive ingredients [9,20], especially GABA and ferulic acid. The mechanism of the hypotensive action of the administered GABA has not yet been fully explained, but GABA showed an anti-hypertensive action as an inhibitory neurotransmitter in the central nervous system [14], while a low dose of a novel cultivar of GABA-rich tomato had an anti-hypertensive effect in SHRs by reducing SBP compared to the control group in both single- and chronic-administration studies [21].

The administration of L-NAME, an L-arginine analog, produces hypertension, vascular resistance, hypertrophy, myocardial remodeling, and vasoconstriction. Angiotensin II (Ang II) influences the phases of the inflammatory response involved in the mechanisms leading to vascular remodeling by stimulating vascular repairs such as transforming growth factor-β1 (TGF-β1), hypertrophy, the accumulation of extracellular matrix, and collagen deposition [17]. Thus, inflammation seems to be closely related and might mediate Ang-II-induced vascular remodeling. Ang II and TGF-β1 stimulate the development of cardiac hypertrophy, myocardial fibrosis, the accumulation of collagen type I (Col I) and collagen type III (Col III), fibrosis, and the structural organization of collagen. The effects of Ang II are either indirect by the upregulation of TGF-β1 expression or direct by Ang II receptors: AT1R [6,20]. All these reports showed relationships among AT1R, TGF-β1, Col I, and Col III that may contribute to myocardial fibrosis in the animal model [22]. This study demonstrated that treatment with PGBR in L-NAME-induced hypertensive rats significantly decreased AT1R, TGF-β1, and Col I gene expression compared to the L-NAME group. The findings indicated that the anti-inflammatory effect of PGBR improved myocardial fibrosis by inhibiting genes involved in the renin–angiotensin system and fibrosis.

Vasoconstriction is also a mechanism leading to hypertension, while the chronic inhibition of nitric oxide synthase (NOS) may lead to vasoconstriction [18,19]. Nitric oxide (NO) is synthesized in endothelial cells from L-arginine and is converted to NO by endothelial nitric oxide synthase (eNOS). Moreover, the inhibition of eNOS activity leads to increased expression of nicotinamide adenine dinucleotide phosphate (NADPH) oxidase (NOX4), a major source of reactive oxygen species (ROS) in vascular tissue. The accumulation of reactive oxygen species decreased antioxidant defense systems and reduced NO bioavailability [23–25]. The L-NAME treatment decreased the expression of the endothelial nitric oxide synthase (eNOS) gene after long-term L-NAME treatment, while many studies reported increased eNOS mRNA levels in the heart and kidney in hypertensive rats [26]. Moreover, L-NAME induced hypertension by involving the renin–angiotensin aldosterone system (RAAS) [27]. It was observed that an L-NAME treatment raised eNOS protein levels [26]. However, in our work, L-NAME had no effect on eNOS mRNA, which might indicate that mRNA is dependable for early responses. Further investigation into the eNOS protein level may be worth pursuing. In this study, treatment with PGBR tended to increase eNOS expression and progressively decreased NOX4 gene expression in the hearts of L-NAME-treated rats, but no significant differences were observed. Therefore, the mechanisms of PGBR that contain bioactive ingredients may be involved in the antioxidant activity and preserve the bioavailability of NO in hypertensive hearts.

The relationship between the development of hypertension, decreased antioxidant capacity, and the increased bioavailability of ROS has been indicated in many experimental models [28]. Major enzymatic antioxidant defenses include superoxide dismutase (SOD), glutathione peroxidase (GPx), and catalase (CAT) as the primary lines of cellular defense against oxidative damage [29]. In this study, only GPX expression was induced after L-NAME treatment, while CAT and SOD expression were not different between the control and L-NAME groups, suggesting that GPx might be the main enzyme quenching oxidative stress in hypertensive rats. This effect displayed a compensatory mechanism to attenuate the excessive ROS production [30]. This result agreed with Kumar et al. (2012) [31], who found improved GPx activity after treatment with syringic acid in L-NAME-induced hypertensive rats. The increased activity of enzymatic antioxidants might be due to free radical scavenging efficacy as a beneficial action against pathophysiological modifications caused by superoxide anions and hydroxyl radicals [21]. These antioxidant genes appeared to fluctuate without being statistically significant. For example, a recent study determined that lead-induced hypertension had no effect on SOD, CAT, or GPx in the hearts of the animals but increase NOX4 [32], and another study showed that SOD, CAT, and GPx were not changed in the kidneys of hypertensive rats (measured at week 16) [33].

As the L-NAME raised the expression of NOX4 in L-NAME-treated rats, resulting in an increase in oxidants, we hypothesized that alternative antioxidant enzymes may play an important role in mitigating the oxidative stress caused by L-NAME.

The inflammatory response and cytokines are important components of the host response to heart injury from hypertension and play a key role in cardiac repair [34], ultimately leading to the replacement of dead myocardium with a collagen-based scar and distorted architecture and function of the heart. Moreover, excess collagen deposition and fibrosis have been linked to myocardial stiffness and systolic and diastolic abnormalities [35,36], suggesting that interstitial myocardial fibrosis may be related to L-NAME-induced vasoconstriction with consequent myocardial ischemia. In this study, hematoxylin and eosin (H&E) staining was used for the test to examine rat tissue for cardiomyopathy. Some studies reported that the histology results related to biochemical parameters and ECG results [37–39]. In this study, rats could be induced to be hypertensive, promoting myocardium inflammation and scar deposition. However, when comparing the microscopic findings (Table 2), collagen involving scars, fibrosis in the myocardium, and collagen deposit in the sub-endocardium were not observed in the three rice groups. Only one lesion was found in the L-NAME group. It is possible that the rats had no severity in their heart pathologies, so the markers, such as gene expression, did not show any differences among the rice groups. Our data showed that the PGBR group seemed to have the lowest downregulation in scar deposition gene expression, such Col I and III. Thus, the administration of PGBR not only inhibited the renin–angiotensin system and fibrosis but also improved cardiac histology.

The results from this study suggested that the ingestion of PGBR might protect the heart against L-NAME-induced hypertension. The molecular mechanism of how PRBR reduced hypertension might be (i) the inhibition of the renin–angiotensin axis, (ii) the inhibition of fibrosis, or (iii) the inhibition of oxidative-stress-generating enzymes (NOX) and the activation of an antioxidant enzyme (GPx). Thus, the consumption of PGBR might reduce heart damage from hypertension pathology in this hypertensive rat model. Some study limitations were observed. The diastolic blood pressure and mean arterial pressure were not recorded, while the protein expression for several genes was not measured. The variations in oxidative stress and antioxidant genes were observed with no statistical findings. A limitation in this study could be that the food was freely accessed by the rats. The exact rice doses were difficult to calculate. Further studies should investigate more biomarkers, such as antioxidant or anti-inflammatory markers, to further clarify the mechanism of PGBR for the treatment of hypertension.

5. Conclusions

This is the first report demonstrating the anti-hypertensive properties of parboiled germinated brown rice (PGBR) in *N*-nitro-l-arginine methyl ester (L-NAME)-induced hypertensive rats. PGBR significantly decreased the angiotensin II type 1 receptor and transforming growth factor-β and significantly increased glutathione peroxidase expression. The gene expression of nicotinamide adenine dinucleotide phosphate oxidase 4 and collagen type I progressively decreased in the PGBR group. Thus, the regular consumption of PGBR could treat hypertension as in this animal model study.

Supplementary Materials: The following supporting information can be downloaded at: https://www.mdpi.com/article/10.3390/foods12010009/s1, Supplementary Table S1: The condition for rice cooking; Supplementary Table S2: The compositions of cooked rice powders; Supplementary Table S3: The composition of the basal formula diet (AIN-76A); Supplementary Table S4: Primers used for amplification in real-time PCR; Supplementary Figure S1: Diet consumption in all animal groups.

Author Contributions: Conceptualization, P.T. and C.C.; data curation, N.O.-N., K.K., P.T., and C.C.; formal analysis, N.O.-N., K.K., P.T., and C.C.; funding acquisition, C.C.; investigation, N.O.-N. and K.K.; methodology, A.K., T.R., P.T., P.P., and C.C.; project administration, C.C.; resources, A.K., T.R., U.S., and C.C.; supervision, P.T., and C.C.; validation, N.O.-N., U.S., P.T., and C.C.; visualization, N.O.-N.; writing—original draft preparation, N.O.-N., U.S., P.T., and C.C.; writing—review and editing, N.O.-N., P.T., and C.C. All authors have read and agreed to the published version of the manuscript.

Funding: This work was supported by the Agricultural Research Development Agency (Public Organization), Thailand, grant No. PRP5705021290.

Institutional Review Board Statement: All experimental protocols were approved for the use of animals by the Siriraj Animal Care and Use Committee (SiACUC) from the Faculty of Medicine of Siriraj Hospital, Mahidol University (SI-ACAP 008/2557, 2014), in compliance with international guidelines for animal research protection, such as the International Guiding Principles for Biochemical Research Involving Animals.

Informed Consent Statement: Not applicable.

Data Availability Statement: All data are contained within this article and the Supplementary Materials.

Acknowledgments: The authors would like to express their gratitude to Anchalee Tungtrongchitr for data collection support. Figure 1 was created with BioRender.com.

Conflicts of Interest: The authors declare no conflict of interest.

References

1. Mittal, B.V.; Singh, A.K. Hypertension in the developing world: Challenges and opportunities. *Am. J. Kidney Dis.* **2010**, *55*, 590–598. [CrossRef] [PubMed]
2. Oparil, S.; Zaman, M.A.; Calhoun, D.A. Pathogenesis of hypertension. *Ann. Intern. Med.* **2003**, *139*, 761–776. [CrossRef] [PubMed]
3. Lim, H.; Zhu, Y.Z. Role of Transforming Growth Factor-Beta in the Progression of Heart Failure. *Cell. Mol. Life Sci.* **2006**, *63*, 2584–2596. [CrossRef] [PubMed]
4. Campbell, S.E.; Katwa, L.C. Angiotensin II Stimulated Expression of Transforming Growth factor-beta1 in Cardiac Fibroblasts and Myofibroblasts. *J. Mol. Cell. Cardiol.* **1997**, *29*, 1947–1958. [CrossRef] [PubMed]
5. Miguel-Carrasco, J.L.; Zambrano, S.; Blanca, A.J.; Mate, A.; Vázquez, C.M. Captopril reduces cardiac inflammatory markers in spontaneously hypertensive rats by inactivation of NF-κB. *J. Inflamm.* **2010**, *7*, 21. [CrossRef] [PubMed]
6. Leask, A. TGFβ, cardiac fibroblasts, and the fibrotic response. *J. Cardiovasc. Dis.* **2007**, *74*, 207–212. [CrossRef]
7. Kelly, D.M.; Ademi, Z.; Doehner, W.; Lip, G.Y.H.; Mark, P.; Toyoda, K.; Wong, C.X.; Sarnak, M.; Cheung, M.; Herzog, C.A.; et al. Chronic Kidney Disease and Cerebrovascular Disease: Consensus and Guidance from a KDIGO Controversies Conference. *Stroke* **2021**, *52*, e328–e346. [CrossRef]
8. Pereira, C.; Lourenço, V.M.; Menezes, R.; Brites, C. Rice Compounds with Impact on Diabetes Control. *Foods* **2021**, *10*, 1992. [CrossRef]
9. Wunjuntuk, K.; Kettawan, A.; Charoenkiatkul, S.; Rungruang, T. Parboiled germinated brown rice protects against CCl4-induced oxidative stress and liver injury in rats. *J. Med. Food* **2015**, *19*, 15–23. [CrossRef]
10. Srichamnong, W.; Thiyajai, P.; Charoenkiatkul, S. Conventional steaming retains tocols and γ-oryzanol better than boiling and frying in the jasmine rice variety Khao dok mali 105. *Food Chem.* **2016**, *191*, 113–119. [CrossRef]
11. Wunjuntuk, K.; Kettawan, A.; Charoenkiatkul, S.; Rungruang, T. Anti-fibrotic and anti-inflammatory effects of parboiled germinated brown rice (Oryza sativa 'KDML 105') in rats with induced liver fibrosis. *J. Funct. Foods* **2016**, *26*, 363–372. [CrossRef]
12. Rattanadee, P.; Naivikul, O. Effect of soaking time, pre-germination time and parboiling processes on chemical and physicochemical properties of paddy and brown rice. In Proceedings of the 49th Kasetsart University Annual Conference: Agro-Industry, Bangkok, Thailand, 1–4 February 2011.
13. Miguel-Carrasco, J.L.; Monserrat, M.T.; Mate, A.; Vázquez, C.M. Comparative effects of captopril and l-carnitine on blood pressure and antioxidant enzyme gene expression in the heart of spontaneously hypertensive rats. *Eur. J. Pharmacol.* **2010**, *632*, 65–72. [CrossRef] [PubMed]

14. Gómez-Amores, L.; Mate, A.; Miguel-Carrasco, J.L.; Jiménez, L.; Jos, A.; Cameán, A.M.; Revilla, E.; Santa-María, C.; Vázquez, C.M. L-carnitine attenuates oxidative stress in hypertensive rats. *J. Nutr. Biochem.* **2007**, *18*, 533–540. [CrossRef] [PubMed]
15. Ontawong, A.; Saowakon, N.; Vivithanaporn, P.; Pongchaidecha, A.; Lailerd, N.; Amornlerdpison, D.; Lungkaphin, A.; Srimaroeng, C. Antioxidant and renoprotective effects of Spirogyra neglecta (Hassall) Kützing extract in experimental type 2 diabetic rats. *BioMed Res Int.* **2013**, *2013*, 820786. [CrossRef]
16. Zambrano, S.; Blanca, A.J.; Ruiz-Armenta, M.V.; Miguel-Carrasco, J.L.; Arévalo, M.; Mate, A.; Vázquez, C.M. L-carnitine attenuates the development of kidney fibrosis in hypertensive rats by upregulating PPAR-γ. *Am. J. Hypertens.* **2014**, *27*, 460–470. [CrossRef]
17. Zambrano, S.; Blanca, A.J.; Ruiz-Armenta, M.V.; Miguel-Carrasco, J.L.; Arévalo, M.; Vázquez, M.J.; Mate, A.; Vázquez, C.M. L-carnitine protects against arterial hypertension-related cardiac fibrosis through modulation of PPAR-γ expression. *Biochem. Pharmacol.* **2013**, *85*, 937–944. [CrossRef]
18. Rajeshwari, T.; Raja, B.; Manivannan, J.; Silambarasan, T. Valproic acid attenuates blood pressure, vascular remodeling and modulates ET-1 expression in L-NAME induced hypertensive rats. *Biomed. Prev. Nutr.* **2014**, *4*, 195–202. [CrossRef]
19. Silambarasan, T.; Manivannan, J.; Priya, M.K.; Suganya, N.; Chatterjee, S.; Raja, B. Sinapic acid prevents hypertension and cardiovascular remodeling in pharmacological model of nitric oxide inhibited rats. *PLoS ONE* **2014**, *9*, e11568. [CrossRef]
20. Ebizuka, H.; Ihara, M.; Arita, M. Antihypertensive effect of pre-germinated brown rice in spontaneously hypertensive rats. *Food Sci. Technol. Res.* **2009**, *15*, 625–630. [CrossRef]
21. Shimada, M.; Hasegawa, T.; Nishimura, C.; Kan, H.; Kanno, T.; Nakamura, T.; Matsubayashi, T. Anti-hypertensive effect of gamma-aminobutyric acid (GABA)-rich Chlorella on high-normal blood pressure and borderline hypertension in placebo-controlled double blind study. *Clin. Exp. Hypertens.* **2009**, *31*, 342–354. [CrossRef]
22. Jane-Lise, S.; Corda, S.; Chassagne, C.; Rappaport, L. The extracellular matrix and the cytoskeleton in heart hypertrophy and failure. *Heart Fail. Rev.* **2000**, *5*, 239–250. [CrossRef] [PubMed]
23. Birben, E.; Umit, M.S.; Cansin, S.; Serpil, E.; Omer, K. Oxidative stress and antioxidant defense. *WAO J.* **2012**, *5*, 9–19. [CrossRef] [PubMed]
24. Colonna, V.D.G.; Rigamonti, A.; Fioretti, S.; Bonomo, S.; Manfredi, B.; Ferrario, P.; Bianchi, M.; Berti, F.; Muller, E.E.; Rossoni, G. Angiotensin-converting enzyme inhibition and angiotensin AT1-receptor antagonism equally improve endothelial vasodilator function in L-NAME-induced hypertensive rats. *Eur. J. Pharmacol.* **2005**, *516*, 253–259. [CrossRef] [PubMed]
25. Litterio, M.C.; Jaggers, G.; Celep, G.S.; Adamo, A.M.; Costa, M.A.; Oteiza, P.I.; Fraga, C.G.; Galleano, M. Blood pressure-lowering effect of dietary (-)-epicatechin administration in L-NAME-treated rats is associated with restored nitric oxide levels. *Free Radic. Biol. Med.* **2012**, *53*, 1894–1902. [CrossRef] [PubMed]
26. Pechanova, O.; Vrankova, S.; Cebova, M. Chronic L-Name-Treatment Produces Hypertension by Different Mechanisms in Peripheral Tissues and Brain: Role of Central eNOS. *Pathophysiology* **2020**, *27*, 46–54. [CrossRef] [PubMed]
27. Villarejo, A.; Prieto, I.; Segarra, A.; Banegas, I.; Wangensteen, R.; Vives, F.; De Gasparo, M.; Ramírez-Sánchez, M. Relationship of Angiotensinase and Vasopressinase Activities Between Hypothalamus, Heart, and Plasma in L-NAME-Treated WKY and SHR. *Horm. Metab. Res.* **2014**, *46*, 561–567. [CrossRef] [PubMed]
28. Briones, A.M.; Touyz, R.M. Oxidative stress and hypertension: Current concepts. *Curr. Hypertens. Rep.* **2010**, *12*, 135–142. [CrossRef]
29. Rodrigo, R.; Libuy, M.; Feliú, F.; Hasson, D. Oxidative stress-related biomarkers in essential hypertension and ischemia-reperfusion myocardial damage. *Dis. Markers* **2013**, *35*, 773–790. [CrossRef]
30. Rincón, J.; Correia, D.; Arcaya, J.L.; Finol, E.; Fernández, A.; Pérez, M.; Yaguas, K.; Talavera, E.; Chávez, M.; Summer, R.; et al. Role of Angiotensin II type 1 receptor on renal NAD(P)H oxidase, oxidative stress and inflammation in nitric oxide inhibition induced-hypertension. *Life Sci.* **2015**, *124*, 81–90. [CrossRef]
31. Vaziri, N.D.; Lin, C.Y.; Farmand, F.; Sindhu, R.K. Superoxide dismutase, catalase, glutathione peroxidase and NADPH oxidase in lead-induced hypertension. *Kidney Int.* **2003**, *63*, 186–194. [CrossRef]
32. Sundaram, A.; Siew Keah, L.; Sirajudeen, K.N.; Singh, H.J. Upregulation of catalase and downregulation of glutathione peroxidase activity in the kidney precede the development of hypertension in pre-hypertensive SHR. *Hypertens. Res.* **2013**, *36*, 213–218. [CrossRef] [PubMed]
33. Kumar, S.; Prahalathan, P.; Raja, B. Syringic acid ameliorates (L)-NAME-induced hypertension by reducing oxidative stress. *Naunyn-Schmiedeberg's Arch. Pharmacol.* **2012**, *385*, 1175–1184. [CrossRef] [PubMed]
34. Nah, D.Y.; Rhee, M.Y. The Inflammatory response and cardiac repair after myocardial infarction. *Korean Circ. J.* **2009**, *39*, 393–398. [CrossRef]
35. Manabe, I.; Shindo, T.; Nagai, R. Gene expression in fibroblasts and fibrosis: Involvement in cardiac hypertrophy. *Circ. Res.* **2002**, *91*, 1103–1113. [CrossRef] [PubMed]
36. Rossi, M.A.; Colombini-Netto, M. Chronic Inhibition of NO Synthesis Per Se Promotes Structural Intimal Remodeling of the Rat Aorta. *J. Hypertens.* **2001**, *19*, 1567–1579. [CrossRef] [PubMed]

37. Jokinen, M.P.; Lieuallen, W.G.; Boyle, M.C.; Johnson, C.L.; Malarkey, D.E.; Nyska, A. Morphologic aspects of rodent cardiotoxicity in a retrospective evaluation of National Toxicology Program studies. *Toxicol. Pathol.* **2011**, *39*, 850–860. [CrossRef]
38. Zaafan, M.A.; Zaki, H.F.; El-Brairy, A.I.; Kenawy, S.A. Protective effects of atorvastatin and quercetin on isoprenaline-induced myocardial infarction in rats. *Bull. Fac. Pharm.* **2013**, *51*, 35–41. [CrossRef]
39. Yang, C.H.; Ting, W.J.; Day, C.H.; Ju, D.T.; Yeh, Y.L.; Chung, L.C.; Tsai, F.J.; Tsai, C.H.; Tsai, Y.; Huang, C.Y. SHSST cyclodextrin complex prevents the fibrosis effect on CCl_4-induced cirrhotic cardiomyopathy in rats through TGF-β pathway inhibition effects. *Int. J. Mol. Sci.* **2014**, *15*, 8037–8048. [CrossRef]

foods

MDPI

Article

Possibility for Prevention of Type 2 Diabetes Mellitus and Dementia Using Three Kinds of Brown Rice Blends after High-Pressure Treatment

Sumiko Nakamura [1], Takeshi Ikeuchi [2,*], Aki Araki [2], Kensaku Kasuga [2], Kenichi Watanabe [3], Masao Hirayama [1], Mitsutoshi Ito [1] and Ken'ichi Ohtsubo [1,*]

1 Faculty of Applied Life Sciences, Niigata University of Pharmacy and Applied Life Sciences, 265-1, Higashijima, Akiha-ku, Niigata 956-8603, Japan; snaka@nupals.ac.jp (S.N.); hirayama@nbrp.co.jp (M.H.); m-ito@nupals.ac.jp (M.I.)

2 Brain Research Institute, Niigata University, 1-757, Asahimachidori, Chuo-ku, Niigata 951-8585, Japan; araki-aki@bri.niigata-u.ac.jp (A.A.); ken39@bri.niigata-u.ac.jp (K.K.)

3 Faculty of Medicine, Niigata University, 1-757, Asahimachidori, Chuo-ku, Niigata 951-8585, Japan; wataken@med.niigata-u.ac.jp

* Correspondence: ikeuchi@bri.niigata-u.ac.jp (T.I.); ohtsubok@nupals.ac.jp (K.O.); Tel./Fax: +81-250-25-1218 (K.O.)

Abstract: As it has been reported that type 2 diabetes mellitus increases the risk of Alzheimer's disease, we investigated how to prevent type 2 diabetes and dementia using biofunctional boiled rice. We adopted unpolished super-hard rice (SHBR) for diabetes and wax-free unpolished black rice (WFBBR) for dementia and blended those with ordinary non-polished rice (KBR) (blending ratio 4:4:2), adding 2.5% waxy black rice bran (WBB) and 0.3% rice oil after high-pressure treatment (HPT) (WFBSK) to improve its palatability. This boiled rice is rich in dietary fiber, anthocyanin, free ferulic acid and β-secretase inhibitory activity. A randomized, parallel-group comparison study was conducted for 12 weeks with 24 subjects, using Cognitrax to evaluate their cognitive function primarily. Furthermore, as the secondary purpose, we performed a single-dose test for postprandial blood glucose and insulin secretion at the end of the human intervention test. After 12 weeks, consumers of the WFBSK rice exhibited significant improvement in language memory by cognitive test battery compared with those who consumed the control white rice ($p < 0.05$). Moreover, subjects who consumed the WFBSK rice had lower insulin secretion levels than those who consumed the control polished rice ($p < 0.05$).

Keywords: wax-free brown rice; waxy black rice bran; n-6/n-3 ratio; RS; insoluble dietary fiber; antioxidative activity; free ferulic acid; anthocyanin

Citation: Nakamura, S.; Ikeuchi, T.; Araki, A.; Kasuga, K.; Watanabe, K.; Hirayama, M.; Ito, M.; Ohtsubo, K. Possibility for Prevention of Type 2 Diabetes Mellitus and Dementia Using Three Kinds of Brown Rice Blends after High-Pressure Treatment. *Foods* **2022**, *11*, 818. https://doi.org/10.3390/foods11060818

Academic Editor: Diego A. Moreno

Received: 9 February 2022
Accepted: 10 March 2022
Published: 12 March 2022

1. Introduction

According to a report by the IDF (International Diabetes Federation), about 537 million people were candidates or patients with diabetes in 2021 [1] and the number of patients with dementia was about 46 million in the world in 2015 [2].

Dementia is a syndrome where there is deterioration in cognitive function that leads to impairment of the activities of daily living (ADL). Dementia may be caused by a variety of diseases and injuries that primarily or secondarily impair the brain function. Among the dementia patients, most have Alzheimer's disease (AD); 60–70% of the patients with dementia suffer from AD [3]. Nowadays, dementia is ranked as the seventh main cause of death among all diseases. Furthermore, it tends to lead to the disability and dependency of aged people all over the world [3]. Cognitive function consists of multiple domains such as memory, attention, language and executive function, all of which are essential for our daily life [4]. Cognitive disabilities caused by dementia negatively affect the quality of life (QOL) of elderly patients [5].

It is well-recognized that the prevalence of dementia is higher in diabetic patients than in non-diabetic subjects [6].

There are many scientific reports about the close relationship between insulin resistance and lifestyle diseases, such as type 2 diabetes and dementia [7–14]. As patients with diabetes have been markedly increasing in number worldwide, the development of a suitable method for its treatment or prevention is very important. The World Health Organization (WHO) and Food and Agriculture Organization of the United Nations (FAO) recommend foods with a low glycemic index (GI) to prevent diabetes [3]. The concept of GI was introduced by Jenkins et al. [15], and low GI and glycemic load diets have become popular to prevent chronic diseases, such as cardiovascular disease, diabetes, cancer and obesity [16], although they all have limited efficacy.

Early intervention for maintaining cognitive function is one of the important factors for successful aging [17]. To investigate dementia patients, the human intervention test is indispensable. Dubois et al. [18] described multi-component interventions targeting several risk factors simultaneously, such as "FINGER", might be needed for optimal preventative effects, to improve and maintain cognitive functioning.

Many researchers reported that the progression of cognitive decline may be considerably affected by various lifestyle factors, such as foods and drinks [4]. For example, it was reported that cognitive decline was prevented by food components such as minerals, polyphenols, flavonoids, vitamins, omega-3 PUFAs, etc. [19].

Findings from prospective studies suggest that greater adherence to the Mediterranean diet may be associated with slower cognitive decline and reduced risk of AD [20,21]. Core et al. [22]. showed that reduced levels or intake of omega-3 fatty acids or fish are associated with increased risk of age-related cognitive decline or dementia such as AD.

Rice (*Oryza sativa* L.) is one of the most important crops, cultivated in over 100 countries around the world, and is a staple food for about half of the world population [23]. As rice consumers, especially Asian people, eat it almost every day, it is very promising that its biofunctionality, such as prevention of diabetes and dementia, would be very effective. Unpolished rice grains contain more nutritional components, such as dietary fibers, phytic acids, flavonoids, tocopherols, γ-oryzanol and E and B vitamins, than ordinary milled rice grains [24]. The germ and bran layers (about 10% of a rice kernel) removed during the milling procedure are rich in proteins, lipids (γ-oryzanol, ferulic acid, sterol, wax, ceramide, phytin and inositol), fiber, minerals, tocotrienols, tocopherols and B-complex vitamins (B1 and B6) [24]. The use of rice bran as food and feed from 1970 to 1998 was recently reviewed, concerning the control of chronic degenerative diseases such as hypercholesterolemia [25]. Juliano showed that stabilized rice bran is mildly crunchy and slightly sweet, and has a mild toasted aroma [25].

The glycemic effect of food depends on numerous factors such as the microstructure of starch, amylose content and amylopectin chain-length distributions [26,27]. In a previous study, we developed a novel method for inhibiting postprandial blood glucose levels in Sprague–Dawley rats by preparing cooked rice grains from amylose extender (*ae*) mutants of rice soaked with functional food ingredients [28,29]. Several studies have reported the development of highly resistant starch rice as well as high-amylose and high-dietary fiber rice via physical or chemical mutation [30,31]. The cooked grains of super-hard rice cultivars are hard and non-sticky because they lack starch branching enzyme IIb and contain many super-long chains (SLCs) [29]. They are promising in terms of their biofunctionality, such as diabetes prevention and reduction of obesity [32]. Pigmented rice contains polyphenol substances, such as anthocyanins and tannins. Black or purple rice gets its color from anthocyanin pigments, which are known to have free-radical scavenging and antioxidant capacities [33]. Black rice may have antiatherogenic activity and may improve certain metabolic pathways associated with diets high in fructose [34,35]. Red rice is known to be rich in minerals, such as iron and zinc, while black and purple rice are especially high in protein, fat and crude fiber [34].

The cooked grains of unpolished rice are too hard and non-sticky for the consumers. For that reason, Watanabe et al. [36] showed that wax-free brown rice (WFBR), which is unpolished rice with only the wax layer removed, keeping other nutrients in the bran layer, is promising for a low-GI and high dietary fibers and vitamins, and its taste is improved. Furthermore, high-pressure treatment (HPT) is very useful in the food industry [37]. The merits of the high pressure are to avoid the destruction of covalent bonding and to keep the natural flavor, taste, and nutrients. HPT is the technological process that has the least effect on heat-labile water-soluble vitamins, thus contributing to the preservation of the nutritional quality of foodstuffs [38]. HPT was reported to be useful for improving the texture of cooked rice without denaturation of enzyme activities, and it led to an increase of free amino acids and change in enzyme activities [39].

The objective of this study was to develop multifunctional boiled rice, which would be useful for preventing type 2 diabetes and dementia, using black rice and super-hard rice with the aid of HPT and wax-free treatment. The antioxidative capacity of black rice and low-GI of super-hard rice would contribute to preventing the onset of diabetes and dementia.

2. Materials and Methods

2.1. Materials

The brown rice of super-hard rice cultivar Niigata 129 go (not registered), the high-quality rice Koshihikari; (registration number in Japan; 8539), waxy black rice Shiho (registration number in Japan; 11846) and black rice Okunomurasaki (registration number in Japan; 11088), were cultivated by the Niigata Prefecture Agriculture Research Institute in 2020, and red rice (Beniroman) and waxy red rice (Yuyake mochi) were purchased from a local market. The wax-free brown rice was manufactured by Toyo Rice (Co., Ltd., Wakayama, Japan) in 2020.

2.2. Preparation of Waxy Black Rice Bran

The bran of waxy black rice (Shihou cultivar) was prepared by polishing using an experimental friction-type rice milling machine (Yamamoto Seisakusyo Co., Tendo, Japan) to a milling yield of 90%.

2.3. Food Processing of Experimental Meal

WFBSK rice (WFBBR:SHBR:KBR = 4:4:2) was combined with 2.5% (*w*/*w*) waxy black rice bran (WBB) and 0.3% (*w*/*w*) rice oil (Tsuno Food Industrial Co., Ltd., Wakayama, Japan), obtained by treating these at 200 MPa for 2 min in a high-pressure machine (Ishikawajima-Harima Heavy Industries Co., Ltd., Tokyo, Japan), and cooked rice was prepared by Echigo Seika, Co., Ltd. (Nagaoka, Japan). Commercial aseptic cooked rice (using polished rice of high-quality rice cultivar, Koshihikari, as material) was prepared with the same condition with the abovementioned procedure, by Echigo Seika, Co., Ltd., and subjected to the analyses, as well as being used as a control meal in the human intervention test.

2.4. Measurement of Moisture Content of Rice Flour

The moisture content of the brown rice and cooked brown rice were measured using an oven-dry method. Accordingly, 2 g of brown rice sample was dried for 1 h at 135 °C, while 2 g cooked rice samples were dried for 3 h at 135 °C.

2.5. Measurement of Water Absorption Rate (WAR) of Wax-Free Brown Rice (WFBR), Brown Rice and White Rice Grains

Each sample grain (5.0 g) was socked in 50 mL distilled water at 25 °C for 15 min, 30 min, 60 min, 90 min, 120 min or 180 min. After draining, the clear supernatant was removed, and we weighed the sample grains (Wa). The water absorption rate (WAR) was calculated as WAR (%) = (Wa − 5)/5 × 100.

2.6. Textural Properties of Cooked Rice

The physical properties of cooked rice grains were measured based on bulk measurement (10 g), using a My Boy System Tensipresser (Taketomo Electric Co., Tokyo, Japan) according to the method described by Okadome et al. [40]. For standard samples, milled rice (10 g) was combined with 14 g distilled water (coefficient [gross water volume/dry matter weight]: 1.77, calculated for each sample) in an aluminum cup. After soaking for 1 h, the samples were cooked. The cooked rice samples were kept in the vessel at 25 °C for 2 h and subjected to the measurements. The bulk measurements were repeated five times, and the mean value was calculated.

2.7. Measurement of the Fatty Acid Composition of Rice Bran

Measurement of the fatty acid composition of rice bran was carried out by the Food Analysis Technology Center (using a gas chromatography method). A rice bran sample (0.2 g) was extracted with 2 mL hexane and mixed well. After that, 2 M potassium hydroxide−methanol solution (0.2 mL) was added and mixed.

2.8. Measurement of RS (Resistant Starch) of Cooked Rice

The resistant starch (RS) was measured according to the AOAC method (2002.02) using an RS assay kit (Megazyme, Ltd., Wicklow, Ireland) except the enzyme reaction time. Freeze-dried rice flours (0.1 g) were treated with a 0.1 M sodium maleate (4 mL) (pH 6.0) buffer solution with enzymes (pancreatin and amyloglucosidase) at 36 °C for 6 h, and then denatured ethanol (99%) (4 mL) was added and the solution was centrifuged. The precipitation was mixed with denatured ethanol (99%) (6 mL) and the reaction mixture was centrifuged, and the process was then repeated. The supernatant was removed, and this was followed by the addition of 2 M potassium hydroxide (2 mL), stirred in ice water for 20 min, and 1.2 M sodium acetate (pH 3.8) (8 mL) and amyloglucosidase (0.1 mL) were added and stirred at 50 °C for 30 min in a water bath. After centrifugation, the supernatant (0.1 mL) was mixed GOPOD (glucose oxidase-peroxidase-aminoantipyrine reagent) (3 mL) and stirred at 50 °C for 20 min in a water bath. The glucose content was measured using a spectrophotometer at 510 nm.

2.9. Measurement of Polyphenol Content of Cooked Rice

The polyphenol content of freeze-dried rice samples was determined using the Folin–Ciocalteu method [41]. Each sample (0.1 g) was extracted with 80% ethanol (4 mL) and then centrifuged. The supernatant (1 mL) was mixed with the same volume of Folin–Ciocalteu solution (1 mL) and incubated for 3 min at room temperature, followed by the addition sodium carbonate (5 mL) and incubation at 50 °C for 5 min. Finally, the sample solution was cooled in ice water for 1 h and then centrifuged. Absorbance was measured at 765 nm. Gallic acid was used for calibration.

2.10. Measurement of Hydrophilic and Lipophilic Oxygen Radical Absorbance Capacity (H-ORAC and L-ORAC)

The hydrophilic and lipophilic oxygen radical absorbance capacities of freeze-dried rice samples were measured as described by Prior et al. [42]. Trolox calibration solutions were made to obtain a standard curve [43,44]. For the hydrophilic antioxidant assay, freeze-dried rice flours (0.1 g) were extracted with hexane (10 mL), and the hexane layer was removed. Residual hexane was evaporated using a water bath at 70 °C, and the residue was then extracted with acetone/water/acetic acid (70:29.5:0.5, $v/v/v$) (10 mL). The solution was sonicated (Ultrasonic cleaner 3510J-MTH, Branson Ultrasonics Co, Ltd., Richmond, VA, USA) at 37 °C for 15 min, and then centrifuged. The supernatant was diluted to a 25-mL total volume. For the lipophilic antioxidant assay, freeze-dried rice flours (0.1 g) were extracted with hexane (1 mL), and then centrifuged. The hexane was evaporated using a water bath at 70 °C, and the dried hexane extract was then dissolved in acetone (250 μL) and diluted with 7% randomly methylated β-cyclodextrin (750 μL) (RMCD; 0.7 g methyl-β-cyclodextrin (Sigma-Aldrich Co. LLC, St. Louis, MO, USA) dissolved in 10 mL 50% acetone). Absorbance values were measured at 485 nm (excitation) and 530 nm (emission) using a fluorescent microplate reader (Grating Based Multimode Reader SH-9000, Corona Electric Co, Ltd., Hitachinaka-shi, Japan).

2.11. *Measurement of the Ferulic Acid Composition of Cooked Rice*

Measurement of the ferulic acid composition of rice bran was carried out by the Japan Food Research Laboratories (using microbiological assays and high-performance liquid chromatography–mass spectrometry).

2.12. *β-Secretase Inhibitory Activity*

The β-secretase (BACE1) inhibitory activity of freeze-dried cooked rice was measured using a BACE1 activity detection kit (Fluorescent; Sigma-Aldrich Co. LLC.). Freeze-dried rice flour (0.1 g) was extracted with 10 mM acetate buffer solution (0.5 mL) (pH 5.0, including 0.1% Triton and 0.05% CHAPS) for 1 h, and then centrifuged. The absorbance values were measured at 320 nm (excitation) and 405 nm (emission) using a fluorescent microplate reader (Grating Based Multimode Reader SH-9000, Corona Electric Co, Ltd.) [45].

2.13. *Measurement of the Dietary Fiber of Cooked Rice*

Measurement of the dietary fiber of cooked rice was carried out by the Japan Food Research Laboratories (using microbiological assays and high-performance liquid chromatography–mass spectrometry).

2.14. *Measurement of the Anthocyanin Content of Cooked Rice*

The anthocyanin content of cooked rice flour was measured based on the pH different method [46]. Freeze-dried rice flours samples (1.0 g) were extracted with 9 mL methanol/deionized water/trifluoroacetic acid (2:3:0.025, *v*/*v*/*v*) at room temperature for 2 min by strong vortexing. The solution was then sonicated (Ultrasonic cleaner 3510J-MTH, Branson Ultrasonics Co, Ltd.) at 37 °C for 5 min and soaked in a water bath at 37 °C for 10 min, then centrifuged for 2 min at 3000× *g*. The supernatant was removed and transferred to a 25-mL volumetric flask. The precipitate was extracted with 8 mL methanol/deionized water/trifluoroacetic acid (2:3:0.025, *v*/*v*/*v*) and the process repeated two times. The supernatant was removed to a volumetric flask and diluted to a 25-mL total volume, and the solution was twofold diluted with pH 1.0 buffer (0.025 M potassium chloride) and another pH 4.5 buffer (0.4 M sodium acetate). The absorbance was measured at 520 nm and 700 nm. The anthocyanin content was expressed as the cyanidin 3-glucoside.

2.15. *Sensory Evaluation*

The sensory test was carried out by the method reported in our previous paper [47]. Seven-grade ranking was used to evaluate the six attributes, such as appearance, aroma, hardness, taste, stickiness, and overall evaluation by 10 trained taste panelists.

2.16. *Study Design of Human Intervention Test*

The study protocol for human test subjects was approved by Niigata Bio-Research Park Inc (Niigata, Japan). and approved by the ethics committee for human tests of the Niigata Bio-Research Park and Niigata University of Pharmacy and Applied Life Sciences, and according to the 2014 guidelines issued by the Ministry of Education, Culture, Sports, Science and Technology and the Ministry of Health, Labor and Welfare, Japan. Informed consent was obtained for experimentation with human subjects, and the test was registered as UMIN000044767.

A randomized, parallel-group comparison study was conducted to examine the effect of WFBSK rice. Inclusion criteria were as follows: (1) subjects aged from 50 to 75 years, (2) not diagnosed as having dementia or diabetes, (3) no consumption of supplements that may affect cognition or blood glucose. We recruited 24 healthy subjects for the study.

The effect of WFBSK rice on cognitive performance was the primary endpoint. Secondary endpoints included inhibition of abrupt increase in postprandial blood glucose level and change in plasma amyloid-β (Aβ) 42/40 ratio.

Components of the test and control meals are shown in Table S1. Participants were randomly assigned to one of two groups. Subjects of each block-consumed the assigned test sample meal (one package of boiled rice) containing 64 g (test meal) or 67 g (control meal) carbohydrate once every day for 12 weeks consecutively. After 12 weeks, test subjects got the single-dose test for BGL measurements, where they consumed the assigned meals within 10 min with frequent mastication (30 times was the guideline) and 200 mL of water.

2.17. Evaluation of Cognitive Function

The Mini-Mental State Examination Japanese version (Nihon Bunka Kagakusha, Tokyo, Japan) was evaluated at the baseline. The Cognitrax test (Health Solution, Inc., Tokyo, Japan) was utilized to assess the cognitive change by intervention. Cognitrax consists of a computerized test battery, which evaluates multiple cognitive domains, including composite memory, verbal memory, visual memory, processing speed, psychomotor speed, executive function, reaction time, complex attention, simple attention, cognitive flexibility and motor speed. Cognitrax scores have been standardized according to the results from large populations of subjects aged from 7 to 90 years [48].

2.18. Measurement of Plasma Concentrations of Aβ1-42 and Aβ1-40

Plasma concentrations of Aβ1-42 and Aβ1-40 were analyzed using a V-PLEX Aβ Peptide Panel 1 (6×10^{10}) Kit (Meso Scale Discovery, Rockville, MD, USA) with MESO QuickPlex SQ 120 (Meso Scale Diagnostics, LLC, Rockville, MD, USA) according to the manufacturer's instructions. The intra-assay and inter-assay coefficients of variation were less than 20% for all assays.

2.19. Blood Examination

Blood was drawn (in November 2021) under fasting conditions from each subject for analysis of Aβ40, Aβ42, HDL cholesterol, LDL cholesterol, insulin sensitivity and HbA1c. Plasma concentrations of Aβ1-42 and Aβ1-40 were analyzed using a V-PLEX Aβ Peptide Panel 1 (6×10^{10}) Kit (Meso Scale Discovery, Rockville, MD, USA) with MESO QuickPlex SQ 120 (Meso Scale Diagnostics, LLC, Rockville, MD, USA) according to the manufacturer's instructions. The intra-assay and inter-assay coefficients of variation were less than 20% for all assays. BGL and insulin levels were measured at 0, 30, 60, 90 and 120 min after two groups of 12 subjects who had eaten two different test meals.

2.20. Statistical Analyses

All results were subjected to t-tests and Dunnett's test using Excel Statistics (version 6, Microsoft Corporation, Tokyo, Japan) and GraphPad Prism V8.4.3 (GraphPad Software, Inc., San Diego, CA, USA). A value of $0.05 < p < 0.10$ was considered to show the tendency, and a value of $p < 0.05$ was considered to be statistically significant.

3. Results and Discussion

3.1. Water Absorption Rate (WAR) of WFBR and Brown Rice

The water absorption rates (WARs) of WFBR and brown rice are shown in Figure 1. The WAR of polished Koshihikari rice is shown as a control. As shown in Figure 1, the WFBRs of Koshihikari, Okunomurasaki (black rice) and Niigata 129 go (super-hard rice) were significantly higher than those of brown rice at every soaking time. Moreover, the WAR of Koshihikari WFBR showed almost the same value as the control polished rice for 1 h soaking. Furthermore, the WAR of Okunomurasaki WFBR was significantly higher than that of control polished rice for 2 h soaking, and that of Niigata 129 go (super-hard rice) showed a similar tendency. Kuwada et al. [49] showed that the WAR of brown rice had almost the same value as polished rice for 17 h soaking, because the wax layer of brown rice interferes with water absorption. In a previous study, we reported that the WAR of brown rice becomes higher due to damage of the cell wall by high-pressure treatment (HPT) [39]. As a result, WFBR could be cooked after 1 h soaking, the same as with white rice.

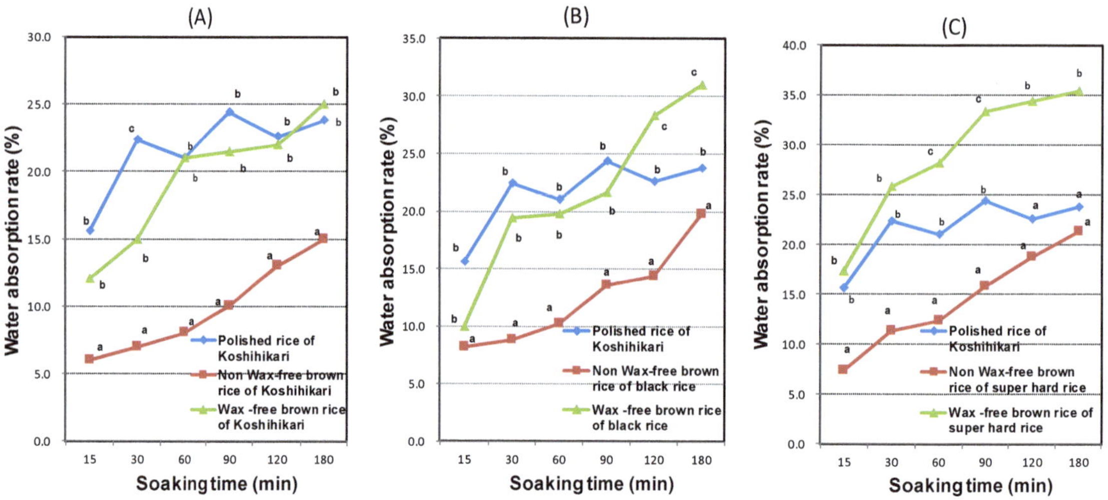

Figure 1. Water absorption rates of three kinds of wax-free brown rice and brown rice. Among the water absorption rates after in same soaking time, different letters (a, b, etc.) denote a statistically significant difference. Water absorption of polished Koshihikari rice is shown as a control (blue line) (**A**); water absorption rates of brown Koshihikari rice and wax-free brown Koshihikari rice (**B**); water absorption rates of brown Okunomurasaki (black) rice and wax-free brown Okunomurasaki rice (**C**); water absorption rates of brown Niigata 129 go (super hard) rice and wax-free brown Niigata 129 go rice.

3.2. *Textural Properties of Cooked Grains of WFBR and Non-Wax-Free Rice*

Hardness (force) and toughness (work) are parameters for rigidness, while adhesiveness (work) and stickiness (force) are indicators of tackiness in the textural measurement.

As shown in Table 1, the hardness of Koshihikari and Niigata 129 go WFBR were significantly lower than those of ordinary brown rice, and toughness showed a similar tendency. Moreover, the stickiness of Koshihikari, Okunomurasaki and Niigata 129 go WFBR were significantly higher than those of ordinary brown rice, and adhesion showed a similar tendency. The cooked rice grains after wax-free treatment were softer and stickier than those of non-wax-free brown rice, while their hardness and toughness were lower. Moreover, the ratio of stickiness and hardness of cooked rice with the wax-free treatment were increased 1.4~1.6 times than with the non-wax-free treatment. We estimate that the reason why WFBR was softer and stickier than non-wax-free brown rice is that the former rice grains absorbed water more rapidly and thoroughly. As a result, wax-free treatment markedly improved the textural properties of cooked brown rice. It is now possible to produce palatable cooked rice grains by wax-free treatment. Furthermore, in terms of the whiteness and brightness of cooked rice, WFBR showed significantly higher values than ordinary brown rice, and on the other hand, in terms of a yellowish color, WFBR showed significantly lower values than brown rice (Table S2). In previous studies, we reported that high-pressure treatment (HPT) improved the physical properties of brown rice [39]. Therefore, wax-free treatment and HPT improved the textural properties of cooked brown rice so that consumers accept them as the table rice they eat every day.

Table 1. Textural properties of various cooked brown rice and wax-free brown rice.

	Hardness $\times10^5$ [N/cm^2]	Toughness $\times10^5$ [N/cm^2]	Adhesion $\times10^5$ [N/cm^2]	Stickiness $\times10^5$ [N/cm^2]
Cooked Koshihikari rice (BR)	0.014 ± 0.002 a	0.032 ± 0.004 a	0.020 ± 0.007 a	0.020 ± 0.003 a
Cooked Koshihikari rice (WFR)	0.009 ± 0.002 b	0.027 ± 0.004 a	0.029 ± 0.005 a	0.024 ± 0.003 b
Cooked black rice (BR)	0.021 ± 0.005 a	0.039 ± 0.005 a	0.015 ± 0.003 a	0.016 ± 0.001 a
Cooked black rice (WFR)	0.020 ± 0.005 a	0.036 ± 0.003 a	0.019 ± 0.004 a	0.022 ± 0.002 b
Cooked super-hard rice (BR)	0.047 ± 0.007 a	0.040 ± 0.004 a	0.000 ± 0.000 a	0.000 ± 0.000 a
Cooked super-hard rice (WFR)	0.025 ± 0.005 b	0.033 ± 0.008 b	0.001 ± 0.000 a	0.005 ± 0.006 b

Brown rice, BR; wax-free brown rice, WFR; within each measure (hardness, toughness, adhesion and stickiness) in the same column and each sample, different letters (a, b, etc.) denote statistically significant differences.

3.3. Fatty Acid Compositions of Six Kinds of Rice Bran

In a previous study, we reported the fatty acid compositions of 30 japonica rice cultivars [50]. We found that a low DP (degree of polymerization) of amylopectin cultivars more easily bonded to linolenic acid (18:3n-3) than linoleic acid (18:2n-6) [50]. As shown in Table 2, the fatty acid compositions of six kinds of rice bran were palmitic acid (16.3–18.2%, mean = 17.3%), oleic acid (39.3–46.0%, mean = 42.7%) and linoleic acid (30.6–36.6%, mean = 34.0%), and the n-6/n-3 ratio was 24.4–40.0 (mean = 29.9). The oleic acid contents of Okunomurasaki (black rice) (46.0%) and Niigata 129 go (super-hard rice) (45.3%) were high; that of Shiho (waxy black rice) (43.9%) was intermediate; those of Koshihikari (41.2%), Yuyake mochi (waxy red rice) (40.5%) and Beniroman (red rice) (39.3%) were low. In contrast, the linoleic acid contents of Koshihikari (36.7%), Beniroman (36.6%) and Yuyake mochi (36.1%) were high, and those of Okunomurasaki (32.0%), Shiho (31.7%) and Niigata 129 go (30.6%) were low. Furthermore, the n-6/n-3 ratios of Okunomurasaki (40.0) and Koshihikari (33.4) were high; those of Niigata 129 go (27.8), Yuyake mochi (27.8) and Beniroman (26.1) were intermediate; that of Shihou (24.4), a waxy black rice cultivar, was very low. Simopoulos [51] showed that increased levels of omega-3 PUFA (a low n-6/n-3 ratio) exert suppressive effects against the pathogenesis of several diseases. Waxy rice lipids tend to be richer in palmitic acid but poorer in oleic, and to a lesser extent, in linoleic acids [52]. Our results showed that the n-6/n-3 ratio was negatively correlated with palmitic acid ($r = -0.85$, $p < 0.05$) (data not shown). As shown in Figure S1, the n-6/n-3 ratios of super-hard rice, waxy black rice, waxy red rice and red rice were low; therefore, they would be biofunctional rice cultivars. As an example of the effects of fatty acid to prevent disease through the diet, eicosapentaenoic acid (EPA) prevented thrombosis and atherosclerosis [53]. Ikemoto and Naganuma [54] reported that the n-6/n-3 ratio of PUFAs would make a very important dietary index because n-3 fatty acids, such as linolenic acid, lead to the generation of eicosapentaenoic acid and docosahexaenoic acid (DHA) in the body.

3.4. Nutrient Intake and Structural Components of Three Kinds of Rice Blends

Shown in Table 3A is a diet of blending super-hard brown rice (SHBR), wax-free black rice (WFBBR) and ordinary brown rice (KBR) (4:4:2) with 2.5% waxy black rice bran (WBB) and 0.3% rice oil added after high-pressure treatment (HPT) (WFBSK). In Table 3B, there is a diet of blending super-hard brown rice (SHBR), black/brown rice (BBR) and ordinary brown rice (KBR) (4:4:2) with 2.5% waxy black rice bran (WBB) and 0.3% rice oil added after high-pressure treatment (HPT) (BSK). In Table 3C, there is a diet of control polished Koshihikari rice. The insoluble dietary fiber (ISDF) of WFBSK (9.5 ± 0.6) (g/100 g) was significantly higher than BSK (7.8 ± 0.2) (g/100 g) and the control diet (0.4 ± 0.4) (g/100 g), whereas the water-soluble dietary fiber (SDF) of BSK (0.8 ± 0.1) (g/100 g) was significantly higher than WFBSK (0.3 ± 0.1) (g/100 g) and the control diet (0.1 ± 0.1) (g/100 g). The protein, lipid and ash contents of WFBSK and BSK showed almost the same values, and those values were significantly higher than the control Koshihikari diet. The effect of the β-glucan and dietary fiber contents on the LDL cholesterol and postprandial blood glucose of the subjects were very similar. Several recent studies, in both hypercholesterolemic and healthy subjects, found that the daily consumption of 5 g of β-glucan significantly lowered the total and LDL cholesterol in serum [55–57]. As a result, it was shown that the ISDF and total dietary fiber contents were significantly higher, by 1.1 times in the case of both wax-free processing and HPT (Table 3A) versus the case of only HPT (Table 3B).

Table 2. Fatty acid compositions of six kinds of japonica rice bran.

		Koshihikari	Black Rice	Waxy Black Rice	Super-Hard Rice	Red Rice	Waxy Red Rice
14:0	Myristic acid (%)	0.3 ± 0.0 a	0.4 ± 0.0 b	0.5 ± 0.0 b	0.3 ± 0.0 a	0.3 ± 0.0 a	0.3 ± 0.0 a
16:0	Palmitic acid (%)	16.7± 0.0 a	16.3 ± 0.0 a	17.6 ± 0.0 b	18.2 ± 0.0 b	17.7 ± 0.0 b	17.3 ± 0.0 b
16:1	Palmitoleic acid (%)	0.1 ± 0.0 a	0.1 ± 0.0 a	0.2 ± 0.0 a	0.2 ± 0.0 a	0.2 ± 0.0 a	0.2 ± 0.0 a
18:0	Stearic acid (%)	1.5 ± 0.0 a	2.0 ± 0.0 b	1.9 ± 0.0 b	2.0 ± 0.0 b	1.7 ± 0.0 b	1.8 ± 0.0 b
18:1	Oleic acid (%)	41.2 ± 0.1 b	46.0 ± 0.0 a	43.9 ± 0.1 a	45.3 ± 0.1 a	39.3 ± 0.0 b	40.5 ± 0.0 b
18:2n-6	Linoleic acid (%)	36.7 ± 0.1 a	32.0 ± 0.0 b	31.7 ± 0.0 b	30.6 ± 0.1 b	36.6 ± 0.0 a	36.1 ± 0.0 a
18:3n-3	α- linolenic acid (%)	1.1 ± 0.0 a	0.8 ± 0.0 b	1.3 ± 0.0 a	1.1 ± 0.0 a	1.4 ± 0.0 a	1.3 ± 0.0 a
20:0	Arachidic acid (%)	0.7 ± 0.0 a	0.8 ± 0.0 a	0.8 ± 0.0 a	0.7 ± 0.0 a	0.6 ± 0.0 a	0.7 ± 0.0 a
20:1	Icosenoic acid (%)	0.5 ± 0.0 a	0.5 ± 0.0 a	0.6 ± 0.0 a	0.5 ± 0.0 a	0.5 ± 0.0 a	0.5 ± 0.0 a
22:0	Behenic acid (%)	0.4 ± 0.0 a	0.4 ± 0.0 a	0.4 ± 0.0 a	0.3 ± 0.0 a	0.4 ± 0.0 a	0.4 ± 0.0 a
24:0	Lignoceric acid (%)	0.8 ± 0.0 a	0.7 ± 0.0 a	0.9 ± 0.0 a	0.7 ± 0.0 a	0.7 ± 0.0 a	0.7 ± 0.0 a

Within each measure (myristic acid, palmitic acid, etc.) in the same column and each cultivar, different letters (a, b, c, etc.) denote statistically significant differences.

Table 3. The principal compositions of three kinds of cooked rice for the human test.

	A	B	C
Energy (kcal/100 g)	143.0 ± 2.8 a	148 ± 1.2 a	147 ± 1.1 a
Protein (g/100 g)	3.5 ± 0.2 a	3.1 ± 0.2 a	2.4 ± 0.1 b
Lipid (g/100 g)	1.5 ± 0.1 a	1.4 ± 0.1 a	0.5 ± 0.1 b
Carbohydrates (g/100 g)	34.0 ± 0.4 a	35.0 ± 0.2 a	33.4 ± 0.1 a
Sugar (g/100 g)	24.2 ±1.1 a	26.4 ± 1.0 a	33.0 ± 1.0 b
Insoluble dietary fiber (g/100 g)	9.5 ± 0.6 a	7.8 ± 0.2 b	0.4 ± 0.4 c
Water soluble dietary fiber (g/100 g)	0.3 ± 0.1 a	0.8 ± 0.1 b	0.1 ± 0.1 c
Moisture (g/100 g)	60.6 ± 0.5 a	59.9 ± 0.4 a	63.6 ± 0.3 a
Ash (g/100 g)	0.6 ± 0.0 a	0.6 ± 0.0 a	0.1 ± 0.0 b

A: Cooked WFBSK rice with 2.5% WBB and 0.3% rice oil added by HPT. B: Cooked BSK rice with 2.5% WBB and 0.3% rice oil added by HPT. C: Cooked polished Koshihikari rice. Waxy black rice bran, WBB; high-pressure treatment, HPT. Blend with wax-free black/brown rice, super-hard brown rice and Koshihikari brown rice (4:4:2), WFBSK. Blend with black/brown rice, super-hard brown rice and Koshihikari brown rice (4:4:2), BSK. Within each measure (energy, protein, etc.) in the same column, different letters (a, b, c, etc.) denote statistically significant differences. Dietary fiber: enzymatic-gravimetric method (Prosky variant). Protein: Kjeldahi method. Lipid: Gas chromatography. Ash: Inductively coupled plasma atomic emission spectrometry. Moisture: Drying method by heating.

3.5. Measurement of Biofunctinal Properties of Three Kinds of Cooked Rice

Many dietary compounds have been proposed to be important antioxidants, and although there is credible evidence that vitamins E and C are important antioxidants, the evidence is weaker for carotenoids and related plant pigment [58]. The H-ORAC values of WFBSK (77.5 ± 3.4) (μmol TE/100 g FW) and BSK (78.0 ± 1.3) (μmol TE/100 g FW) were significantly higher than the control diet (0.0 ± 0.0) (μmol TE/100 g FW), and the L-ORAC values trended in similar directions to the H-ORAC values. As a result, the total ORAC values of WFBSK and BSK proved to be higher than the control diet by about

80 times, and similarly, the anthocyanin and polyphenol contents of WFBSK and BSK were significantly higher than the control diet, by about eight times. Pigmented rice contains naturally occurring pigmented flavonoids known as anthocyanins. Positive health effects have been reported for the pigments in the bran layer of rice [50]. In a previous study, we reported that the H-ORAC values and available polyphenol content of black rice were increased by HPT [39].

RS is starch that eludes digestion in the small intestine and may ferment in the large intestine. Several studies have reported that long-term consumption of RS might reduce the fasting cholesterol and triglyceride concentrations [59,60]. Yang et al. reported the starch properties of mutant rice, which is rich in resistant starch [30]. Furthermore, Noro et al. showed that Niigata 129 go (Chou 2418) contains long-chain glucans of amylopectin and that the RS content is higher than in high-amylose rice cultivars [61]. Maeda et al. [62] showed that the RS content of mutant rice became slightly higher than in that untreated by HPT, although their physical properties were improved. As shown in Table 4, the RS contents of WFBSK (5.1 ± 0.1%) and BSK (5.1 ± 0.1%) were significantly higher than with the control diet (0.7 ± 0.0%).

Table 4. Biofunctional properties of three kinds of cooked rice for human test.

	H-ORAC μmol TE/100 gFW	L-ORAC μmol TE/100 gFW	Total ORAC μmol TE/100 gFW	Anthocyanin (mg/g)	Polyphenol GAEmg/ 100 gFW	RS (%)	Free Ferulic Acid (mg/100 g)	β-Secretase Inhibition Rate (%) (0.27μg-eq/μL)
A	77.5 ± 3.4 b	5.2 ± 0.4 b	82.7 ± 3.8 b	8.03 ± 0.03 b	8.30 ± 0.20 b	5.1 ± 0.1 b	0.60 ± 0.00 b	9.2 ± 0.3 b
B	78.0 ± 1.3 b	4.9 ±0.0 b	82.9 ± 1.3 b	5.84 ± 0.02 b	7.87 ± 0.20 b	5.1 ± 0.1 b	0.57 ± 0.01 b	8.8 ± 0.3 b
C	0.0 ± 0.0 a	1.1 ± 0.1 a	1.1 ± 0.2 a	0.00 ± 0.00 a	0.00 ± 0.00 a	0.7 ± 0.0 a	0.30 ± 0.00 a	0.0 ± 0.0 a

A: Cooked WFBSK rice with 2.5% WBB and 0.3% rice oil added by HPT. B: Cooked BSK rice with 2.5% WBB and 0.3% rice oil added by HPT. C: Cooked polished Koshihikari rice. Waxy black rice bran, WBB; high-pressure treatment, HPT. Blend with wax-free black/brown rice, super-hard brown rice and Koshihikari brown rice (4:4:2), WFBSK. Blend with black/brown rice, super-hard brown rice and Koshihikari brown rice (4:4:2), BSK. Within each measure (H-ORAC, L-ORAC, etc.) in the same column, different letters (a, b, c, etc.) denote statistically significant differences.

BACE1 is the first protease involved in the process of converting APP to Aβ in the brain. As shown in Table 4, the BACE1 inhibitory activities for WFBSK (9.2 ± 0.3%) and BSK (8.8 ± 0.3%) were significantly higher than for the control diet (0.0%). The BACE1 inhibitory activity of pigmented rice bran is much stronger than the inhibitory activities of ordinary rice bran, as reported in our previous paper [45].

Ferulic acid is used as an antioxidant and antimicrobial agent. It is also recognized that ferulic acid exhibits a preventive effect on discoloration in various food products, and a variety of physiological functions such as suppression of Alzheimer's disease, prevention of muscular fatigue, improvement of hypertension and antitumor activity in the breast, liver and colon [24]. As shown in Table 4, the free ferulic acid levels of WFBSK (0.60 ± 0.00) (mg/100 g) and BSK (0.57 ± 0.01) (mg/100 g) were significantly higher than for the control diet (0.30 ± 0.00) (mg/100 g). These results were presumed to be due to HPT.

As a result, the biofunctional properties of WFBSK tended to be slightly higher than the BSK ones. Therefore, we selected WFBSK as the test diet for the human intervention test.

3.6. Textural Properties of Three Kinds of Cooked Rice

As shown in Table 5, the hardness and adhesion of three kinds of cooked rice showed similar values, but the toughness and stickiness of WFBSK and BSK were significantly higher than those of the control diet. As a result, the textures of the cooked WFBSK and BSK rice became acceptable for consumers, in terms of palatability, by HPT.

The biofunctional and physical properties of WFBSK and BSK showed a similar tendency. Therefore, we evaluated the palatability of blended meals by sensory evaluation. We found the "taste" and "overall evaluation", in sensory analysis of WFBSK meal, were significantly higher than those of BSK meal at the level of 5%, as shown in Table 6. Furthermore, the "taste" and "overall evaluation" of WFBSK meal were improved by adding 0.3%

rice oil according to the method for improving the eating quality of cooked rice by adding rice oil [63].

Table 5. Textural properties of three kinds of cooked rice for human test.

	Hardness $\times 10^5$ [N/cm^2]	Toughness $\times 10^5$ [N/cm^2]	Adhesion $\times 10^5$ [N/cm^2]	Stickiness $\times 10^5$ [N/cm^2]
A	0.002 ± 0.00 a	0.025 ± 0.00 b	0.007 ± 0.00 a	0.010 ± 0.00 b
B	0.002 ± 0.00 a	0.024 ± 0.01 b	0.008 ± 0.00 a	0.012 ± 0.01 b
C	0.002 ± 0.00 a	0.018 ± 0.00 a	0.007 ± 0.00 a	0.006 ± 0.00 a

A: Cooked WFBSK rice with 2.5% WBB and 0.3% rice oil added by HPT. B: Cooked BSK rice with 2.5% WBB and 0.3% rice oil added by HPT. C: Cooked polished Koshihikari rice. Waxy black rice bran, WBB; high-pressure treatment, HPT. Blend with wax-free black/brown rice, super-hard brown rice and Koshihikari brown rice (4:4:2), WFBSK. Blend with black/brown rice, super-hard brown rice and Koshihikari brown rice (4:4:2), BSK. Within each measure (hardness, toughness, etc.) in the same column, different letters (a, b, c, etc.) denote statistically significant differences.

Table 6. Result of sensory test of three kinds of the blended meals.

	Appearance	Aroma	Hardness	Taste	Stickiness	Overall Evaluation
A	3.13 ± 0.85 a	3.13 ± 0.25 a	2.38 ± 0.48 a	3.85 ± 0.02 a	3.35 ± 0.02 a	3.85 ± 0.01 a
B	3.25 ± 0.29 a	3.13 ± 0.25 a	2.88 ± 0.48 a	3.30 ± 0.01 b	3.25 ± 0.01 a	3.31 ± 0.02 b
C	3.00 ± 0.00 a	3.00 ± 0.00 a	3.00 ± 0.00 a	3.00 ± 0.00 b	3.00 ± 0.00 a	3.00 ± 0.00 b

A: Cooked WFBSK rice with 2.5% WBB and 0.3% rice oil added by HPT. B: Cooked BSK rice with 2.5% WBB and 0.3% rice oil added by HPT. C: Cooked WFBSK rice with 2.5% WBB added by HPT. Waxy black rice bran, WBB; high-pressure treatment, HPT. Blend with wax-free black/brown rice, super-hard brown rice and Koshihikari brown rice (4:4:2), WFBSK. Blend with black/brown rice, super-hard brown rice and Koshihikari brown rice (4:4:2), BSK. Within each measure (appearance, aroma, etc.) in the same column, different letters (a, b, c, etc.) denote statistically significant differences. a, b < 0.05, $n = 10$.

3.7. Human Intervention Test for Cognitive Function

We examined the effects of WFBSK meal on the cognitive function of healthy Japanese adults. Twenty-four subjects were randomly assigned to the WFBSK meal or control meal based on the MMSE score. There was no significant difference in subject age, BMI, MMSE score, fasting blood glucose or HbA1c between the WFBSK and control meal groups (Table 7). One pack of WFBSK or the control of white rice was taken every day for a 12-week intervention period (Table S1). No adverse effects were observed for 12 weeks and all 24 subjects completed the intervention test in good health (Figure S2). Cognitrax was used as a test battery to assess the cognitive function of the participants. Evaluation was performed before and after 12 weeks of intervention. At the baseline, none of the cognitive domains showed significantly different scores between the WFBSK and control meal. As shown in Table 8 and Figure 2, Cognitrax demonstrated that subjects taking WFSFK meal showed significant improvement in the cognitive domain of language memory after 12 weeks ($p < 0.05$).

As biofunctional components to improve the cognitive ability, astaxanthin and sesamin [64], green tea catechins [65], rosemary extracts [66], chlorogenic acids [4], L-theanine [67], astaxanthin and tocotrienol [68] have been reported. It seems that the main component in our sample meal to improve cognitive ability would be anthocyanin because many papers reported the important role of its anti-oxidative capacity, the same as with propolis [69,70], astaxanthin [67,68], green tea catechins [65] and chlorogenic acids [4]. In the present paper, we showed that the WFSFK meal contained a high amount of anthocyanin, a kind of polyphenol.

Although other components, such as gamma-oryzanol, ferulic acid, and tocopherol, in the brown rice and rice oil, could be the candidates for the improvement of cognitive ability, we consider that anthocyanin would be most important because of the stronger ORAC values in the black/brown rice, as shown in Table 4. We will continue our investigation to clarify this it in the next report.

It seems that our results are one of the first examples of effects shown for a foodstuff itself like cooked rice, not for a specified functional component or its extracts.

Table 7. Comparison between test and control meals for human intervention test items.

	Test Meal (n = 12)	SD	Control Meal (n = 12)	SD	Paired-t p-Value
Age	60.1	6.3	58.0	4.8	0.4345
BMI (kg/m^2)	21.9	2.4	21.0	2.3	0.6763
FBG (mg/mL)	94.9	5.2	95.2	9.3	0.9306
HbA1c	5.6	0.2	5.6	0.3	0.7525
MMSE	29.0	1.0	29.3	0.8	0.4649

BMI, body mass index; FBG, fasting blood glucose; HbA1c, hemoglobin A1c; MMSE, mini-mental state examination.

Table 8. Results of Cognitrax.

	Test Meal	Control Meal
Comprehensive memory	7.25 ± 13.87 a	−0.83 ± 12.19 a
Language memory	16.00 ± 18.35 a	1.00 ± 15.08 b
Visual memory	−3.67 ± 12.28 a	−2.75 ± 14.05 a
Cognitive function speed	4.08 ± 7.18 a	0.17 ± 7.33 a
Reaction time	1.83 ± 7.90 a	7.33 ± 8.90 a
Comprehensive attention	6.42 ± 11.98 a	1.17 ± 17.83 a
Cognitive flexibility	7.17 ± 13.07 a	4.50 ± 17.48 a
Processing speed	5.50 ± 12.90 a	5.33 ± 5.77 a
Execution mechanism	7.33 ± 12.76 a	4.75 ± 15.71 a
Simple attention	2.75 ± 13.80 a	−8.50 ± 41.58 a
Movement speed	1.25 ± 6.54 a	−2.92 ± 7.91 a
Finger tapping (left hand)	0.50 ± 4.06 a	−6.00 ± 9.72 b
Symbol digit coding	−7.50 ± 18.30 a	2.83 ± 8.39 a
Stroop test	−6.70 ± 27.2 a	11.42 ± 18.58 a

Within each measure (comprehensive memory, language memory, etc.) in the same column, different letters (a, b, etc.) denote statistically significant differences. a, b < 0.05.

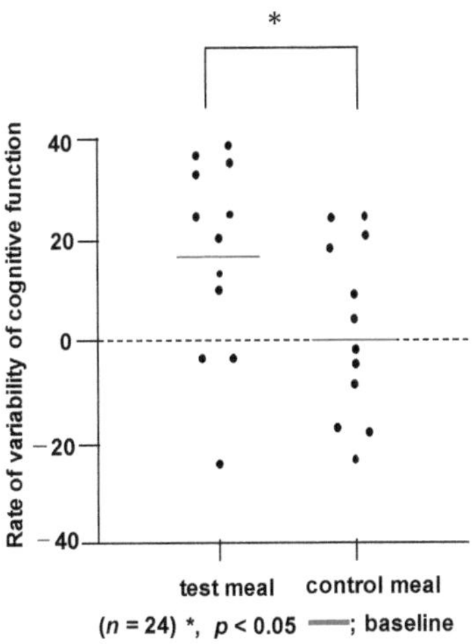

Figure 2. Language memory test.

3.8. Changes in Plasma Aβ42/40 Ratio

Blood samples were taken from the 24 subjects after meals before the diets began (baseline) and after 12 weeks. Plasma Aβ40 and Aβ42 were measured, and Aβ42/40 ratio was evaluated as a secondary endpoint. A change in Aβ42/40 ratio from the baseline to the end of the test (after 12 weeks) did not show a significant difference between the subjects who consumed the WFBSK and those who consumed the control meal (Figure 3).

Aβ spices are important components of plaque in the AD brain. The Aβ42/40 ratios in plasma may decline early in the course of AD [71]. High plasma concentrations of Aβ40, especially when combined with low concentrations of Aβ42, indicate an increased risk of dementia [71,72]. The reason why the Aβ 42/40 ratios did not change with the WFBSK meal may be explained by the relatively short period of intervention in this study.

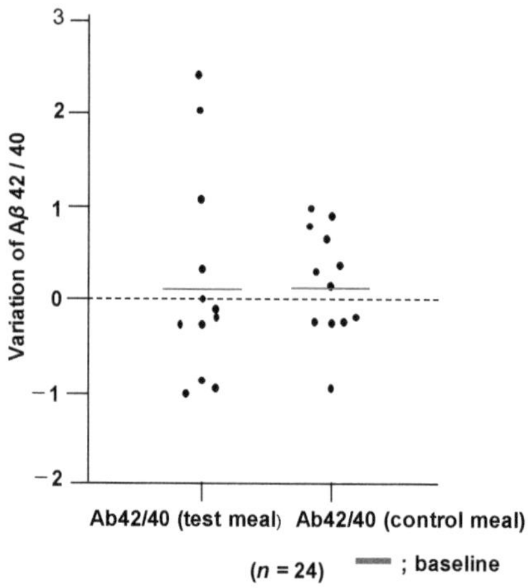

Figure 3. Changes of variation of amyloid β-42/4 peptide ratio by ELISA after 12 weeks.

3.9. Results of Postprandial Blood Glucose and Insulin Secretion

Panlasigui and Thompson [73] showed that unpolished rice is healthier and more beneficial than polished rice for diabetics and hyperglycemic individuals, and that the glycemic area and GI were 35.4% lower in unpolished rice than in polished rice.

Jung et al. [74] showed that ferulic acid might be beneficial in the treatment of type 2 diabetes because it regulates blood glucose levels (BGLs) by elevating glucokinase activity and producing glycogen in the liver.

The BGL of test subjects at 90 min and 120 min after ingesting the WFBSK meal showed no significantly lower tendency after 12 weeks on the diets than that in subjects who consumed the control meal, as shown in Figure 4A.

Tokutake et al. [10] reported that the disturbed insulin signaling cascade may be implicated in the pathways through which soluble Aβ induces Tau phosphorylation, and lent further support to the notion that correcting the insulin signal dysregulation in AD may offer a potential therapeutic approach.

After 12 weeks on treatment meals, the insulin levels of test subjects that ingested the WFBSK meal were significantly lower at 120 min after the meal than the insulin levels in subjects consuming the control meal ($p < 0.05$) (Figure 4B).

Test meal: wax-free black/brown rice, "Okunomurasaki", super-hard brown rice "Niigata 129 gou" and ordinary brown rice "Koshihikari" blended in a ratio of 4:4:2 with 2.5% waxy black rice bran (WBB) "Shihou" and 0.3% rice oil added, cooked after high-pressure-treatment (HPT) by Echigo Seika, Co., Ltd. Control meal: "Koshihikari" cooked polished rice was prepared by Echigo Seika, Co., Ltd.

Although we measured HDL cholesterol, LDL cholesterol, insulin sensitivity and HbA1c, they did not show significant differences between subjects who consumed the test meal or the control meal.

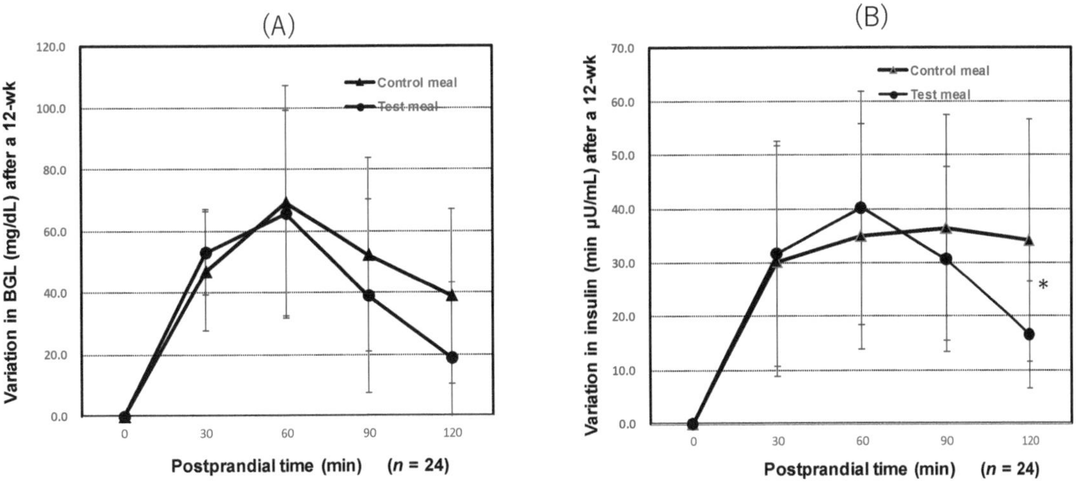

Figure 4. Variation in the BGL and insulin in 24 subjects after eating blending cooked rice of test meal (WFBSK) and polished rice of control after 12 weeks. (**A**) variation in postprandial BGL. (**B**) variation in postprandial insulin secretion. * means significant difference ($p < 0.05$).

4. Conclusions

In this study, as the texture of cooked brown rice is too hard and non-sticky to be table rice, we adopted novel processing technology to create "wax-free brown rice" (WFBR) and compared it with ordinary brown rice in terms of textural properties and functional ingredients. A human intervention test, in a randomized, parallel-group comparison study, was conducted by using the abovementioned meals as samples for 12 weeks, to be consumed by 24 subjects.

(1) Wax-free unpolished rice showed higher water absorption, which led to the improvement of eating quality and hydrophilic ORAC of black unpolished rice.
(2) In addition to the wax-free processing, HPT (high-pressure treatment) and the addition of rice oil were effective at improving the eating quality of our test meal, which led to the completion of the human intervention test without dropout in terms of palatability.
(3) Wax-free processing, HPT and addition of WBB (waxy black rice bran) strengthened the biofunctionality of our test meal, WFBSK.
(4) After 12 weeks of human intervention testing, the postprandial blood insulin levels of the test subjects were significantly lower at 120 min than those of subjects who consumed the control diet ($p < 0.05$), which would prevent hyperinsulinemia, one of the causes of the shortage of insulin in the brain, and contribute to improvement in the cognitive domain of language memory.
(5) The results of Cognitrax demonstrated that consuming WFSFK meal was significantly effective at improving the cognitive domain of "language memory" after 12 weeks ($p < 0.05$).

In conclusion, our test meal showed the possibility of controlling dementia and type 2 diabetes mellitus in the 12-week human intervention test. This result would lead to the prevention of dementia through a diet based on boiled rice. As rice is a staple food for many people in the world, this would support not only satiety but also a healthy life without dementia.

Supplementary Materials: The following supporting information can be downloaded at: https://www.mdpi.com/article/10.3390/foods11060818/s1, Table S1: The principal ingredients of 2 kinds of cooked rice for Human test; Table S2: Difference in appearance among of cooked rice of ordinary polished rice, unpolished rice and wax-free brown rice; Figure S1: n6/n3 ratio of PUFAs of 6 kinds of japonica rice bran; Figure S2: Flow diagram of the progress through the phases of a parallel randomized trial of two groups.

Author Contributions: S.N. and K.O. designed this research, S.N., K.K., A.A., K.W., M.H. and M.I. conducted the experiments, S.N., T.I. and K.O. wrote the manuscript. All authors have read and agreed to the published version of the manuscript.

Funding: Part of this research was supported by a grant from the Project of the NARO Bio-Oriented Technology Research Advancement Institution (research program on development of innovative technology, 01026C).

Institutional Review Board Statement: The human intervention study was conducted in accordance with the Declaration of Helsinki, and approved by the Institutional Ethics Committee of Niigata Bioresearch Park Inc. (IRB2021-BPD-001, 8/19, 2021).

Informed Consent Statement: Informed consent was obtained from all subjects involved in the human intervention study.

Data Availability Statement: The datasets generated for this study are available on request to the corresponding author.

Acknowledgments: We are deeply grateful to Masatoyo Nishizawa of Niigata University of Health and Welfare, for his enlightening discussion. We express our gratitude to the Niigata Bio-Research Park Inc. for its analysis of the human intervention test. We express our thanks to Toyo Rice Co., Ltd. and Echigo Seika Co., Ltd. for production of the sample meals for the human test, and to the Niigata Prefectural Agriculture Research Institute and Ogata National Countries for their kind gift of rice samples of the super-hard rice cultivar Niigata 129 go, the high-quality rice Koshihikari, waxy black rice Shiho and black rice Okunomurasaki. We express our gratitude to Tsuno Group Co., Ltd. for the gift of rice oil.

Conflicts of Interest: The authors declare no conflict of interest.

Abbreviations

Aβ	amyloid β protein
Ratio of (n6/n3)	ratio of omega-6/omega-3
β-secretase	BACE1
APP	amyloid precursor protein
MMSE	mini-mental state examination
WFBR	wax-free brown rice
WFBBR	wax-free brown rice of black rice
WBB	waxy black rice bran
SHBR	super-hard brown rice
KBR	Koshihikari brown rice
BBR	black/brown rice
WFBSK	rice blend of WFBBR, SHBR and KBR added with WBB and rice oil after HPT
BSK	rice blend of BBR, SHBR, and KBR added with WBB and rice oil after HPT
HPT	high-pressure treatment
RS	resistant starch
WAR	water absorption rate

References

1. The IDF (International Diabetes Federation) Diabetes Atlas 10th. Available online: https://diabetesatlas.org/date/en/ (accessed on 7 February 2022).
2. ADI (Alzheimer's Disease International) Report. 2015. Available online: https://www.alzint.org/resource/world-alzheimer-report-2015/ (accessed on 5 February 2022).
3. Nantel, G. Glycemic carbohydrate: An international perspective. *Nutr. Rev.* **2003**, *61*, S34–S39. [CrossRef]

4. Saitou, K.; Ochiai, R.; Kozuma, K.; Sato, H.; Koikeda, T.; Osaki, N.; Katsuragi, Y. Effect of chlorogenic acids on cognitive function: A randomized, double-blind, placebo-controlled trial. *Nutrients* **2018**, *10*, 1337. [CrossRef] [PubMed]
5. Sarach, O.; Akca, A.S.D.; Atasoy, N.; Onder, O.; Senormanci, O.; Kaygisiz, I.; Atik, L. The relationship between quality of life and cognitice functions, anxiety and depression among hospitalized elderly patients. *Clin. Psychopharmacol. Neurosci.* **2015**, *13*, 194–200. [CrossRef] [PubMed]
6. Kawamura, K.; Umemura, T.; Hotta, N. Cognitive impairment in diabetic patients: Can diabetic control prevent cognitive decline? *J. Diabetes Investig.* **2012**, *18*, 413–423. [CrossRef] [PubMed]
7. Ohara, T.; Doi, Y.; Ninomiya, T.; Hirakawa, Y.; Hata, J.; Iwaki, T.; Kanba, S.; Kiyohara, Y. Glucose tolerance status and risk of dementia in the community. The Hisayama Study. *Neurology* **2011**, *77*, 1126–1134. [CrossRef] [PubMed]
8. Biessels, G.J.; Staekenborg, S.; Brunner, E.; Brayne, C.; Scheltens, P. Risk of dementia in diabetes mellitus: A systematic review. *Lancet Neurol.* **2006**, *5*, 64–74. [CrossRef]
9. Kasuga, K.; Kaneko, H.; Nishizawa, M.; Onodera, O.; Ikeuchi, T. Generation of intracellular domain of insulin receptor tyrosine kinase by gamma-secretase. *Biochem. Biophys. Res. Commun.* **2007**, *360*, 90–96. [CrossRef] [PubMed]
10. Tokutake, T.; Kasuga, K.; Yajima, R.; Sekine, Y.; Tezuka, T.; Nishizawa, M.; Ikeuchi, T. Hyperphosphorylation of Tau induced by naturally secreted amyloid-β at nanomolar concentrations is modulated by insulin-dependent AKt-GSK3β signaling pathway. *J. Biol. Chem.* **2012**, *287*, 35222–35233. [CrossRef]
11. Gudala, K.; Bansal, D.; Schifano, F.; Bhansali, A. Diabetes mellitus and risk of dementia: A meta-analysis of prospective observational studies. *J. Diabetes Investig.* **2013**, *4*, 640–650. [CrossRef]
12. Moran, C.; Bears, R.; Phan, T.G.; Bruce, G.G.; Callisaya, M.L.; Srikauth, V. Type 2 diabetes mellitus and biomarkers of neurodegeneration. *Neurology* **2015**, *85*, 1123–1130. [CrossRef]
13. Willette, A.A.; Bendlin, B.B.; Starks, E.J.; Birdsill, A.C.; Johnson, S.C.; Christian, B.T.; Okonkwo, O.; Rue, A.L.; Hermann, B.P.; Koscik, R.L.; et al. Association of insulin resistance with cerebral glucose uptake in late middle-aged adults at risk for Alzheimer's disease. *Arch. Neurol.* **2015**, *72*, 1013–1020.
14. Macauley, S.; Stanley, M.; Caesar, E.; Yamada, S.A.; Raichle, M.E.; Perez, R.; Mahan, T.E.; Sutphen, C.L.; Holtzman, D.M. Hyperglycemia modulates extracellular amyloid-β concentrations and neuronal activity in vivo. *J. Clin. Investig.* **2015**, *125*, 2463–2467. [CrossRef] [PubMed]
15. Jenkins, D.J.; Wolever, T.M.; Taylor, R.H.; Barker, H.; Fielden, H.; Baldwin, J.M.; Bowling, A.C.; Newman, H.C.; Jenkins, A.L.; Goff, D.V. Glycemic index of foods: A physiological basis for carbohydrate exchange. *Am. J. Clin. Nutr.* **1981**, *34*, 362–366. [CrossRef] [PubMed]
16. Venn, B.J.; Green, T.J. Glycemic index and glycemic load: Measurement issues and their effect on diet-disease relationships. *Eur. J. Clin. Nutr.* **2007**, *61* (Suppl. 1), S122–S131. [CrossRef] [PubMed]
17. Depp, C.A.; Harmell, A.; Vahia, I.V. Successful cognitive aging. *Curr. Top. Behav. Neurosci.* **2012**, *10*, 35–50. [CrossRef]
18. Dubois, B.; Hampel, H.; Feldman, H.H.; Scheltens, P.; Aisen, P.; Andrieu, S.; Bakardjian, H.; Benali, H.; Bertram, L.; Blennow, K.; et al. Preclinical alzheimer's disease: Definition, natural history, and diagnostic criteria. *Alzheimers Dement.* **2016**, *12*, 292–323. [CrossRef]
19. Vauzour, D.; Camprubi-Robles, M.; Miquel-Kergoat, S.; Andres-Lacueva, C.; Banati, D.; Barberger-Gateau, P.; Bowman, G.L.; Caberlotto, L.; Marie Minihane, A.; Siobban Mitchell, E.; et al. Nutrition for the ageing brain: Towards evidence for an optimal diet. *Ageing Res. Rev.* **2017**, *35*, 222–240. [CrossRef]
20. Lourida, I.; Soni, M.; Thompson-Coon, J.; Purandare, N.; Lang, I.A.; Ukoumunne, O.C.; Llewellyn, D.J. Mediterranean diet, cognitive function, and dementia. *Epidemiology* **2013**, *24*, 479–489. [CrossRef]
21. Sofi, F.; Macchi, C.; Abbate, R.; Gensini, G.F.; Casini, A. Effectiveness of the Mediterranean diet: Can it help delay or prevent Alzheimer's disease? *J. Alzheimers Dis.* **2010**, *20*, 795–801. [CrossRef]
22. Cole, G.M.; Ma, Q.L.; Frautschy, S.A. Omega-3 fatty acids and dementia. *Prostaglandins Leukot. Essent. Fat. Acids* **2009**, *81*, 213–221. [CrossRef]
23. Childs, N.W. Production and Utilization of Rice. In *Rice-Chemistry and Technology*, 3rd ed.; Champagne, E.T., Ed.; AACC Inc.: St. Paul, MN, USA, 2004; pp. 1–24.
24. Taniguchi, H.; Hashimoto, H.; Hosoda, A.; Kometani, T.; Tsuno, T.; Adachi, S. Functionality of compounds contained in rice bran and their improvement. *Nippon Shokuhin Kagaku Kogaku Kaishi* **2012**, *59*, 301–318. (In Japanese) [CrossRef]
25. Juliano, B.O. Chapter 9: Bran. In *Rice Chemistry and Quality*; Juliano, B.O., Ed.; The Philippine Rice Research Institute (PhilRice): Munos, Philippines, 2003; pp. 357–367.
26. Nishi, A.; Nakamura, Y.; Tanaka, N.; Satoh, H. Biochemical and genetic analysis of the effects of amylase-extender mutation in rice endosperm. *Plant Physiol.* **2001**, *127*, 459–472. [CrossRef] [PubMed]
27. Englyst, H.N.; Cummings, J.H. Digestion of the polysaccharides of some cereal foods in the human small intestine. *Am. J. Clin. Nutr.* **1985**, *42*, 778–787. [CrossRef] [PubMed]
28. Nakamura, S.; Nakano, Y.; Satoh, H.; Ohtsubo, K. Improved palatability and bio-functionality of super-hard rice by soaking in a barley-*koji miso* suspension. *Biosci. Biotechnol. Biochem.* **2013**, *77*, 2419–2429. [CrossRef] [PubMed]
29. Nakamura, S.; Ohtsubo, K. Improvement of palatability and inhibition of abrupt increase in postprandial blood glucose level by the boiled rice after soaking with functional food ingredients. *J. Appl. Glycosci.* **2015**, *62*, 53–63. [CrossRef]

30. Yang, C.Z.; Shu, X.L.; Zhang, L.L.; Wang, X.Y.; Zhao, H.J.; Ma, C.X.; Wu, D.X. Starch properties of mutant rice high in resistant starch. *J. Agric. Food Chem.* **2006**, *54*, 523–528. [CrossRef]
31. Goddard, M.; Young, G.; Marcus, R. The effect of amylose content on insulin and glucose responses to ingested rice. *Am. J. Clin. Nutr.* **1984**, *39*, 388–392. [CrossRef] [PubMed]
32. Kang, H.J.; Hwang, I.K.H.; Kim, K.S.; Choi, H.C. Comparative structure and physicochemical properties of Ilpumbyeo, a high-quality *Japonica* rice, and its mutant, Suwon 464. *J. Agric. Food Chem.* **2003**, *51*, 6598–6603. [CrossRef]
33. Sutharut, J.; Sudarat, J. Total anthocyanin content and antioxidant activity of germinated colored rice. *Int. Food Res. J.* **2012**, *19*, 215–221.
34. Muntana, N.; Prasong, S. Study on total phenolic contents and their antioxidant activities of Thai white, red and black rice bran extracts. *Pak. J. Biol. Sci.* **2010**, *13*, 170–174. [CrossRef]
35. Ichikawa, H.; Ichiyanagi, T.; Xu, B.; Yoshii, Y.; Nakajima, M.; Konishi, T. Antioxidant activity of anthocyanin extract from purple black rice. *J. Med. Food* **2001**, *4*, 211–218. [CrossRef]
36. Watanabe, S.; Hirakawa, A.; Takei, N.; Beppu, S.; Hashimoto, H.; Saika, K. Medical Rice: A new wax-free brown rice and its protein reduced rice. *Adv. Food Technol. Nutr. Sci.* **2018**, *4*, 10–16. [CrossRef]
37. Hayashi, R.; Hayashida, A. Increased amylase digestibility of pressure-treated starch. *Agric. Biol. Chem.* **1989**, *53*, 2543–2544. [CrossRef]
38. Hayashi, R. Application of high pressure to food processing and preservation: Philosophy and development. In *Engineering and Food*, 2nd ed.; Spiess, W.E.L., Schubert, H., Eds.; Applied Science: London, UK, 1989; pp. 815–826.
39. Yamakura, M.; Haraguchi, K.; Okadome, H.; Suzuki, K.; Tran, U.T.; Horigane, A.K.; Yoshida, M.; Homma, S.; Sasagawa, S.; Yamazaki, A.; et al. Effects of soaking and high-pressure treatment on the qualities of cooked rice. *J. Appl. Glycosci.* **2005**, *52*, 85–93. [CrossRef]
40. Okadome, H.; Toyoshima, H.; Ohtsubo, K. Multiple measurements of physical properties of individual cooked rice grains with a single apparatus. *Cereal Chem.* **1999**, *76*, 855–860. [CrossRef]
41. Folin, O.; Denis, W. A colorimetric method for the determination of phenols (and phenol derivatives) in purine. *J. Biol. Chem.* **1915**, *22*, 305–308. [CrossRef]
42. Prior, R.L.; Hoang, H.; Gu, L.; Wu, X.; Bacchiocca, M.; Howard, L.; Hampsch-Woodill, M.; Huang, D.; Ou, B.; Jacob, R. Assay for hydrophilic and lipophilic antioxidant capacity (oxygen radical absorbance capacity (ORACFL) of plasma and other biological and food samples. *J. Agric. Food Chem.* **2003**, *51*, 3273–3279. [CrossRef] [PubMed]
43. Watanabe, J.; Oki, T.; Takebayashi, J.; Yamasaki, K.; Takano-Ishikawa, Y.; Hino, A.; Yasui, A. Method validation by interlaboratory studies of improved hydrophilic oxygen radical absorbance capacity methods for the determination of antioxidant capacities of antioxidant solutions and food extracts. *Anal. Sci.* **2012**, *28*, 159–165. [CrossRef] [PubMed]
44. Watanabe, J.; Oki, T.; Takebayashi, J.; Yamasaki, K.; Takano-Ishikawa, Y.; Hino, A.; Yasui, A. Improvement of the lipophilic-oxygen radical absorbance capacity (L-ORAC) method and single-laboratory validation. *Biosci. Biotechnol. Biochem.* **2013**, *77*, 857–859. [CrossRef] [PubMed]
45. Nakamura, S.; Hara, T.; Joh, T.; Kobayashi, A.; Yamazaki, A.; Kasuga, K.; Ikeuchi, T.; Ohtsubo, K. Effects of super-hard rice bread blended with black rice bran on amyloid β peptide production and abrupt increase in postprandial blood glucose levels in mice. *Biosci. Biotechnol. Biochem.* **2017**, *81*, 323–334. [CrossRef]
46. Sawai, Y.; Sugawara, T.; Oki, T.; Nishiba, Y.; Ujihara, K.; Suda, I. Comparison of high-performance liquid chroma tography and pH differential method: Correlation for quantitation of anthocyanins in black pearl rice and soybean. *Nippon Shokuhin Kagaku Kogaku Kaishi* **2021**, *59*, 104–108. [CrossRef]
47. Nakamura, S.; Suzuki, D.; Kitazume, R.; Ohtsubo, K. Quality evaluation of rice crackers based on physicochemical measurements. *Biosci. Biotechnol. Biochem.* **2012**, *76*, 794–804. [CrossRef]
48. Gualtieri, C.T.; Johnson, L.G. Reliability and validity of a computerized neurocognitive test battery, CNS Vital Signs. *Arch. Clin. Neuropsychol.* **2006**, *21*, 623–643. [CrossRef]
49. Kuwada, H.; Teramoto, A.; Jibu, Y.; Tabuchi, M.; Fuchigami, M. Texture and fine structure of cooked brown rice. *J. Cook. Sci. Jpn.* **2011**, *44*, 137–144.
50. Nakamura, S.; Katsura, J.; Maruyama, Y.; Ohtsubo, K. Relationship between fatty acid composition and starch properties of 30 japonica rice cultivars. *Cereal Chem.* **2019**, *96*, 228–242. [CrossRef]
51. Simopoulos, A.P. The importance of the ratio of omega-6/omega-3 essential fatty acids. *Biomed. Pharm.* **2002**, *56*, 365–379. [CrossRef]
52. Taira, H. Lipid content and fatty acid opposition of rice. *JARQ* **1983**, *16*, 273–280.
53. Ohnishi, H.; Saito, Y. Eicosapentaenoic acid (EPA) reduces cardiovascular events: Relationship with the EPA/Arachidonic acid ratio. *J. Atheroscler. Thromb.* **2013**, *20*, 861–877. [CrossRef] [PubMed]
54. Ikemoto, A.; Naganuma, S. Varietal differences in lipid content and fatty acid composition of milled rice and their influence on dietary balance of lipid nutrition (in Japanese). *Mem. Fac. Integr. Arts Akita Univ.* **2007**, *62*, 19–28.
55. Aoe, S. Physiological function of β-glucan in barley. *J. Integr. Study Diet. Habits* **2015**, *1*, 3–6. [CrossRef]
56. Theuwissen, E.; Mensink, R.P. Simultaneous intake of beta-glucan and plant stanol esters affects lipid metabolism in slightly hypercholesterolemic subjects. *J. Nutr.* **2007**, *137*, 583–588. [CrossRef]

57. Yokoyama, W.; Hudson, C.A.; Knuckles, B.E.; Chiu, M.M.; Sayre, R.N.; Turnlund, J.R.; Schneeman, B.O. Effect of barley β-glucan in durum wheat pasta on human glycemic response. *Cereal Chem.* **1997**, *74*, 293–296. [CrossRef]
58. Ling, W.H.; Cheng, Q.X.; Ma, J.; Wang, T. Red and black rice decrease atherosclerotic plaque formation and increase antioxidant status in rabbits. *J. Nutr.* **2001**, *131*, 1421–1426. [CrossRef] [PubMed]
59. Behall, K.M.; Scholfield, D.J.; Hallfrisch, J.G.; Liljeberg-Elmstahl, H.G. Consumption of both resistant starch and beta-glucan improves postprandial plasma glucose and insulin in women. *Diabetes Care* **2006**, *29*, 976–981. [CrossRef] [PubMed]
60. Robertson, M.D.; Bickerton, A.S.; Dennis, A.L.; Vidal, H.; Frayn, K.N. Insulin-sensitizing effects of dietary re sistant starch and effects on skeletal muscle and adipose tissue metabolism. *Am. J. Clin. Nutr.* **2005**, *82*, 559–567. [CrossRef] [PubMed]
61. Noro, W.; Itayanagi, S.; Hosono, R.; Matsui, T.; Takahashi, M.; Nakamura, S.; Ishizaki, K.; Watanabe, S.; Ohtsubo, K. Characterization of the new amylopectin long-chain rice cultivar Chou 2418 and its boiled rice grains. *Food Sci. Technol. Res.* **2019**, *25*, 227–235. [CrossRef]
62. Maeda, S.; Kazama, Y.; Kobayashi, A.; Yamazaki, A.; Nakamura, S.; Yamaguchi, M.; Maeda, H.; Ohtsubo, K. Improvement of palatability and prevention of abrupt increases in postprandial blood glucose levels by Hokurikukona243 after high pressure treatment. *J. Appl. Glycosci.* **2015**, *62*, 127–134. [CrossRef]
63. Fujimura, M. Oils and Fats Constituent for the Cooking Rice. Japanese Patent Application 2008-118987, 2008. (In Japanese).
64. Ito, N.; Saito, H.; Seki, S.; Ueda, F.; Asada, T. Effects of composite supplement containing astaxanthin and sesamin on cognitive functions in people with mild cognitive impairment: A randomized, double-blind, placebo-controlled trial. *J. Alzheimers Dis.* **2018**, *62*, 1767–1775. [CrossRef]
65. Baba, Y.; Inagaki, S.; Nakagawa, S.; Kaneko, T.; Kobayashi, M.; Takihara, T. Effect of daily intake of green tea catechins on cogni tive function in middle-aged and older subjects: A randomized, placebo-controlled study. *Molecules* **2020**, *25*, 4265. [CrossRef]
66. Araki, R.; Sasaki, K.; Onda, H.; Nakamura, S.; Kassai, M.; Kaneko, T.; Isoda, H.; Hashimoto, K. Effects of continu ous intake of rosemary extracts on mental health in working generation healthy Japanese men: Post-hoc testing of a randomized controlled trial. *Nutrients* **2020**, *12*, 3551. [CrossRef]
67. Baba, Y.; Inagaki, S.; Nakagawa, S.; Kaneko, T.; Kobayashi, M.; Takihara, T. Effects of L-Theanine on cognitive func tion in middle-aged and older subjects: A randomized placebo-controlled study. *J. Med. Food* **2021**, *24*, 333–341. [CrossRef]
68. Sekikawa, T.; Kizawa, Y.; Li, Y.; Takara, T. Cognitive function improvement with astaxanthin and tocotrienol intake: A random ized, double-blind, placebo-controlled study. *Clin. Biochem. Nutr.* **2020**, *67*, 307–316. [CrossRef] [PubMed]
69. Asama, T.; Hiraoka, T.; Ohkuma, A.; Okumura, N.; Yamaki, A.; Urakami, K. Cognitive improvement and safety assessment of a dietary supplement containing propolis extract in elderly Japanese: A placebo-controlled, randomized, parallel-group, double-blind human clinical study. *Evid.-Based Complementary Altern. Med.* **2021**, *2021*, 6664217. [CrossRef]
70. Nakajima, Y.; Shimazawa, M.; Mishima, S.; Hara, H. Water extract of propolis and its main constituents, caffeoylquinic acid derivatives, exert neuroprotective effects via antioxidant actions. *Life Sci.* **2007**, *80*, 370–377. [CrossRef] [PubMed]
71. Graff-Radford, N.R.; Crook, J.E.; Lucas, J.; Boeve, B.F.; Knopman, D.S.; Ivnik, R.J.; Smith, G.E.; Younkin, L.H.; Petersen, R.C.; Younkin, S.G. Association of low plasma abeta42/abeta40 ratios with increased imminent risk for mild cognitive impairment and Alzheimer disease. *Arch. Neurol.* **2007**, *64*, 354–362. [CrossRef] [PubMed]
72. Van, O.; Hofman, A.; Soares, H.D.; Korudstaal, P.J.; Breteler, M.M. Plasma abeta (1-40) and abeta (1-42) and the risk of dementia: A prospective case-cohort study. *Lancet Neurol.* **2006**, *5*, 655–660. [CrossRef]
73. Panlasigui, L.; Thompson, L.U. Blood glucose lowering effects of brown rice in normal and diabetic subjects. *Int. J. Food Sci. Nutr.* **2006**, *57*, 151–158. [CrossRef] [PubMed]
74. Jung, E.H.; Kim, S.R.; Hwang, I.K.; Ha, T.Y. Hypoglycemic effects of a phenolic acid fraction of rice bran and ferulic acid in C57 BL/KsJ-db/db mice. *J. Agric. Food Chem.* **2007**, *55*, 9800–9804. [CrossRef] [PubMed]

foods

MDPI

Article

The Effect of Rice vs. Wheat Ingestion on Postprandial Gastroesophageal Reflux (GER) Symptoms in Patients with Overlapping GERD-Irritable Bowel Syndrome (IBS)

Tanisa Patcharatrakul [1,2], Sittikorn Linlawan [3], Suppawatsa Plaidum [1] and Sutep Gonlachanvit [1,2,*]

1 Division of Gastroenterology, Department of Medicine, King Chulalongkorn Memorial Hospital, The Thai Red Cross Society, Bangkok 10330, Thailand; dr_tanisa@yahoo.com (T.P.); vavai_vv@hotmail.com (S.P.)
2 Center of Excellence on Neurogastroenterology and Motility, Faculty of Medicine, Chulalongkorn University, Bangkok 10330, Thailand
3 Department of Medicine, Phrachomklao Hospital, Phetchaburi 76000, Thailand; sittilin@yahoo.com
* Correspondence: sutep.g@chula.ac.th; Tel.: +66-2256-4265

Abstract: A randomized crossover study in twenty-one patients (18F, age 50 ± 13 years) with overlapping GERD-IBS was conducted to evaluate the effects of rice noodles (low FODMAPs) vs. wheat noodles (high FODMAPs) on typical GER symptoms, and the correlation between GERD symptoms and intestinal gas production. Results: Heartburn and regurgitation scores were highest in most patients (19/21) during the first 15 min after meals. At 15 min after lunch, wheat was significantly associated with more regurgitation and heartburn than rice. Also, at 15 min after breakfast, wheat aggravated more regurgitation than rice. Wheat ingestion was significantly associated with higher H_2 and CH_4 levels after lunch compared to rice, whereas gas levels before lunch were similar ($p > 0.05$). The area under the curve of H_2 and CH_4 concentration 15 min after a lunch of wheat moderately correlated with the regurgitation severity at 15-min ($r = 0.56$, $p < 0.05$). Conclusion: Wheat induced more GERD symptoms than rice in patients with overlapping GERD-IBS. This effect, immediately developed after lunch, was associated with more intestinal gas production. Thus, a low FODMAPs diet may relieve postprandial GERD symptoms in GERD patients with overlapping IBS. Wheat inducing more regurgitation than rice after breakfast suggests other mechanism(s) besides gut fermentation.

Keywords: rice; wheat; fermentable oligosaccharides, disaccharides, monosaccharides, and polyols (FODMAPs); irritable bowel syndrome; GERD; intestinal gas; fermentation

Citation: Patcharatrakul, T.; Linlawan, S.; Plaidum, S.; Gonlachanvit, S. The Effect of Rice vs. Wheat Ingestion on Postprandial Gastroesophageal Reflux (GER) Symptoms in Patients with Overlapping GERD-Irritable Bowel Syndrome (IBS). *Foods* **2022**, *11*, 26. https://doi.org/10.3390/foods11010026

Academic Editor: Hanne Christine Bertram

Received: 27 November 2021
Accepted: 15 December 2021
Published: 23 December 2021

1. Introduction

Food can exacerbate gastrointestinal (GI) symptoms in patients with functional gastrointestinal disorders as a consequence of several mechanisms, such as changes in intestinal motility, visceral sensations, microbiome, intestinal permeability, immune activation, and brain–gut interactions [1,2]. Carbohydrates, a major component of each meal, may cause bloating, flatulence, abdominal pain and discomfort, and diarrhea if the small intestine's absorption process is not complete. The products of bacterial fermentation in the colon, such as hydrogen (H_2), methane (CH_4), and short-chain fatty acids (SCFAs), can modulate colonic propagated contraction, and colonic transit [3,4]. Studies showed that colonic infusion of SCFAs affects upper GI tract motor functions, including relaxation of the proximal stomach [5], and induces transient lower esophageal sphincter relaxations (TLESRs) [6] in healthy individuals. Research in animals also suggested the role of SCFAs on gut–brain communication and brain function directly or indirectly through immune, endocrine, vagal, and other humoral pathways [7].

Overlapping functional gastrointestinal disorders are common, and associated with poor treatment outcomes and lower quality of life [8]. IBS and GERD have been reported in about one-third of patients with each disease [9,10]. Meta-analysis of randomized controlled studies supported the benefit of consuming a low FODMAPs diet on gastrointestinal symptoms and quality of life in patients with IBS [11,12]. However, the role of carbohydrates on GERD has not been well understood.

Rice and wheat are the common food staples for humans worldwide. Besides starch, wheat also contains different proteins, such as gluten, albumin, globulin, and polypeptides [13]. Among them, gluten is the major protein component. In contrast, albumin, globulin, and glutelin are the major protein in rice [14]. Likewise, starch components in rice and wheat are also different. Rice and wheat are representative of a low and high FODMAPs diet, respectively. Rice has been proposed as a good carbohydrate source for patients with functional gastrointestinal disorders [15]. Studies showed that hydrogen gas, a colonic bacterial fermentation product, in breath samples after rice ingestion is not significantly increased from the fasting period. In contrast, wheat ingestion produced more breath hydrogen gas in healthy humans [16] and patients with IBS who had negative tests for celiac disease [17]. This suggests that wheat is not completely absorbed in the small intestine, and it may produce GI symptoms independent of gluten hypersensitivity.

Therefore, this study aimed to evaluate the effects of low vs. high FODMAPs meals on typical GER symptoms using rice vs. wheat noodles, and the correlation between GERD symptoms and intestinal gas production in patients with overlapping GERD-IBS. We hypothesized that ingestion of wheat, a high FODMAP food, will increase colonic fermentation, intestinal gas, and SCFAs, leading to more TLESRs and GERD symptoms than rice, a low FODMAP food.

2. Materials and Methods

2.1. Study Subjects

Adult patients (aged 18–65 years old) who had typical GERD symptoms (bothersome heartburn and/or regurgitation) and non-constipation type IBS according to the Rome III criteria were recruited from the gastroenterology clinic in King Chulalongkorn Memorial Hospital, The Thai Red Cross Society, Bangkok, Thailand. A symptom questionnaire with a Bristol Stool Form Scale (BSFS) was used to exclude the IBS constipation-predominant type. We excluded patients with a history of allergies to the test meals; history of abdominal surgery, except for appendectomy and hemorrhoidectomy; pregnancy; major psychological disorders; and comorbid pulmonary conditions, such as chronic obstructive pulmonary disease (COPD). A serologic test for celiac disease (serum immunoglobulin A, anti-tissue transglutaminase; tTG antibody), and a specific skin prick test for gluten and wheat allergy (ALK Abello Pharm., Inc., Mississauga, ON, Canada) were performed. All participants needed to have stable medical treatment at least four weeks before the study enrollment; stop probiotics, antibiotics, prokinetics, laxatives, or medications that affect GI functions and symptoms during a 2-week run-in and the study period; and record their food diary for three days before the study. Patients who still had overall GI symptoms severity during the last week of a run-in period more than 5 of 10 cm of visual analog scale (VAS) were enrolled.

2.2. Study Design

All subjects were randomly assigned by block randomization to the different test meals (rice noodles or wheat noodles), then crossover with a one-week washout period. This period was judged as a sufficient time for the meals in the previous period to have washed out, as the participants had a stool frequency of at least three times per week. After at least 8 h fasting, baseline GI symptom scores and BSFS were assessed. All subjects ingested a standard 250 g rice noodles or 250 g wheat noodles at 8.00 AM (breakfast) and noon (lunch). Exhaled breath hydrogen (H_2) and methane (CH_4) gas were acquired from all subjects at fasting and after breakfast every 15 min for 8 h. Although the rice and wheat noodles looked different, all patients were not informed about the major component of the noodles, and the term "FODMAPs" was not mentioned in the patient's information sheet. The investigators who evaluated GI symptoms and measured the breath H_2 and CH_4 gas levels were blinded to the test meals. The postprandial GI symptoms, including typical

GERD symptoms (heartburn, regurgitation), gas-related symptoms (bloating, belching, flatulence), satiety, nausea, abdominal pain, and stool urgency, were evaluated using 10 cm visual analog scales (VAS), in which 0 indicated no symptom and 10 indicated the most severe symptoms. All subjects gave their informed consent before they participated in the study. The study was conducted following the Declaration of Helsinki, and the protocol was approved by the Institutional Review Board of Faculty of Medicine, Chulalongkorn University (project identification code 027/56 and 678/62).

2.3. Interventional Meals

The study meals were made from 90-g-dry weight of rice or wheat noodles, and cooked as 250-g-cooked weight, as described in our previous study [17]. No vegetable or other fermentable ingredients, such as garlic or soy sauce, were added, to avoid intestinal gas production from other sources. In each study, participants took the test meal for breakfast and lunch. A glass of water (250 mL) was allowed with the test meals, and subjects were asked to finish their meal within 15 min. No food, drink, or medication were allowed during the study. We recorded the amount of food taken in every meal. According to the USDA Food Composition Database (https://ndb.nal.usda.gov/ndb/search/list (accessed on 30 November 2021)) [18], the total energy in each serving size for wheat and rice noodles in this study was 440 kcal and 450 kcal, respectively. The carbohydrate, protein, fat, and fiber contents in wheat and rice noodles serving were 42:32:14:2 g and 50:30:14:<1 g, respectively. The oligosaccharides (fructans and galacto-oligosaccharides) for a grain cutoff value of less than 0.3 g per serving was classified as a low FODMAPs diet, which was well tolerated and did not trigger symptoms in patients with IBS in clinical studies [19–21]. A study reported a serving size (cooked, 165 g) of wheat-based pasta containing 2.5 g fructans, which is considerably high in FODMAPs content [22]. In contrast, 100 g cooked rice noodles did not contain fructans and short-chain carbohydrates [23]. In this study, we used 250 g rice noodles, representing a low FODMAPs meal; and 250 g wheat noodles, representing a high FODMAPs meal.

2.4. Breath Tests

In the evening before the study day, we advised all subjects to avoid poorly absorbable carbohydrates affecting intestinal gas production on the study day. Patients ensured their good oral hygiene during the breath testing by brushing their teeth before taking the first breath sample [24]. The baseline gas sample was a fasting sample taken before breakfast, then every 15 min for 8 h. Each breath sample was collected using a 250 mL sample holding bag (Quintron Instrument Co., Inc., Milwaukee, WI, USA). We used a Quintron Microlyzer Model DP Plus (Quintron Instrument Co., Inc., Milwaukee, WI, USA) for measuring H_2 and CH_4, and reported in parts per million (ppm).

2.5. Statistical Analysis

The primary outcome was the postprandial typical GERD symptoms (heartburn, regurgitation) severity scores compared between the test meals. Secondary outcomes were the gas-related symptoms (bloating, belching, flatulence) and other GI symptom scores after breakfast and lunch, and exhaled H_2 and CH_4 over an 8 h study comparing the interventional meals. The sample size was calculated to determine at least a 30% difference of GERD symptom severity score between rice and wheat [25] with 90% power at $\alpha = 0.05$, and at least 20 subjects were needed.

A comparison of GI symptoms, H_2, and CH_4 gas levels between two groups were analyzed using the paired T-test and Wilcoxon sign rank test, depending on data distribution. A p-value of less than 0.05 was defined as statistical significance. Data were expressed as mean $\pm$ SD or median (interquartile range) as appropriate. The repeated measures analysis of variance was also performed to determine whether period, sequence, and carryover effects can arose in a crossover trial. The data were analyzed using SPSS software version 26.0 for Windows.

3. Results

Twenty-one patients (18F, age 50 ± 13 years, BMI 24.1 ± 4.3 kg/m^2) with bothersome typical GERD symptoms overlapping non-constipation type IBS were included. The duration of GERD and IBS symptoms was 6 (5–12) and 8 (6–13) months, respectively. Baseline global symptoms severity (VAS 0–10) was 7.6 ± 1.5, with a typical GERD symptoms severity score of 6.1 ± 2.3 at the study enrollment. The median BSFS during the past month before study enrollment was 5(4–5), with a stool frequency of 7(6–10) times per week. Thirteen patients had serologic tests for celiac disease (serum immunoglobulin A, anti-tissue transglutaminase; tTG antibody), and a specific skin prick test for gluten and wheat allergy, and all the results were negative. Nineteen patients underwent upper endoscopy after GERD onset, and all of them had no reflux esophagitis. The other two patients did not undergo upper endoscopy as they were under 50 years and had no alarm features. Three-day food diaries before the study day showed comparable food items between arms. The fasting GI symptom scores in the morning before each study meal ingestion were not significantly different compared between test meals ($p > 0.05$) (Table 1).

Table 1. Fasting gastrointestinal symptom scores in the morning before each study meal ingestion.

	Wheat Noodles ($n = 21$)	Rice Noodles ($n = 21$)
GERD symptoms (Visual analog scale 0–10)		
Heartburn	0(0–0)	0(0–0)
Regurgitation	1(0–2.5)	0(0–0)
Other GI symptoms (Visual analog scale 0–10)		
Bloating	3(0–5.5)	3(0–3.5)
Belching	2(0–6)	2(0–3.5)
Flatulence	2(0–6)	1(0–4)
Nausea	0(0–1.5)	0(0–1)
Satiety	5(3.5–5.5)	5(2–6)
Chest discomfort	0(0–0)	0(0–0)
Abdominal pain	0(0–2.5)	0(0–3)
Abdominal burn	0(0–2.5)	0(0–2.5)
Stool urgency	0(0–0)	0(0–0)

Data expressed as median (interquartile ranges; $p > 0.05$, wheat noodles vs. rice noodle for all symptoms.

3.1. *Effects of Wheat vs. Rice Ingestion on Typical GERD Symptoms*

All patients finished the assigned wheat and rice noodles at a similar amount, and completed the studies without serious adverse events. Regarding the symptoms after breakfast, regurgitation severity after the wheat noodles arm was significantly higher than rice noodles only at 15 min ($p < 0.05$). Then, the symptom gradually decreased, and was not significantly different between wheat and rice noodles after that. Heartburn symptoms 2 h after wheat noodles for breakfast were not significantly different from rice noodles ($p > 0.05$). In contrast, at 15 min after lunch, heartburn and regurgitation severity in the wheat noodles arm were significantly higher than the rice noodles arm ($p < 0.05$). Also, at 2 h after lunch, regurgitation severity in the wheat noodles arm was significantly higher than the rice noodles arm ($p < 0.05$) (Figure 1A,B).

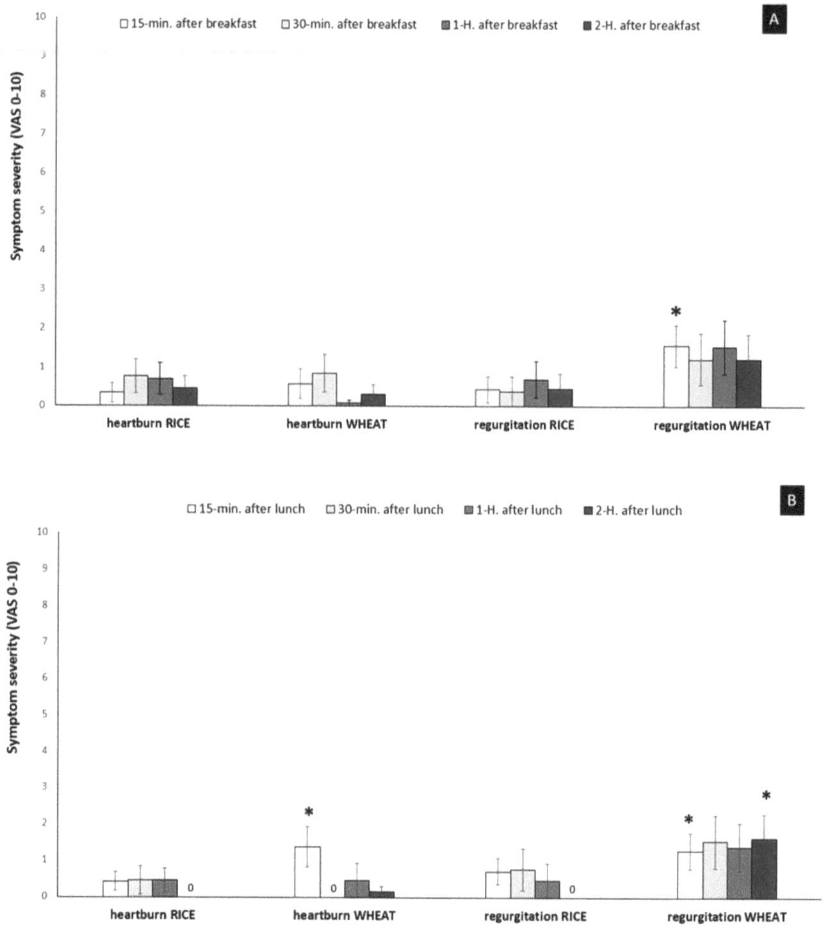

Figure 1. The typical gastroesophageal reflux disease (GERD) symptoms scores at postprandial period comparing between wheat and rice noodles ingestion. (**A**) shows symptoms after breakfast, and (**B**) shows symptoms after lunch. Heartburn and regurgitation were evaluated by visual analog scale (VAS) 0–10. * $p < 0.05$ wheat noodles vs. rice noodles. Data expressed as mean and SEM.

3.2. Effects of Wheat vs. Rice Ingestion on Other GI Symptoms

During 2 h after breakfast, wheat noodles ingestion was significantly associated with higher maximal satiety symptom scores than rice noodles ingestion ($p < 0.05$), whereas bloating, belching, flatulence, nausea, chest discomfort, abdominal pain, abdominal burn, and stool urgency severity scores were not significantly different compared to rice noodles ($p > 0.05$). In contrast, wheat noodles ingestion significantly aggravated more severe bloating, satiety, and chest discomfort than rice noodles ingestion during 2 h after lunch ($p < 0.05$). Belching, flatulence, nausea, abdominal pain, abdominal burn, and stool urgency severity scores were not significantly different after the wheat and rice test meals. Wheat noodles for lunch were associated with more 2 h postprandial bloating symptoms than wheat noodles for breakfast ($p < 0.05$) (Figure 2).

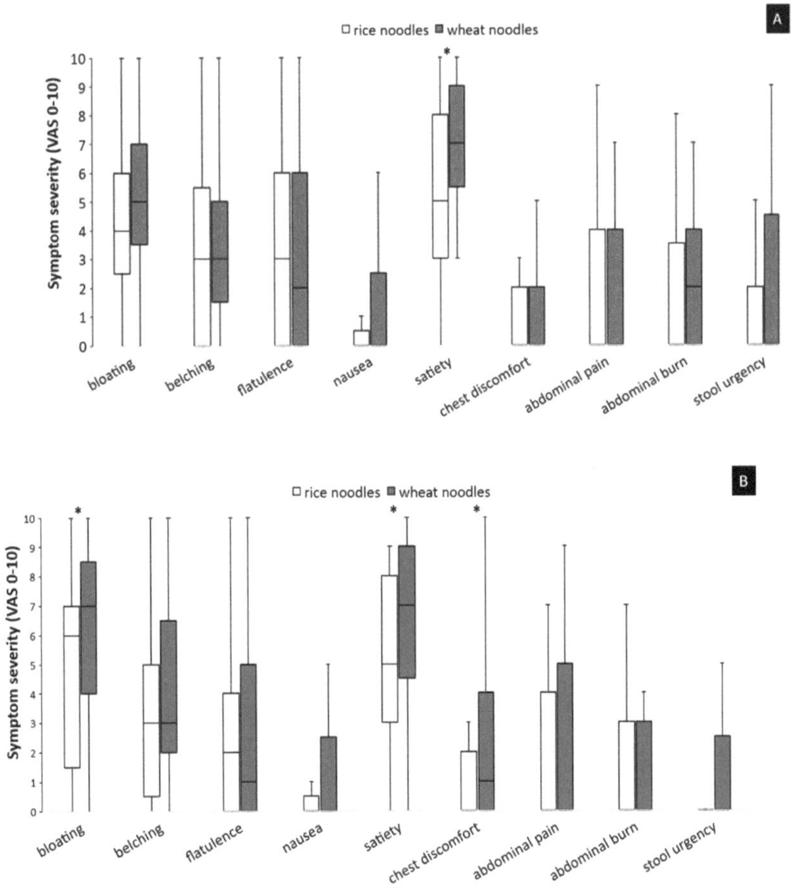

Figure 2. The maximal gastrointestinal symptom scores during 4 h after breakfast (**A**) and lunch (**B**), comparing between wheat and rice noodles ingestion. Symptoms were evaluated by visual analog scale (VAS) 0–10. A box and whisker plot represents the median, interquartile range, minimal, and maximal values of symptom severity. * $p < 0.05$ wheat noodles vs. rice noodles.

3.3. H_2 and CH_4 Production after Rice vs. Wheat Ingestion

The area under the curve (AUC) of H_2 and CH_4 concentration over 8 h after wheat noodles ingestion were significantly greater than the levels after rice noodles ingestion (wheat vs. rice; AUC: H_2 = 2925(1710–5334) vs. 1523(944–2446) ppm-min; CH_4 = 833(581–1628) ppm-min, $p < 0.001$). During 4 h after lunch, the AUC of H_2 and CH_4 concentration in the wheat study arm were significantly higher than the rice study arm (AUC: H_2 = 1208(300–2145) vs. 600(278–840) ppm-min; CH_4 = 390(296–1050) vs. 330(214–566) ppm-min, $p < 0.05$). In the wheat noodles arm, the AUC of H_2 and CH_4 concentrations after lunch were significantly higher than the concentrations after breakfast ($p < 0.05$), but not rice ($p > 0.05$). This effect was demonstrated immediately after lunchtime (Figure 3). During 4 h after breakfast, the AUC of H_2 and CH_4 concentration were not significantly different between wheat and rice noodles ingestion (AUC: H_2 = 533(266–999) vs. 485(248–739) ppm-min; CH_4 = 350(243–494) ppm-min, $p > 0.05$). Also, H_2 and CH_4 concentration levels at every time point during this period were similar.

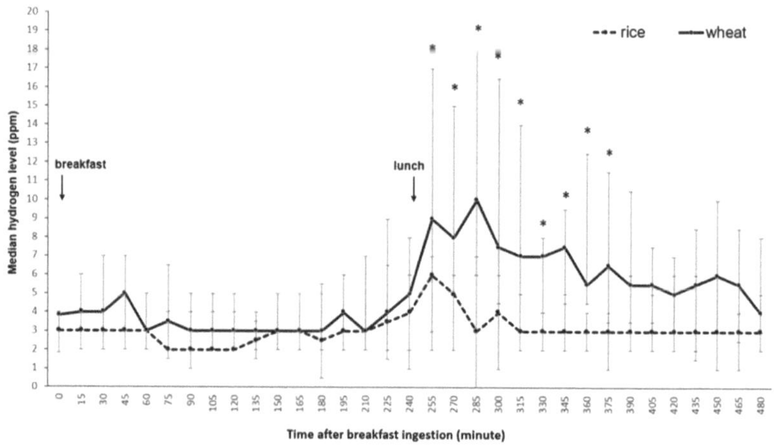

Figure 3. Exhaled hydrogen gas concentrations after ingestion of different test meals. * $p < 0.05$ wheat noodles vs. rice noodles.

3.4. The Correlation between Postprandial Symptoms and H_2 and CH_4 Production

The maximum regurgitation symptom was developed at the first 15 min after lunch in nearly all (19/21) patients. Thus, we correlated the regurgitation severity scores with H_2 and CH_4 concentrations during 15 min after lunch. The regurgitation severity score at 15 min after wheat noodles for lunch significantly correlated with the area under the curve of exhaled H_2 concentration during 15 min after lunch ($r = 0.56$, $p = 0.009$) and the area under the curve of exhaled CH_4 concentration during 15 min after lunch ($r = 0.55$, $p = 0.009$).

In the analysis of variance, there was no indication of a sequence effect, a period effect, or a carryover effect on the GI symptoms, and gas production ($p > 0.05$)

4. Discussion

This study demonstrated wheat noodles, a high FODMAPs meal, were significantly associated with higher postprandial heartburn and regurgitation than rice noodles, a low FODMAPs meal, in patients with GERD overlapping non-constipation type IBS. This effect was clearly observed after lunch. In addition, wheat produced higher exhaled H_2 and CH_4 concentrations than rice noodles, demonstrated immediately after lunchtime onward, but not after breakfast. Moreover, after lunch, the amount of H_2 and CH_4 production with wheat noodles significantly correlated with postprandial GERD symptom severity. This finding suggests that intestinal gas production, or colonic fermentation of incomplete carbohydrate absorption might play roles in the postprandial GERD symptoms after lunch. This study also demonstrated that regurgitation symptom scores, but not heartburn scores, after wheat noodles for breakfast were significantly higher than rice noodles, independent of intestinal gas production. All thirteen patients who underwent serologic tests for celiac disease, and specific skin prick tests for gluten and wheat allergy had negative results. Wheat noodles still have different components than rice noodles, such as gluten [13,14], and may trigger more regurgitation symptoms without increased intestinal gas in the morning. Furthermore, this effect was augmented by gas production in the afternoon, as both heartburn and regurgitation symptoms scores after wheat noodles were significantly higher than rice noodles. This finding suggests that mechanisms other than colonic fermentation might also play a role in the pathogenesis of wheat-induced regurgitation symptoms.

The rising rate of exhaled intestinal gas being the highest at the first 15 min after wheat lunch suggests a clearance of incomplete absorbed carbohydrates from the ileum into the large bowel stimulated by lunch ingestion. Studies showed that the bacterial fermentation products in the colon not only stimulate colonic propagated contraction, and accelerate colonic transit [3,4], but also decrease proximal gastric tone in a dose-dependent manner [5], and induce transient lower esophageal sphincter relaxations (TLESRs) [6], and TLESR-associated acid reflux episodes in healthy individuals [6]. The maximal difference in regurgitation and heartburn severity scores between wheat and rice in this study occurred at the first 15 min after lunch. As a randomized crossover design, wheat-induced regurgitation and heartburn symptoms after lunch likely link to TLESRs via colonic fermentation.

The participants in our study also had a more increased regurgitation after breakfast with wheat noodles than rice noodles, which were not associated with the intestinal gas production amount. After breakfast, the difference in GERD symptoms was only regurgitation, but not heartburn, and symptoms were rapidly improved rather than prolonged, and associated with other gas-related symptoms in the afternoon. The mechanisms independent of colonic fermentation that explain wheat-induced regurgitation symptoms in the morning remain elusive. One hypothesis is the impact of different diets on reflux perception. A recent randomized study comparing between a 4-week low FODMAPs diet (< 3 g/d) and a diet following usual dietary advice (low-fat diet; avoid alcohol, caffeine, and overeating) in the PPI-refractory GERD patients who had normal total acid exposure time ((AET), (median AET 1.1% (0.2–2.6)) demonstrated a similar GERD symptoms improvement [26]. Almost all patients in both groups displayed unchanged pH-impedance parameters (reflux number, acid exposure, bolus exposure time) at the end of the diet. In contrast, the number of patients with positive symptom association numerically decreased. So, different food components or dietary modifications might affect GERD symptoms perception in patients with PPI-refractory GERD. A physiologic study in healthy subjects showed that food could activate gastric accommodation, an underlying mechanism of GER [27], since it is in the mouth without swallowing (oropharyngeal phase) to the duodenal phase [28]. A study in NERD patients from our center showed that capsaicin, a TRPV1 receptor agonist, enhanced gastric accommodation more than placebo in GERD patients, and this effect appeared at 20 min postprandial period [29]. Whether different chemical stimuli from wheat, rice, or their components may enhance gastric accommodation in GERD patients differently is not known. A pre-clinical study showed that hexane extract of wheat flour could activate capsaicin (TRPV1) receptors, but do not have a pungency effect [30]. This finding may explain why only regurgitation, but not heartburn, symptoms were triggered after wheat ingestion in the morning in the present study.

Most starch in the western diet usually comes from refined wheat grains. The Monash University FODMAPs Diet Application® classified wheat as a high FODMAPs food, and rice as a low FODMAPs food. This study uses 250 g rice noodles, representing a low FODMAPs meal; and 250 g wheat noodles, representing a high FODMAPs meal. Although, this study suggests the possible role of a low FODMAPs diet as a non-pharmacologic treatment for GERD with overlapping IBS patients. It is impossible to exclude the effect of gluten or other components in wheat noodles on GI symptoms and intestinal gas production. Previous studies in healthy volunteers showed that gluten-containing wheat meals produce a significantly higher cumulative breath H_2 excretion than gluten-free wheat meals and gluten-free wheat meals with added gluten [31,32]. A previous study from our group also showed that mung bean noodles, considered as gluten-free noodles, produce intestinal gas comparable to rice noodles, although mung beans are classified as a high FODMAPs food [17]. It is possible that the procedures for gluten extraction from the flour altered the carbohydrates, making them more absorbable. Further studies are needed to elucidate whether FODMAPs, gluten, or other components in wheat induce GERD symptoms.

The effectiveness of a gluten-free diet and low FODMAPs for GERD management has been reported [26,33–35]. A prospective study in untreated celiac disease patients showed that GERD is six-fold more common than healthy individuals. After treatment with a gluten-free diet, GERD symptoms rapidly improved during the first three months, and there was a persistent improvement, despite PPIs being discontinued [33]. Thus, the mechanism of GERD in celiac disease patients might be different from GERD in the general population, and diet might play an important role. A linkage between GERD symptoms and gluten/wheat ingestion has been reported in non-celiac disease patients. A study among 498 patients without celiac disease who were referred to a digestive endoscopy showed that 20% of them reported GERD symptoms after gluten/wheat ingestion at least once per week, and symptom disappearance on a gluten/wheat-free diet. The GERD prevalence in patients with self-reported wheat sensitivity was more than in the control group [36]. The beneficial effect of a low FODMAPs diet other than a gluten-free diet on GERD has yet to be explored.

Limitations of our study are: (1) We did not perform an esophageal pH study in all patients. Although almost all the participants had no esophagitis, the study results could not be generalized to all GERD phenotypes and GERD patients without IBS. Future studies on GERD patients with esophagitis or significant esophageal acid exposure may provide different results. (2) Although the enrolled patients reported bothersome heartburn and/or regurgitation during the last week of a run-in period, the postprandial GERD symptoms severity after tested meals was low. The tested food may be different from patients' usual food. It may not induce GERD symptoms as severe as the patients' previous experience, even though the difference in symptom severity was statistically significant between study arms. (3) It is difficult to make two different study meals identical looking. We tried to minimize these limitations by avoiding informing the patients about the carbohydrates sources and FODMAPs. Previous experience of the patients with food might cause an expected symptoms bias. However, we demonstrated the impact not only on the patient symptoms, but on objective parameters, which are the intestinal gases and blinding the study meals to the assessors. (4) Other high FODMAP starch might cause different effects from wheat noodles. A longer period of ingestion might also cause different effects from the two meals ingestion in this study. So, this needs to be evaluated before a universal recommendation of low FODMAPs diets in overlapping GERD-IBS to treat GERD and upper GI symptoms.

5. Conclusions

This study demonstrated that wheat noodles, a high FODMAP meal, produce more typical GERD symptom scores postprandially, especially early or the first 15 min after lunch, than rice noodles, a low FODMAPs meal. The effect of wheat noodles on GERD symptoms is associated with increased intestinal gas production after lunch, confirming the linkage between GERD symptoms and colonic fermentation. Other mechanisms of wheat-induced GERD symptoms after breakfast independent of gas production must be further explored. This study suggests that rice is a better source of carbohydrates for patients with overlapping GERD and non-constipation type IBS than wheat, and provides insight into the role of a FODMAPs dietary modification for treating these patients.

Author Contributions: Conceptualization, T.P., S.L., S.P. and S.G.; methodology, T.P. and S.G.; software, T.P., S.L. and S.P.; Validation, T.P. and S.G.; formal analysis, T.P., S.L. and S.P.; investigation, S.L. and S.P.; resources, T.P., S.L., S.P., and S.G.; data curation, T.P., S.L., S.P. and S.G.; writing—T.P.; review and editing, S.G.; visualization, T.P.; supervision, T.P. and S.G.; project administration, S.L. and S.P.; funding acquisition, T.P., S.L., S.P. and S.G. All authors have read and agreed to the published version of the manuscript.

Funding: This study was supported in part by the Ratchadapiseksompotch Fund, Chulalongkorn University, and a grant from The Gastroenterological Association of Thailand.

Institutional Review Board Statement: The study was conducted according to the guidelines of the Declaration of Helsinki, and approved by the Institutional Review Board of Faculty of Medicine, Chulalongkorn University (project identification code 027/56 and 678/62 date of approval 12 December 2019).

Informed Consent Statement: Informed consent was obtained from all subjects involved in the study.

Acknowledgments: Authors are thankful to Wachinee Promjampa, a scientist and research coordinator, for her superb support.

Conflicts of Interest: The authors declare no conflict of interest. The funders had no role in the design of the study; in the collection, analyses, or interpretation of data; in the writing of the manuscript, or in the decision to publish the results.

References

1. Farré, R.; Tack, J. Food and symptom generation in functional gastrointestinal disorders: Physiological aspects. *Am. J. Gastroenterol.* **2013**, *108*, 698–706. [CrossRef] [PubMed]
2. Spencer, M.; Chey, W.D.; Eswaran, S. Dietary Renaissance in IBS: Has Food Replaced Medications as a Primary Treatment Strategy? *Curr. Treat. Options Gastroenterol.* **2014**, *12*, 424–440. [CrossRef] [PubMed]
3. Barrow, L.; Steed, K.P.; Spiller, R.C.; Watts, P.J.; Melia, C.D.; Davies, M.C.; Wilson, C.G. Scintigraphic demonstration of lactulose-induced accelerated proximal colon transit. *Gastroenterology* **1992**, *103*, 1167–1173. [CrossRef]
4. Jouët, P.; Sabaté, J.M.; Coffin, B.; Lémann, M.; Jian, R.; Flourié, B. Fermentation of starch stimulates propagated contractions in the human colon. *Neurogastroenterol. Motil.* **2011**, *23*, 450–456 e176. [CrossRef]
5. Ropert, A.; Cherbut, C.; Rozé, C.; Le Quellec, A.; Holst, J.J.; Fu-Cheng, X.; des Varannes, S.B.; Galmiche, J.P. Colonic fermentation and proximal gastric tone in humans. *Gastroenterology* **1996**, *111*, 289–296. [CrossRef]
6. Piche, T.; Zerbib, F.; Varannes, S.B.; Cherbut, C.; Anini, Y.; Roze, C.; le Quellec, A.; Galmiche, J.P. Modulation by colonic fermentation of LES function in humans. *Am. J. Physiol. Gastrointest. Liver Physiol.* **2000**, *278*, G578–G584. [CrossRef] [PubMed]
7. Dalile, B.; Van Oudenhove, L.; Vervliet, B.; Verbeke, K. The role of short-chain fatty acids in microbiota-gut-brain communication. *Nat. Rev. Gastroenterol. Hepatol.* **2019**, *16*, 461–478. [CrossRef]
8. Mönnikes, H.; Heading, R.C.; Schmitt, H.; Doerfler, H. Influence of irritable bowel syndrome on treatment outcome in gastroesophageal reflux disease. *World J. Gastroenterol.* **2011**, *17*, 3235–3241.
9. Lovell, R.M.; Ford, A.C. Prevalence of gastro-esophageal reflux-type symptoms in individuals with irritable bowel syndrome in the community: A meta-analysis. *Am. J. Gastroenterol.* **2012**, *107*, 1793–1801. [CrossRef]
10. Nastaskin, I.; Mehdikhani, E.; Conklin, J.; Park, S.; Pimentel, M. Studying the overlap between IBS and GERD: A systematic review of the literature. *Dig. Dis. Sci.* **2006**, *51*, 2113–2120. [CrossRef]
11. van Lanen, A.S.; de Bree, A.; Greyling, A. Efficacy of a low-FODMAP diet in adult irritable bowel syndrome: A systematic review and meta-analysis. *Eur. J. Nutr.* **2021**, *60*, 3505–3522. [CrossRef] [PubMed]
12. Hahn, J.; Choi, J.; Chang, M.J. Effect of Low FODMAPs Diet on Irritable Bowel Syndromes: A Systematic Review and Meta-Analysis of Clinical Trials. *Nutrients* **2021**, *13*, 2460. [CrossRef]
13. Zilić, S.; Barać, M.; Pešić, M.; Dodig, D.; Ignjatović-Micić, D. Characterization of proteins from grain of different bread and durum wheat genotypes. *Int. J. Mol. Sci.* **2011**, *12*, 5878–5894. [CrossRef] [PubMed]
14. Fabian, C.; Ju, Y.H. A review on rice bran protein: Its properties and extraction methods. *Crit. Rev. Food Sci. Nutr.* **2011**, *51*, 816–827. [CrossRef]
15. Gonlachanvit, S. Are rice and spicy diet good for functional gastrointestinal disorders? *J. Neurogastroenterol. Motil.* **2010**, *16*, 131–138. [CrossRef]
16. Levitt, M.D.; Hirsh, P.; Fetzer, C.A.; Sheahan, M.; Levine, A.S. H2 excretion after ingestion of complex carbohydrates. *Gastroenterology* **1987**, *92*, 383–389. [CrossRef]
17. Linlawan, S.; Patcharatrakul, T.; Somlaw, N.; Gonlachanvit, S. Effect of Rice, Wheat, and Mung Bean Ingestion on Intestinal Gas Production and Postprandial Gastrointestinal Symptoms in Non-Constipation Irritable Bowel Syndrome Patients. *Nutrients* **2019**, *11*, 2061. [CrossRef] [PubMed]
18. USDA. Food Data Central. Available online: https://ndb.nal.usda.gov/ndb/search/list (accessed on 30 November 2021).
19. Varney, J.; Barrett, J.; Scarlata, K.; Catsos, P.; Gibson, P.R.; Muir, J.G. FODMAPs: Food composition, defining cutoff values and international application. *J. Gastroenterol. Hepatol.* **2017**, *32* (Suppl. 1), 53–61. [CrossRef] [PubMed]
20. Halmos, E.P.; Power, V.A.; Shepherd, S.J.; Gibson, P.R.; Muir, J.G. A diet low in FODMAPs reduces symptoms of irritable bowel syndrome. *Gastroenterology* **2014**, *146*, 67–75.e5. [CrossRef]
21. Ong, D.K.; Mitchell, S.B.; Barrett, J.S.; Shepherd, S.J.; Irving, P.M.; Biesiekierski, J.R.; Smith, S.; Gibson, P.R.; Muir, J.G. Manipulation of dietary short chain carbohydrates alters the pattern of gas production and genesis of symptoms in irritable bowel syndrome. *J. Gastroenterol. Hepatol.* **2010**, *25*, 1366–1373. [CrossRef]
22. Shepherd, S.J.; Gibson, R.P. Fructose malabsorption and symptoms of irritable bowelsyndrome: Guidelines for effective dietary management. *Am. Diet. Assoc.* **2006**, *106*, 1631–1639. [CrossRef] [PubMed]
23. Biesiekierski, J.R.; Rosella, O.; Rose, R.; Liels, K.; Barrett, J.S.; Shepherd, S.J.; Gibson, P.R.; Muir, J.G. Quantification of fructans, galacto-oligosacharides and other short-chain carbohydrates in processed grains and cereals. *J. Hum. Nutr. Diet.* **2011**, *24*, 154–176. [CrossRef]
24. Simren, M.; Stotzer, P.O. Use and abuse of hydrogen breath tests. *Gut* **2006**, *55*, 297–303. [CrossRef] [PubMed]
25. Camilleri, M.; Mangel, A.W.; Fehnel, S.E.; Drossman, D.A.; Mayer, E.A.; Talley, N.J. Primary endpoints for irritable bowel syndrome trials: A review of performance of endpoints. *Clin. Gastroenterol. Hepatol.* **2007**, *5*, 534–540. [CrossRef]
26. Rivière, P.; Vauquelin, B.; Rolland, E.; Melchior, C.; Roman, S.; Bruley des Varannes, S.; Mion, F.; Gourcerol, G.; Sacher-Huvelin, S.; Zerbib, F. Low FODMAPs diet or usual dietary advice for the treatment of refractory gastroesophageal reflux disease: An open-labeled randomized trial. *Neurogastroenterol. Motil.* **2021**, *33*, e14181. [CrossRef]

27. Gonlachanvit, S.; Maurer, A.H.; Fisher, R.S.; Parkman, H.P. Regional gastric emptying abnormalities in functional dyspepsia and gastro-oesophageal reflux disease. *Neurogastroenterol. Motil.* **2006**, *18*, 894–904. [CrossRef]
28. Vanden Berghe, P.; Janssen, P.; Kindt, S.; Vos, R.; Tack, J. Contribution of different triggers to the gastric accommodation reflex in humans. *Am. J. Physiol. Gastrointest. Liver Physiol.* **2009**, *297*, G902–G906. [CrossRef]
29. Patcharatrakul, T.; Kriengkirakul, C.; Chaiwatanarat, T.; Gonlachanvit, S. Acute Effects of Red Chili, a Natural Capsaicin Receptor Agonist, on Gastric Accommodation and Upper Gastrointestinal Symptoms in Healthy Volunteers and Gastroesophageal Reflux Disease Patients. *Nutrients* **2020**, *12*, 3740. [CrossRef]
30. Iwasaki, Y.; Saito, O.; Tanabe, M.; Inayoshi, K.; Kobata, K.; Uno, S.; Morita, A.; Watanabe, T. Monoacylglycerols activate capsaicin receptor, TRPV1. *Lipids* **2008**, *43*, 471–483. [CrossRef] [PubMed]
31. Di Stefano, M.; Carnevale Maffè, G.; Bergonzi, M.; Mengoli, C.; Formagnana, P.; Di Sabatino, A.; Corazza, G.R. The effect of gluten on intestinal fermentation, gastric and gallbladder emptying in healthy volunteers. *Dig. Liver Dis.* **2015**, *47*, 751–756. [CrossRef]
32. Anderson, I.H.; Levine, A.S.; Levitt, M.D. Incomplete absorption of the carbohydrate in all-purpose wheat flour. *N. Engl. J. Med.* **1981**, *304*, 891–892. [CrossRef] [PubMed]
33. Nachman, F.; Vázquez, H.; González, A.; Andrenacci, P.; Compagni, L.; Reyes, H.; Sugai, E.; Moreno, M.L.; Smecuol, E.; Hwang, H.J.; et al. Gastroesophageal reflux symptoms in patients with celiac disease and the effects of a gluten-free diet. *Clin. Gastroenterol. Hepatol.* **2011**, *9*, 214–219. [CrossRef] [PubMed]
34. Usai, P.; Manca, R.; Cuomo, R.; Lai, M.A.; Russo, L.; Boi, M.F. Effect of gluten-free diet on preventing recurrence of gastroesophageal reflux disease-related symptoms in adult celiac patients with nonerosive reflux disease. *J. Gastroenterol. Hepatol.* **2008**, *23*, 1368–1372. [CrossRef]
35. Cuomo, A.; Romano, M.; Rocco, A.; Budillon, G.; Del Vecchio Blanco, C.; Nardone, G. Reflux oesophagitis in adult coeliac disease: Beneficial effect of a gluten free diet. *Gut* **2003**, *52*, 514–517. [CrossRef]
36. Mansueto, P.; Soresi, M.; Peralta, S.; Perricone, S.; La Blasca, F.; Sichera, R.; Giambalvo, O.; Carroccio, A. Self-reported nonceliac wheat sensitivity in an outpatient digestive endoscopy center: High frequency but insufficient medical approach. *Eur. J. Gastroenterol. Hepatol.* **2021**. [CrossRef] [PubMed]

foods

MDPI

Review

Stabilization of Rice Bran: A Review

Neşe Yılmaz Tuncel

Department of Food Technology, Faculty of Applied Sciences, Onsekiz Mart University, Çanakkale 17100, Turkey; neseyilmaz@comu.edu.tr; Tel.: +90-2862180018 (ext. 16097)

Abstract: One of the major problems in food science is meeting the demand of the world's growing population, despite environmental limitations such as climate change, water scarcity, land degradation, marine pollution, and desertification. Preventing food from going to waste and utilizing nutritive by-products as food rather than feed are easy and powerful strategies for overcoming this problem. Rice is an important staple food crop for more than half of the world's population and substantial quantities of rice bran emerge as the main by-product of rice grain milling. Usually, rice bran is used as animal feed or discarded as waste. Although it is highly nutritious and comprises many bioactive compounds with considerable health benefits, the rapid deterioration of bran limits the exploitation of the full potential of rice bran. Hydrolytic rancidity is the main obstacle to using rice bran as food, and the enzyme inactivation process, which is termed stabilization, is the only way to prevent it. This study reviews the methods of stabilizing rice bran and other rice-milling by-products comprising rice bran in the context of the efficiency of the process upon storage. The effect of the process on the components of rice bran is also discussed.

Keywords: rice bran; stabilization; by-product; utilization; sustainability; extrusion; microwave; infrared; lipase activity; free fatty acid

Citation: Yılmaz Tuncel, N. Stabilization of Rice Bran: A Review. *Foods* **2023**, *12*, 1924. https://doi.org/10.3390/foods12091924

Academic Editors: Piebiep Goufo, John C. Beaulieu, Ming Hsuan Chen and Ida Wenefrida

Received: 31 March 2023
Revised: 25 April 2023
Accepted: 6 May 2023
Published: 8 May 2023

1. Introduction

Scientists are searching for innovative strategies to increase food production without jeopardizing food security and biodiversity. Within the framework of sustainability, reducing food waste and utilizing by-products generated during processing are among the main objectives, considering population growth and limited resources. Rice is a major crop worldwide and rice bran (RB) is one of the most underutilized by-products of the rice milling process, despite its high nutritive, functional, and bioactive properties.

The hulled form of rice harvested from the field is called paddy. Milling of paddy yields around 60% white rice as the main product and approximately 10% broken rice, 20% husk, and 10% RB as the by-products [1]. The first step in rice milling is the removal of the hull or husk (Figure 1). Rice husk or hull is the coating on the seed and contains approximately 50% cellulose, 25–30% lignin, and 15–20% silica [2]. Therefore, it has no value as food. It is mostly used for energy production. Biochar from carbonized rice husk is also used as a soil amendment, activated carbon, processing fertilizer, etc. [3].

Brown rice, also called cargo, is unpolished whole-grain rice with the inedible rice husk or hull removed. The second step in milling involves removing bran layers to obtain white rice. Although it is known that the nutritional quality of brown rice is higher, white rice consumption is preferred all over the world. Brown rice is not well accepted mainly because of its flavor and taste. Moreover, brown rice has a coarse texture and a "dirty" look, requires longer cooking time, is more expensive, is not conveniently available, has a shorter shelf life, and gets rancid in the long term if it is not processed using a suitable stabilization process [4–6].

RB is the by-product produced as a result of milling brown rice into white rice. This is a multi-stage process employed by sophisticated milling systems at industrial scale (Figure 1). Multi-stage milling reduces the mechanical stress and heat buildup in the grain, thereby minimizing grain breakage and increasing head rice yield. The yield of head rice, unbroken white rice, is the basis for grading rice quality and establishing the market value. The bran layer is removed by whitening and polishing mills. Whitening machines remove the bran layers and rice germ and polishing machines remove the remaining bran by polishing the exterior of the milled kernel, which improves the appearance of milled or white rice. Composite RB consists of coarse bran fraction from the whitening steps and fine bran fraction or polish (the part of the bran fractions containing the most endosperm) from the polishing step. Composite RB may also contain rice germ and tiny fractions of rice hull (Figure 1) [7].

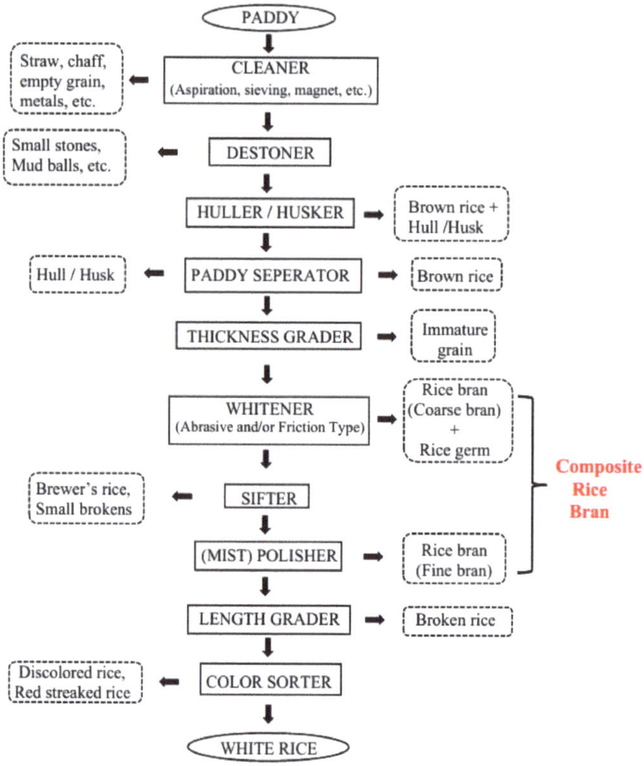

Figure 1. A typical multi-stage commercial rice milling system.

Fresh RB has great potential as a food product. It is highly nutritious and has a characteristic bland flavor and fine texture. However, unprocessed RB becomes rancid very quickly after the milling process, which limits its use as food. The rancidity of RB is predominantly triggered by lipophilic enzymes, mainly lipases. In intact paddy, lipases are primarily found in the seed coat and most of the oil is localized in the aleurone layer and rice germ. In other words, lipolytic enzymes and their substrates are physically separated in unmilled paddy rice. However, the milling process disrupts this individual localization and lipase enzymes come into contact with fat, causing hydrolysis of fat into free fatty acids (FFA) and glycerol [8]. High concentration of FFA causes a rancid soapy taste and induces oxidation and even lipotoxicity.

RB contains several types of lipases as well as glycolipases, phospholipases, and esterases [9]. So far, two types of lipases have been purified from RB. Lipase I has a molecular mass of 40 kD and an optimum pH of about 7.5. Lipase I preferentially cleaves fatty acids from the *sn*-1 and *sn*-3 positions of triacylglycerols and is activated by calcium. On the other hand, lipase II has a molecular mass of 32 kD, a pI of 9.1, and an optimum pH of about 7.5 [10]. In addition to endogenous lipases, lipases of microbial origin can also initiate hydrolytic deterioration. Dehulling causes surface damage and disrupts the aleurone and germ (RB-oil-concentrated parts of the kernel), and lipase-producing mold and bacteria found on kernel surfaces interact with bran oil, which results in an increase in FFA [9].

The world production of rice was 787 million metric tons in 2021 [11]. Considering rice production worldwide, approximately 80 million tons of RB is produced as a by-product and is mainly utilized as feed as many food by-products. However, processed RB has great potential as a value-added commodity in the food industry. RB has been used in bakery products such as bread [12–16], noodles or pasta [14,17], crackers [18], biscuits [19,20], extruded snacks [21], breakfast cereals, muffins, pancakes, cookies, cakes, pies, and wafers [14,19,22], as well as a protein supplement [20,22], a binder or fat substitute in meats and sausages [22,23], ingredient in plant-based meat analogs [24], and as a beverage base [25]. RB also has great potential in applications as an ingredient in infant formulas and gluten-free products due to it being highly soluble, hypoallergenic, and gluten-free [26,27]. In addition to direct utilization, RB is also used as a source of RB protein concentrate [26–31], RB oil [32,33], RB fiber [34–36], RB wax [37,38], γ-oryzanol [39], and a phytochemical-rich ingredient [40]. Furthermore, RB is not only used for culinary purposes but also in the pharmaceutical and cosmetic industries. RB-derived ingredients are used in hair or skin care products, sunscreen formulations (due to natural sun protection factors), shampoos, bath oils, foundations, and various cosmetics [41].

2. Rice Bran Stabilization Methods

In order to utilize RB as food instead of feed, it is essential to apply a process that will stop lipolytic activity, which is called "stabilization". In other words, stabilization is an enzyme inactivation process that enables RB to be incorporated into the human diet. Stabilization refers to the prevention of hydrolytic degradation. Therefore, oxidative deterioration was not covered in the context of stabilization. Lipase activity and FFA formation are the main measures of hydrolytic degradation in RB. To a lesser extent, lipoxygenase and peroxidase activities were also used as RB deterioration indicators. However, FFA is the most widely used indicator due to its ease of determination. In general, RB with an excess of 10% FFA is accepted as unsuitable for human consumption [42]. An increase in FFA occurs very rapidly in freshly milled RB without proper stabilization treatment and FFA levels can reach 10% within hours depending on the post-harvest conditions. Hydrolytic rancidity development can also be avoided by rapid oil extraction soon after the rice milling process [43,44]. However, in practice, milling of rice and extraction of RB oil are not performed consecutively and the time lapse between milling and extraction results in excess amounts of FFA. However, delaying hydrolytic degradation using a stabilization process may save time for good-quality oil extraction economically from RB.

A review of the literature on methods whose main objective is RB stabilization, clearly defined stabilization conditions, and studies of storage following stabilization is presented in Table 1. Stabilization studies carried out using chemicals, acids, or similar substances that are not edible or may be inconvenient for human and animal consumption, and studies with plant-derived extracts of unknown purity are excluded. Although they have the effect of delaying rancidity to a certain extent, studies employing physical stabilization strategies such as low-temperature storage or refrigeration were also excluded since these methods have serious limitations such as restoration of enzyme activity at room temperature and oxidative rancidity.

It was observed that extrusion, microwave (MW) heating, hot air heating, autoclaving, infrared (IR) heating, ohmic heating, radio frequency (RF) heating, ultraviolet (UV) treatment, ultrasound treatment, γ-irradiation, antioxidant addition, enzyme addition, phenolics addition, parboiling, toasting, roasting, and steaming are the RB methods for stabilization studied in the literature.

Some of the studies on RB stabilization examined solely the effect of the stabilization process on nutritional and/or bioactive components, and did not examine the effectiveness of the stabilization itself, nor did they analyze any of the stabilization markers such as FFA or lipase activity [45–48].

Table 1. Studies on rice bran stabilization (2000–2023) following the effect of stabilization during storage.

Method and Conditions of Stabilization	Measure of Stabilization	Effect of Stabilization upon Storage	Co-Effect of Stabilization	Reference
Hot air drying at 110 °C for 2 h, Solar drying (8 h per day, max 42 °C), Steaming at 100 °C for 30 min	FFA	FFA levels of hot-air-dried, solar-dried, and steamed-RB were 9%, 39%, and 6%, respectively, while the FFA content of unstabilized RB was 78% at the end of 50 days.	Steaming showed the best results, with only 3% reduction in oil yield after 50 days in comparison with 89% oil reduction forunstabilized bran.	[49]
Extrusion at 125–130 °C for 30 s and held in the auger at 97–99 °C for 3 min	FFA	FFA levels of extruded rice bran increased from 2.8% to 3.2% and 3.3% in vacuum packs and zipper-top bags, respectively, at the end of 8 weeks when stored at 4–5 °C.	Water absorption capacity increased while fat absorption capacity decreased after the treatment compared with untreated raw bran. Emulsification and foaming capacities of extruded brans were significantly lower than that of raw bran.	[9]
Extrusion with a co-rotating twin-screw extruder (screw speed 140 rpm) at 130 °C for 20 s	Acid value	No significant change in the acid value of extruded RB for at least 20 days of storage at 25 °C.	Extrusion reduced the extractability of phytic acid.	[50]
Dry heat stabilization at 120 °C for 30 min Extrusion under the following conditions: Water flow rate: 0.000038 m^3/s, feed rate: 27 kg/h, steam supply: 275.80 kPa, die opening: 0.0078 m, temperature: 135–140 °C. Extrudates were air dried at 50 °C for 24 h and then ground	FFA	FFA content of raw bran increased from 4.05% to 64.60%, while FFA content of dry-heated RB increased from 3.66% to 9.15%. Variation in FFA content (from 3.85% to 4.10%) of extruded bran during storage at ambient temperature in polyethylene bags for 60 days was insignificant ($p > 0.05$).	Slight increase in crude fat content, no significant change in protein content and calorific value, reducing, non-reducing, and total sugar contents, significant loss of lysine bioavailability, significant reduction in phytic acid content, significant increase in bulk density, water absorption, damaged starch content, significant decrease in fat absorption, protein solubility, and darker color were observed in stabilized brans compared with raw bran.	[51]
Extrusion (at 135 °C, 6% moisture addition, 5 s) Roasting (RB heated to 70, 90, 100, and 105 °C in a steam jacket roaster with four chambers) Pelleting (RB was conditioned with steam to increase the temperature to 90 °C, passed through a die of 5 mm, and cooled) Antioxidant addition (at 125, 250, and 375 ppm levels)	FFA Peroxide value Iodine value	Raw and pelleted RB behaved similarly in terms of storage stability. Addition of antioxidant to rice bran was not effective for stabilizing FFA, peroxide, and iodine values at any level. Extrusion was partly effective, resulting in 49.5% FFA as an average over 345 days of storage under ambient conditions.		[52]
Extrusion with pH modification (addition of HCI or $Ca(OH)_2$ at the levels of 1, 5, and 10%; extrusion temperature 110–140 °C; feed moisture 20, 30, and 40%	Spectrophotometric determination of FFA and copper acetate complex	10% of HCI addition provided the lowest FFA increase at the end of 98 days of storage at room temperature (25 ± 3 °C).		[53]

Table 1. *Cont.*

Method and Conditions of Stabilization	Measure of Stabilization	Effect of Stabilization upon Storage	Co-Effect of Stabilization	Reference
Extrusion (four zones: 70 °C, 90 °C, 110 °C and 130 °C, with screw speed set at 300 rpm) Microwave heating (800 W, 75 s) Dry heating (130 °C, 120 min) Steaming (30 min) High-pressure steam (121 °C, 0.1 MPa, 15 min)	Lipase activity Lipoxygenase activity Acid value Peroxide value	After 28 days of accelerated storage at 37 °C, the acid value of untreated RB was about 6 mg KOH/100 g, while the acid values of stabilized samples were around 1.7 KOH/100 g, except for atmospheric steam-treated RB.	High-pressure steam-treated, steam-treated, and extruded RB samples were more suitable for long-term storage and better maintained the stability of flavor.	[54]
Microwave heating at 850 W for 3 min (moisture content was adjusted to 21% before the treatment)	FFA	No significant change in FFA content of MW-treated RB stored at 4–5 °C at the end of 16 weeks.	No dramatic change in proximate composition and fatty acid composition as a result of MW stabilization.	[55,56]
Microwave heating at 550 W, 2450 MHz for 3 min (moisture content was adjusted to 21% before the treatment). The temperature of the bran was 107 °C after the process.	FFA	FFA content of MW-treated RB increased from 3.2% to 3.9% in both vacuum and zipper-top packs at the end of 8 weeks when stored at 4–5 °C.	Fat absorption, emulsification, and foaming capacities of MW-treated RB were significantly lower than that of raw bran. However, MW treatment increased water absorption capacity.	[9]
Microwave heating at 700 W, 2450 MHz for 1, 3, and 5 min (moisture content was adjusted to 21% before the treatment). MW power densities were 2, 4, and 6 W/g	FFA Peroxide value	FFA content of RB treated with MW at a power density of 6 W/g for 5 min was 1.12%, while FFA of untreated control was 58.5% at the end of 28 days of storage at ambient temperature.	Protein content and oil yield significantly decreased under the noted effective MW stabilization condition.	[57]
Microwave heating Flowing MW drum heater with a max power of 6 kW (6 magnetrons × 1000 W, 2450 MHz)	Lipase activity Lipoxygenase activity FFA Peroxide value	Residual lipase activity was 13.21 U/kg for untreated control and only 1.25 U/kg for MW-treated RB at 5 kW for 13 min. Lipoxygenase activity was completely destroyed after 7 min of MW processing.	No significant decrease in fatty acid, tocopherols, and γ-oryzanol content of bran treated under optimum conditions (4 kW, 10 min).	[58]
Microwave heating at 850, 925, and 1000 W for 3, 4.5, and 6 min	FFA Acid value Peroxide value	Stabilization was investigated in 3 RB milling fractions and composite RB. Stabilization at 925 W for 3 min was recommended as the suitable condition for stabilization of RB milling fractions.	–	[59]
Infrared heating Short-wave IR radiation at 200, 300, 400, 500, 600, 700, 800, and 900 W for 1, 2, 3, 4, 5.2, 6.3, 7.2, 8.3, and 10 min	FFA	No significant increase was observed in FFA content of RB treated at IR powers of 600 W (for 4 and 5 min) and 700 W (for 2 and 3 min), while FFA content of raw RB increased from 4.32% to 43.08% at the end of 6 months at 25 °C.	The effect of IR stabilization was insignificant on γ-oryzanol content and fatty acid composition of RB. However, a significant decrease in tocopherol content of RB of up to 50% was determined.	[60]
Infrared heating Medium wave IR radiation at 500, 600, and 700 W IR power for 3.0, 5.5, and 7.0 min	FFA	Stabilization was investigated in 3 RB milling fractions. Stabilization at 700 W IR power for 7.0 min provided 90 days of shelf life without a notable change in FFA content of RB fraction obtained from the first whitening step.	Total tocopherol and γ-oryzanol contents of stabilized RB fractions were higher than in their crude counterparts.	[61]

Table 1. *Cont.*

Method and Conditions of Stabilization	Measure of Stabilization	Effect of Stabilization upon Storage	Co-Effect of Stabilization	Reference
Infrared heating (Simultaneous drying and stabilization) RB was heated for 55s to reach a surface temperature of 60 °C during each drying pass and tempered in an incubator set at 60 °C for various durations (1, 2, 3, 4, and 5 h).	FFA	It took 7 days for the FFA level of control RB to reach 10%. This period has been extended to 38 days for IR-treated RB.	IR heating at the employed conditions did not have a negative effect on milling quality.	[62]
Infrared heating with a max power of 2400 W, unspecified radiation intensity at 100, 120, and 140 °C for 5, 10, 15, and 20 min.	FFA Peroxide value	FFA content of raw RB increased from 4.4% to 62.8% after 6 months of storage at 25–30 °C, while RB treated with IR at 140 °C for 15 and 20 min maintained FFA content at around 7%.	IR stabilization at 140 °C for 15 min did not cause a significant decrease in γ-oryzanol content and fatty acid composition, but significantly decreased E vitamins.	[63]
Infrared heating with a laboratory-scale ceramic infrared drying device until the surface temperature of the rice bran reached 85 °C.	Lipase activity Lipoxygenase activity Peroxidase activity FFA Peroxide value	FFA content of untreated RB increased from 5.41% to 49.87%, and from 5.21% to 14.11% for IR-treated RB at the end of 20 days of storage at 20 °C. Lipase, lipoxygenase, and peroxidase activities in IR-treated samples were significantly higher at the end of the storage.	The contents of palmitic acid (C16:0), oleic acid (C18:1), and linoleic acid (C18:2) were well maintained by IR treatment during the whole storage period. The aroma of fresh RB was preserved.	[64]
Infrared heating either with medium or shortwave IR emitters at 700 W for 3 min	FFA	Hydrolytic lipid degradation occurred more likely in samples stored in bran form; however, samples stored in oil form were more prone to oxidative degradation. Medium-wave IR radiation was more effective in terms of retarding FFA increase compared with short-wave IR radiation.	Peroxide values of either raw or IR stabilized samples were <10 meq/kg at the end of 6 months of storage at 25 °C. Conjugated dienoic acids and *p*-anisidine values increased in all samples during storage. Storage in oil form resulted in a higher loss of total tocopherol and γ-oryzanol compared with storage in bran form.	[43]
Ohmic heating using an alternating current of 1 or 60 Hz and an electric field strength of 100 V/cm.	FFA	FFA content of RB (moisture content adjusted to 21%) stabilized with ohmic heating at 60 Hz increased from 3.25% to 5.47%, while FFA content of raw RB increased from 3.96% to 18.03% when stored in Ziploc bags at 4 °C for 6 weeks. The moisture content of the sample had a very decisive effect.	Increase in rice bran oil extraction yield.	[65]
Ohmic heating Using an alternating current of 50 Hz and an electric field strength of 75, 150, and 225 V.cm^{-1}. Moisture content of RB was adjusted to 20, 30, and 40%.	FFA Lipase activity	Lower FFA levels and lipase activity were observed in ohmically treated RB compared with raw RB stored at 4 °C for 21 days. The moisture content of the sample had a very decisive effect.	A very slight increase was observed in total phenolic, α-tocopherol, and γ-oryzanol levels of ohmic-heated RB compared with raw RB.	[66]
Ohmic heating at 20, 30, and 40% initial moisture content, 132, 150, 168, 189, 216 V, and 50 Hz. Electrical field strengths were 44, 50, 56, 63, and 72 V/cm.	FFA Peroxide value	FFA of ohmically heated bran was 4.77% after 75 days of storage, whereas it was 41.84% for raw RB.	The peroxide value of ohmically heated samples after 75 days of storage was 4.7 meq/kg.	[67]

Table 1. *Cont.*

Method and Conditions of Stabilization	Measure of Stabilization	Effect of Stabilization upon Storage	Co-Effect of Stabilization	Reference
Hot air-assisted radio frequency (HA-RF) heating using a 6 kW, 27.12 MHz pilot-scale free-running oscillator RF system combined with a hot air oven	FFA Peroxide value Lipase activity Lipoxygenase activity	RB treated by HA-RF heating to 100 °C with 15 min holding can be stored at 35 °C and remain below acceptable thresholds for a period of 60 days without adverse effects on product quality.	Optimum HA-RF heating treatment led to a significant increase in tocopherol content, but had no significant effect on γ-oryzanol.	[68]
Radiofrequency heating at 5 kW, 40.68 MHz for 2 min	FFA Acid value Peroxide value Lipase activity	Lipase activity retention was close to zero after 2 min of RF heating. FFA content of untreated bran stored at 37 °C was almost 4-fold higher than that of RF-treated bran after 8 weeks. Peroxide values of either raw or processed brans were below 10 meq O_2/kg for 8 weeks and at all storage temperatures (4, 25, 37 °C).	No significant difference in total phenolic and flavonoid contents, DPPH scavenging activity, reducing power, and color between the untreated control and radio frequency-treated RB ($p > 0.05$)	[69]

2.1. Extrusion

Extrusion is one of the oldest and possibly the most widespread method of RB stabilization. Generally, temperatures between 120 and 130 °C were sufficient to inactivate RB lipase. Randall et al. (1985) reported that no notable increase was observed in FFA content of RB extruded at 130 °C during storage at 32 °C and 85% relative humidity (RH) for 28 days [70]. The authors stated that although a consistent temperature of 120 °C was usually suitable, stabilization was always sufficient at 130 °C. However, FFA content of raw bran that was stored at 32 °C increased to over 80% [70]. Additionally, Kim et al. (1987) reported a complete inactivation in lipase activity at temperatures above 128 °C regardless of the moisture content of the RB fed to a single screw extruder [71]. Similarly, Fuh and Chiang (2001) indicated that extrusion at 130 °C for 20 s with a screw speed of 140 rpm was sufficient to inactivate lipase [50]. Shin et al. (1997) extruded RB at 110, 120, 130, and 140 °C with post-extrusion times of 0, 3, and 6 min and stored RB at ambient temperature for 375 days [72]. FFA content of raw RB reached over 70% at the end of the year. Although the FFA level of the processed RB was below 7% at all extrusion temperatures after 1 year, it was observed that the amount of FFA content increased as the extrusion temperature decreased during storage. In addition, it was reported that post-extrusion holding time had no effect on FFA levels in extruded RB [72].

Malekian et al. (2000) carried out extrusion (125–130 °C for 30 s) with the aim of RB stabilization. FFA levels of raw (untreated) and extruded RB samples were 26.7% and 3.2% when stored in vacuum packs, and 22.2% and 3.3% when stored in zipper-top packs, respectively, at the end of 8 weeks of storage at 4–5 °C [9]. It was observed that for untreated RB, vacuum-packed samples had a higher increase in FFA levels compared with samples in zipper-top bags. The authors attributed this result to anaerobic microorganisms present in RB [9]. Sharma et al. (2004) stabilized RB using dry heating (120 °C for 30 min) and extrusion cooking (135–140 °C) [51]. The authors stated that they observed an incomplete destruction of lipase in dry heat-treated RB since FFA content increased from 3.66% to 9.15% at the end of 60 days of storage at ambient temperature. However, extruded RB showed no significant increase during storage under the noted conditions [51]. Escamillo–Castillo et al. (2005) also studied extrusion stabilization of RB with pH modification [53]. The authors added HCI or $Ca(OH)_2$ to the RB samples at the levels of 1, 5, and 10% before extrusion. Although extrusion alone or in combination with any of the chemicals resulted in lower FFA when compared with the unprocessed RB, the authors concluded that the addition of $Ca(OH)_2$ promoted the activity of lipases and led to higher FFA concentrations during storage. The lowest FFA increase was observed in RB samples treated with 10% HCI, regardless of the initial moisture content of the bran [53]. Rafe and Sadeghian (2017) extruded RB at temperatures between 100 and 130 °C and the lowest lipase and peroxidase activities were obtained at 123 °C die temperature, 354 rpm screw speed, and 10.8% initial moisture content [73].

2.2. Microwave Heating

Many researchers proposed MW heating as an efficient RB stabilization process [42,55–59,74,75]. However, it should be noted that household MW ovens were used in almost all these studies. Although attempts to increase homogeneity are made by rotating the sample, it is well recognized that domestic scale MW ovens can suffer from non-uniform heating, resulting in cold and hot spots. However, industrial scale MW ovens may provide more uniform heating since they can have different designs, magnetron placements, modes, and dimensions of cavity.

Tao et al. (1993) stabilized RB with MW heating at 2450 MHz for 3 min and found that the FFA content of MW-treated long grain RB increased from 4.0% to 4.9% and that of medium grain RB increased from 4.6% to 6.25% at the end of 4 weeks of storage at 33 °C and 75% RH. On the other hand, FFA content of untreated raw RB ranged from 4.6% to 56.8% and 4.0% to 68.3% in medium and long-grain bran, respectively [42]. Ramezanzadeh et al. (1999) adjusted the moisture content of RB (150 g per batch) to 21% and heated the moistened RB in plastic zipper-top bags at 850 W for 3 min in a household MW oven [55]. The temperature of the heated RB was 107 ± 2 °C after the process. The FFA content of raw RB increased from 2.5% to 54.9% and 48.1% at 25 °C to 25.4% and 19.5% at 4–5 °C at the end of 16 weeks of storage in vacuum bags and zipper-top bags, respectively. However, FFA content of MW-heated RB increased from 2.8% to 6.9% and 5.2% at the end of the storage period when stored in vacuum and zipper-top bags, respectively, at 25 °C, while no significant change was observed in FFA content of MW-heated RB when stored at 4–5 °C [55].

Patil et al. (2016) reported that the FFA content of RB treated with MW at a power density of 6 W/g for 5 min was 1.12%, while the FFA content of untreated control was 58.5% at the end of 28 days of storage at ambient temperature [57]. Ertürk and Meral (2019) compared MW and conventional heating with regard to RB stabilization [75]. Although the researchers did not conduct a storage study, it was reported that lipase activity was significantly decreased by both processing methods in proportion to MW power or oven temperature, although MW treatment had a greater effect [75]. Li et al. (2018) stabilized RB using a flowing MW drum heater under different conditions of MW power, duration time, and ventilation rate [58]. FFA contents of RB samples treated with MW at 4 and 5 kW for 10 min were below 3% at the end of 60 days of storage at 35 °C. Significant and negative correlations were reported either between lipase activity and final bran temperature at the end of the process or between lipase activity and process time. Researchers stated that LOX activity was completely destroyed after 7 min of the MW process, even at 3 kW, and residual lipase activity was only 1.25 U/kg in RB treated with MW at 5 kW for 13 min, while it was 13.21 U/kg for the untreated control [58]. Process durations were notably longer than those reported in RB stabilization studies carried out using household MW ovens and it was clearly shown that longer process time is more effective for stabilization. Therefore, it can be interpreted that even with the same method, different results may be obtained in studies where the processing equipment is different.

2.3. Dry or Moist Heat Treatments

Stabilization of RB using dry or moist heat treatments such as hot air heating, toasting, roasting, steaming, and autoclaving is also one of the most widely employed techniques. Thanonkaew et al. (2012) reported that hot air heating at 150 °C for 10 min, roasting at 150 °C for 2 min, and steaming at 130 °C for 2 min with domestic kitchen equipments resulted in lower FFA, acid value, and peroxide value compared with raw RB [76]. However, the researchers did not perform any storage study and measured the noted parameters only after the process [76]. Amarashinge et al. (2009) reported that FFA contents of unstabilized, solar-dried (8 h per day, max temperature 42 °C), hot-air-dried (110 °C, 2 h), and steamed (100 °C, 30 min) RB were 78%, 39%, 9%, and 6%, respectively, at the end of 50 days of storage [49]. The researchers concluded that steaming showed the best results, with only 3% reduction in oil yield and the lowest FFA content after 50 days. Many studies have reported that moist heating treatments (i.e., steaming or pre-moisturization) provide a more effective stabilization than dry heating [49,77,78]. Brunschwiler et al. (2013) showed that heating RB with a moisture content of 20% at 110 °C for 5 min almost completely inactivated lipase/esterase activity (0.3% of the activity in raw RB). However, the treatment conditions required to achieve the same inactivation rate in RB with 10% moisture content was heating at 120 °C for nearly 40 min [78].

Li et al. (2023) analyzed the efficiency of different stabilization treatments and their effects on the flavor of RB based on GC–MS, E-nose, and E-tongue analyses [54]. The researchers reported that MW and high-pressure steam treatments were effective in terms of retarding FFA increase and passivating lipase activity, while atmospheric steaming had the worst effect. The content of total volatile compounds in extruded and MW-treated RB was lower than that of untreated RB, whereas untreated and steam-treated RB had almost the same content of total volatile compounds. Bitterness increased slightly during storage. It was concluded that flavors of high-pressure steam-treated, steam-treated, and extruded RB were similar and can better maintain the stability of flavor, which makes these stabilization methods more suitable for long term RB storage [54].

2.4. Infrared Heating

The first study on infrared stabilization of RB was reported by Yılmaz et al. (2014) [60]. The researchers found no significant increase in the FFA content of RB treated with IR radiation (using short-wave IR emitters) at 600 W IR power for 4 and 5 min and 700 W IR power for 2 and 3 min at 25 °C over 6 months, while the FFA content of raw RB increased from 4.32% to 43.08% at the end of storage [60]. Irakli et al. (2018), who also employed IR radiation (unspecified radiation intensity), reported very similar results [63]. It was found that the FFA content of RB treated with IR radiation at 140 °C for 15 and 20 min was around 6–7%, while FFA concentration in raw RB increased from 4.4% to 62.8% during 6 months of storage at 25–30 °C [63].

Yan et al. (2020) used a laboratory-scale ceramic IR drying device to stabilize RB [64]. The researchers employed IR heating until the temperature of the bran surface reached 85 °C; however, they did not justify why they chose this temperature. It was found that lipase activity of the IR-treated RB was lower than that of the untreated RB. Nonetheless, lipase activity of the IR-treated bran also significantly increased during the storage period (20 days at 20 °C). Although they showed a fluctuating trend, both lipoxygenase and peroxidase activities of the IR-treated RB significantly increased during the storage period. Additionally, FFA content of the IR-treated samples significantly increased during storage, albeit with a lower acceleration compared with control samples [64]. Wang et al. (2017) carried out a simultaneous rough rice drying and RB stabilization study using IR radiation [62]. The authors used a catalytic IR emitter and reached a rice surface temperature of 60 °C and then tempered the samples in an incubator at 60 °C for different durations ranging from 1 to 5 h. Although FFA levels increased during the course of storage (38 days at a temperature of 20 ± 1 °C and RH of 46 ± 3%) under all tested conditions, it was reported that the time until the FFA reached 10% was extended to 38 days compared with 7 days for the untreated control sample [62].

It should be noted that the intensity and penetration power of IR radiation is strongly dependent on the IR emitter used. Yılmaz Tuncel and Yılmaz Korkmaz (2021) reported that medium-wave IR heating was more effective at retarding the increase in FFA compared with short-wave IR heating when employed at the same IR emitter power and process time (700 W, 3 min) [43]. It was found that the FFA content of short-wave IR-stabilized samples was significantly higher than that of medium-wave IR-stabilized counterparts during 6 months of storage at ambient temperature [43]. Similar findings were also reported by Atungulu and Pan (2011) who indicated that foodstuffs absorb medium-wave IR energy more efficiently; however, absorptivity decreases and transmissivity increases in short-wave IR processes, especially for thin materials [79].

2.5. Ohmic Heating

Ohmic heating generates heat through passage of electrical current through food which resists the flow of electricity and can be used in enzyme inactivation [80]. Lakkakula et al. (2004) used ohmic heating to stabilize RB and to improve the yield of RB oil [65]. It was noted that FFA contents of raw and electrically heated RB were 18.0% and 5.5%, respectively, at the end of 6 weeks when stored at 4 °C. Researchers reported that although the process had a limited effect on FFA levels—probably due to the electroporation phenomenon—ohmically heating RB without moisture adjustment (adding water) was not effective due to the low electrical conductivity of the bran, which has an intial moisture content of around 10% [65].

Stabilization of RB with ohmic heating was also studied by Loypimai et al. (2009) [66]. The authors used an alternating current of 50 Hz and applied 75, 150, and 225 $V.cm^{-1}$ electrical field strengths to RB with moisture contents of 20, 30, and 40%. Although it was shown that ohmically heated RB had a lower FFA level and lipase activity, the effect of different ohmic processing conditions on these parameters was not so clear since the authors stored samples at 4 °C for only 21 days. However, it was shown that lipase activity of ohmically treated samples with a moisture content of 20% was higher than that of samples with 30% and 40% moisture content [66]. Dhingra et al. (2012) also stated that the FFA content of ohmically heated RB was 4.77% after 75 days of storage, whereas it was 41.84% for raw bran [67]. However, the authors did not indicate the conditions of storage. In all the aforementioned studies, it was observed that the moisture content of the sample had a critical effect on the efficiency of the process.

2.6. Radio Frequency Heating

Ling et al. (2018) used hot air-assisted RF heating (6 kW, 27.12 MHz) combined with a hot air oven to stabilize RB [68]. In this study, significant decreases in residual lipase and lipoxygenase activities after treatment under specific conditions were observed; however, it was found that the activities of the noted enzymes increased again during storage, probably due to water absorption. The researchers concluded that stabilization methods using dry heating may not be successful at irreversibly inactivating lipase and lipoxygenase activities, especially when the moisture content of the bran increased to atmospheric equilibrium during storage. Nevertheless, hot air-assisted RF heating at 90 °C for 30 min resulted in 4.01% of FFA, while the FFA content of the untreated control sample was 50.67% at the end of 60 days of storage at 35 °C and 70% RH [68].

Chen et al. (2021) also employed RF heating (5 kW, 40.68 MHz) to stabilize RB and reported that lipase activity retention was close to zero after 2 min [69]. However, the increase in FFA content during storage could also not be prevented, even at 4 °C [69]. Liao et al. (2020) did not perform FFA analysis or storage study; however, they showed that relative lipase, polyphenol oxidase, and peroxidase activities of RB treated with high-temperature hot air-assisted radio frequency (110–115 °C, 6 min) were 20.1%, 22.9%, and 7.6% that of raw bran, respectively [81].

2.7. Irradiation

Shin and Godber (1996) irradiated RB at 5, 10, and 15 kGy doses using a Cobalt-60 source [82]. The authors found higher levels of FFA in irradiated RB compared with untreated bran after the process, and FFA levels reached almost 90% at the end of 52 weeks of storage at ambient temperature (22–26 °C). It was concluded that γ-irradiation of RB did not decrease lipolytic enzyme activity in the range used. Significant losses were also reported for E vitamins and γ-oryzanol in irradiated RB [82]. However, positive effects of γ-irradiation have also been reported in other studies whose aim was not stabilization.

For instance, Masamran et al. (2023) applied γ-irradiation to defatted RB before protein extraction and reported that the extraction yield and protein recovery increased with the treatment. The authors also concluded that γ-irradiation changed the structure of RB and increased the release of bioactive compounds such as phenolic compounds and resulted in enhanced antioxidant activity in irradiated extracts [83].

2.8. Other Stabilization Approaches

Pourali et al. (2009) employed subcritical water extraction as an environmentally friendly technique to stabilize and extract RB oil simultaneously [84]. Researchers also conducted conventional solid–liquid extraction (hexane–bran, hexane–water–bran, and ethanol–bran) for comparison. No increase was observed in the content of FFA in subcritical-water-extracted and ethanol-extracted (60 °C) RB oil, while FFA concentration increased from 5.0% to 5.6% and from 6.5% to 7.0% in hexane-extracted (25 °C) and hexane–water-extracted (25 °C) RB oils, respectively, after 12 weeks of storage [84]. Another interesting approach was performed by Raghavendra et al. (2017) [85]. It is known that polyphenols can bind proteins and enzymes and alter their structural properties and biological activities. Raghavendra et al. (2017) showed that the activity of isolated and purified RB lipase decreased in the presence of chlorogenic and prominently caffeic acids. Researchers found 56% loss of lipase activity at 60 µM caffeic acid concentration. The loss of enzymatic activity increased with increasing concentration of the noted ligands [85].

Gopinger et al. (2015) treated RB with a mixture of acetic and propionic acids (1:1, m/m) with the aim of stabilization. The organic acid mixture (2% based on bran weight) was applied via spraying and the bran samples were stored at +18 °C for 120 days. Although the authors did not analyze typical stabilization indicators such as lipase activity and FFA content, lower lipid acidity (titratable acidity) increase and less lipid oxidation product formation were reported in organic acid-treated RB [86]. Yu et al. (2020) compared various stabilization methods such as MW (700 W for 2, 4, 6 min), steam heating (for 20, 40, 60 min), dry heating (at 105 °C for 30, 60, 90 min), IR heating (at 105 °C for 30, 60, 90 min), autoclaving (at 121 °C for 20 min), extrusion (at 60, 65, 115, 120 °C subsequent heating, 400–500 rpm screw speed), enzyme treatment (pepsin and papain), low-temperature storage (at 4, −18, and −80 °C for 72 h), ultraviolet irradiation (at 254 nm for 6, 12, 18 h), and ultrasound (28 kHz and 300 W for 30, 60, 90 min) for acid value, lipase, and peroxidase activities [87]. The authors reported that autoclaving is the most effective method for lipase inactivation at the noted conditions. Significant decreases in acid values were found after MW, autoclaving, steam heating, low-temperature storage, IR heating, and extrusion treatments ($p < 0.05$). However, non-thermal methods were not effective in terms of lipase inactivation. Residual lipase activities of RB treated with UV radiation even for 18 h and RB stored at extreme low temperature (−80 °C for 72 h) were 57% and 58%, respectively. Moreover, ultrasound markedly increased the peroxide value of RB oil [87].

Parboiling is another practice employed for RB stabilization. Although the term "parboiling" does not define a specific condition, it generally refers to soaking the paddy in water (at varying temperatures) followed by a short steaming procedure and drying (solar drying in most cases). In general, the bran obtained from milling of parboiled paddy is used for extraction of RB oil [49].

3. Stabilization of Individual Rice Bran Fractions

In sophisticated multi-brake systems, brown rice is whitened in several stages in order to avoid heat generation, which causes higher rates of cracking and broken grains. Therefore, RB is obtained step-by-step and all of these fractions are accumulated as composite RB at the end of the process. These fractions do not directly represent the outer, middle, and inner bran layers of the grain, since RB is not removed layer-by-layer, but rather, more like scraped away piece-by-piece. Nevertheless, it is not incorrect to indicate that these fractions contain more pieces from the outermost to innermost layers of the bran as the milling progresses.

Yılmaz (2016) showed that the FFA content of the unprocessed bran fraction obtained from the last whitening machine and water-mist polisher (BF3), which contain mostly the innermost bran layer, was significantly higher than that of the unprocessed bran fractions obtained from the first (BF1) and second (BF2) whitening machines, which contain mostly the outer layers of the bran ($p < 0.05$) [61]. This result was attributed to the water-mist polishing which may have favored the lipolytic enzyme activity due to the addition of a small amount of water and an increase in temperature due to the friction forces during this milling step. Moreover, it was found that the lowest increase in the FFA content (from 4.88% to 6.00%) was observed in IR-treated (at 700 W for 3 min) BF1 during 3 months of storage at ambient temperature [61]. Similarly, Lavanya et al. (2019) found that the rancidity levels were significantly different in bran fractions, and FFA, acid value, and peroxide values significantly increased from the first fraction to the third fraction (from outer to inner RB layers) [59]. This trend continued throughout the storage of MW-stabilized fractions as well. The increase in FFA was also greater in the BF3 sample compared with the BF1 sample, which was exposed to the same MW power for the same process time [59]. Yu et al. (2022) classified RB into four fractions using a specific surface abrasion apparatus and indicated that fractions one and two mainly consisted of pericarp, testa, and nucellus, while fractions three and four mainly consisted of aleurone layer. Controversially, the researchers found that lipase activity decreased significantly from the outer fractions to the inner fractions [88].

4. Stabilization of Other Rice Milling by–Products Comprising Rice Bran

Brown rice or cargo is a whole grain and can be defined as the unpolished version of the rice grain consisting of RB, rice germ, and endosperm. Immature rice is also a whole grain and a form of brown rice. Immature rice is defined as either chalky, thin kernels or kernels that have green seed coats. Immature rice is produced as a by-product during the rice milling process in around 5% of the paddy milled [89]. Like the RB itself, brown rice and immature rice are also unstable in nature due to the presence of RB and require a stabilization process for long-term storage.

A very recent study investigated the effect of cold plasma, an innovative non-thermal processing technology, on the storage stability of brown rice [90]. Researchers found that peroxidase activity was completely inactivated when cold plasma (at 160 kV for 15 min) was applied to brown rice in the form of brown rice powder. However, 61% of residual peroxidase activity was found after treatment under the same conditions when cold plasma was applied to whole brown rice grain [90]. In other words, cold plasma had a very limited effect on peroxidase activity when applied to the grain itself instead of the powder form, probably due to the low depth of penetration.

Qian et al. (2014) applied pulsed electric field (PEF) treatment to monolayer brown rice grains and reported that lipase activity could be significantly inactivated by PEF [91]. The authors found that the voltage was the most important parameter for the inactivation efficiency of the process, followed by frequency and pulse width [91].

Bergonio et al. (2016) investigated the effect of dry heating (at 60 °C for 15, 20, and 25 min), MW heating (800 W, 2450 MHz, for 30, 60, and 90 s), and steam (for 30, 60, and 90 s) on lipase activity and FFA content of brown rice and found that all treated brown rice samples showed significantly lower FFA content and lipase activity after the process. However, a steady and significant increase in FFA content of each treated and untreated brown rice sample was noted during storage, while the increase was significantly lower for treated samples [92].

Yılmaz et al. (2018) stabilized immature rice grain using either medium or short-wave IR radiation at 1000–1600 W for various process times and showed that the FFA content of untreated immature rice increased from 5.59% to 35.71% at the end of 3 months of storage at room temperature, while the increase in FFA was statistically insignificant in the samples treated with both medium and short-wave IR radiation at 1600 W for longer than 4 min [89].

5. Factors Affecting Hydrolytic Stability of Unprocessed Rice Bran

Hydrolytic stability of raw RB is affected by various factors, including variety, milling conditions, storage conditions, and the analytical procedure employed for the stability markers.

Rice variety remarkably affects the hydrolytic stability of RB and therefore, it is hard to propose a single condition for effective stabilization. Different rice varieties milled within the same milling system under the same conditions and stored at the same temperature and RH may have different levels of FFA [43]. Considering the remarkable effect of variety, the use of breeding techniques is reported to have a potential to increase RB stability against lipid hydrolysis [93,94]. Goffman and Bergman (2003a) stated that hydrolytic rancidity was highly and significantly correlated with esterase activity but not with RB oil concentration [93]. The authors reported about two-fold higher hydrolytic rancidity and esterase activity in RB samples with low oil content (16–20%) compared with samples with high oil content (24–28%). Researchers found 58% higher FFA content in the Cypress variety, which had 26% higher lipase activity than the Earl variety at the end of 5 months of storage at room temperature [94]. It was concluded that the intensity of the hydrolytic rancidity is primarily related to lipase activity of RB and, therefore, selection of varieties with low lipase activity could be useful for increasing RB stability. Rattanathanan et al. (2022) investigated the hydrolytic stability of either regular non-pigmented RB or black RB and reported similar lipase activities (269 and 241 U/kg, respectively) [95]. Controversially, Goffman and Bergman (2003a) reported that red RB had lower values for both hydrolytic rancidity and esterase activity and attributed this result to the high tannin content that may inhibit the lipase activity [93].

Higher rates of FFA accumulation in the first stages of storage were reported in many studies [43,55,60]. Goffman and Bergman (2003b) indicated that depletion of RB triacylglycerols is concentration-dependent. The lipolytic activity is maximal at the beginning of storage when the triacylglycerol content is highest and decreases gradually with decreasing concentration [94]. Additionally, it was clearly shown that the food matrix is considerably decisive with regard to RB oil degradation mechanism [43]. Almost 4-fold higher FFA occurred in raw RB when stored in bran matrix (68.27%) compared with oil matrix (17.94%) at the end of 6 months of storage at 25 °C. This result was attributed to the fact that lipase exhibited enzyme activity when the substrate oil is dispersed in a medium containing considerable amounts of water [43]. Similarly, Pourali et al. (2009) showed that the FFA content of stored (at 25 °C) raw RB increased from 5.6% to 36.0%, while the FFA content of RBO extracted using hexane increased from 5.0% to 5.6% at the end of 12 weeks [84].

Determination of FFA content using different analytical procedures is another factor that complicates the comparison of hydrolytic stability of RB. For instance, Escamilla-Castillo et al. reported 5.25% increase in FFA (from 0.14% to 5.39%) at the end of 98 days (spectrophotometric determination), while Mujahid et al. found 87.3% increase in FFA (from 9.5% to 96.8%) after 345 days (titrimetric determination) in raw RB stored at room temperature [52,53].

Degree of milling, which defines the amount of bran removed from brown rice, affects the composition of the bran since it affects the bran layers or fractions included in the composite RB. Thus, the hydrolytic stabilities of RB obtained from rice grains that were milled to different degrees may be different. Moreover, some studies obtained the RB sample using laboratory-scale portable rice debranners that have completely different milling dynamics and therefore, it should be noted that studies using this type of sampling may not reflect the behavior of the commercial RB sample.

6. Nutritive Value of Rice Bran and the Effect of Stabilization

RB contains significant amounts of macronutrients such as proteins, fats, and fiber and bioactive micronutrients such as vitamins, especially B group vitamins and tocols, minerals, γ-oryzanol, phytosterols, and squalene [96]. The composition of RB is presented in Table 2. Although the amount of nutrients can vary depending on factors such as rice variety, climatic conditions, agricultural practices, rice milling system, maturity level of the grain, method of analysis, type of solvents used, etc., Table 2 presents an average range of values in the literature. Stabilization methods also affect the nutritional attributes of RB, as discussed in a recent review [97].

Crude protein content of RB ranged between 11% and 16% [50,51,53,68,73,98,99]. The potential of RB as a protein source should be emphasized since plant-based protein sources are one of the largest growing product groups in the food market. Studies on RB protein concentrates and their properties have dramatically increased in recent years. RB protein has high nutritional value and is efficiently digested; it has a protein efficiency ratio of 1.6 [1]. Although there are discrepancies in the amino acid composition of RB protein in the literature, probably due to the variation in rice cultivars, all essential amino acids are present in varying proportions in RB protein [22,40]. It has been reported that the concentrates of RB protein have an efficiency ratio of 2.0–2.2, which is comparable to milk protein casein (2.5) [1,22]. Prakash and Ramaswamy (1996) found that threonine and isoleucine are limiting amino acids in RB protein concentrates [22]. On the other hand, Wang et al. (2015) reported that the contents of essential amino acids such as phenylalanine, valine, methionine, leucine, and histidine were higher in RB protein concentrates compared with soy protein isolates [100]. Moreover, antioxidant, anti-inflammatory, antimicrobial, antihypertensive, antidiabetic, and anticancer activities were reported for hydrolysates and peptides derived from RB protein [27].

Table 2. Nutritional composition of rice bran.

Component	Average Range	Reference
Crude fat (%)	18–23	[50,68,73,98,99,101]
Crude protein (%)	11–16	[50,51,53,68,73,98,99]
Ash (%)	8–12	[50,68,73,98,99]
Soluble dietary fiber (%)	2–5	[50,73,98,99]
Insoluble dietary fiber (%)	20–27	
Total dietary fiber (%)	22–32	
γ-oryzanol (g/kg)	0.5–5.5	[46,50,60,82,95]
	Vitamins	
Total Tocopherols (mg/kg)	100–150	[60,82,101]
α-T (mg/kg)	50–130	
β-T (mg/kg)	2–10	
γ-T (mg/kg)	10–50	
δ-T (mg/kg)	0–2	
Total Tocotrienols (mg/kg)	130–170	[82,101]
α-T3 (mg/kg)	38	
β-T3 (mg/kg)	–	
γ-T3 (mg/kg)	120–140	
δ-T3 (mg/kg)	0–10	
Vitamin B1 (Thiamin) (mg/kg)	12–40	[5,50,99,102]
Vitamin B2 (Riboflavin) (mg/kg)	1–4	[5,50,99,102]
Vitamin B3 (Niacin) (mg/kg)	300–800	[5,50,99,102]
Vitamin B5 (Pantothenic acid) (mg/kg)	74	[102]
Vitamin B6 (mg/kg) (Pyridoxamine, pyridoxal, pyridoxine)	20–40	[99,102]

Table 2. *Cont.*

Component	Average Range	Reference
	Minerals	
Ca (mg/kg)	300–1200	[5,16]
K (mg/kg)	5992	[16]
Fe (mg/kg)	86–430	[5,16]
Zn (mg/kg)	50–250	[5,16]
P (mg/kg)	6278	[16]

Although the crude fat content of RB has been found to be in the range of 18–23% in many studies [50,68,73,99,101], very low values such as 9% [53] and very high values such as 30% [1] have also been reported. Goffman et al. (2003) investigated the lipid content and fatty acid profiles of RB obtained from 204 genetically diverse rice accessions and found that genotype effects were highly statistically significant ($p < 0.001$) [103]. It was stated that the effect of season (year) on oil content and fatty acids was also significant, except for palmitic acid. In general, saturated and unsaturated fatty acid contents of RB oil ranged between 15 and 30% and 70 and 85%, respectively. Palmitic acid is the most abundant saturated fatty acid, while oleic and linoleic acids are the dominant fatty acids among unsaturated fatty acids [1,60,82,104]. Similar fatty acid distributions were reported for RB oil in the literature. Besides, it was reported that stabilization processes such as IR and MW did not significantly alter the fatty acid composition of RB [56,60].

RB oil contains over 4% unsaponifiable matter mainly consisting of plant sterols (mostly campesterol, stigmasterol, and β–sitosterol), triterpene alcohols (24–methylene cycloartenol and cycloartenol), tocopherols, and tocotrienols [105]. Among them, γ-oryzanol, which is a mixture of ferulic acid esters of triterpenoid alcohols, composes 20–30% of the unsaponifiable matter and 1–3% of RB oil and is one of the most important bioactive components of RB oil [1]. Gamma-oryzanol has great potential in nutraceutical, pharmaceutical, and cosmeceutical applications due to its numerous health benefits. It has been reported that γ-oryzanol has notable antioxidant (at least four times higher than tocopherols), anti–inflammatory, anticancer, and cholesterol-lowering properties. In addition, it helps to reduce platelet aggregation, fight menopausal distress, increase bile acid excretion, inhibit and tumor growth, and also has protective effects against UV light [105–111]. Furthermore, γ-oryzanol is a very stable compound. It has been demonstrated in many studies that there is no decrease in the amount of γ-oryzanol as a result of various stabilization processes such as IR heating [60,61,63,87], MW heating [46,58,87], ohmic heating [66], hot air heating [76] RF heating [68], autoclaving, extrusion, steam heating, ultrasound, and UV radiation [87].

Tocopherols are another important bioactive component found in RB oil. Among the tocopherol analogues, α-tocopherol is the most abundant analogue in RB oil, followed by γ-tocopherol [46,58,60,61,82,112]. Unlike γ-oryzanol, tocopherols are more susceptible to heat and showed significant decreases as a result of stabilization processes such as extrusion [72], IR [60,113], MW [58,113], and steaming [95].

The contents of both γ-oryzanol and tocopherols are affected by both variety and environmental factors. However, Bergman and Xu (2003) showed that growing environment has a greater effect on γ-oryzanol and tocopherol levels than genotype [112]. Many researchers found that the γ-oryzanol content of RB is many times higher than tocopherols [60,61,112]. Moreover, individual RB fractions have different levels of γ-oryzanol and tocopherols [61,114]. Yılmaz (2016) reported that the fraction obtained from the first mill, which contains mostly the outermost bran layer, is the richest in γ-oryzanol and this amount decreases towards the inner layers. However, tocopherols show the opposite trend [61]. Similarly, Britz et al. (2007) stated that γ-oryzanol is concentrated in the outermost pericarp and seed coat, while tocols are well presented in the deeper aleurone [114].

Furthermore, RB oil is the only readily available oil that contains significant levels of tocotrienols, with the exception of palm oil [41]

In addition, RB is a notable source of B group vitamins, especially niacin. Yılmaz and Tuncel (2015) found 36.92 mg/kg of thiamin (B_1), 0.91 mg/kg of riboflavin (B_2), 338.50 mg/kg of niacin (B_3), 8.46 mg/kg of pyridoxamine, 1.90 mg/kg of pyridoxal, and 12.86 mg/kg of pyridoxine (23.23 mg/kg Vitamin B_6 in total) in raw RB [99]. The authors observed a significant decrease of up to 42% in thiamin content, while no changes in the contents of riboflavin, niacin, and total B6 vitamin were observed as a result of IR stabilization [99]. Moreover, Rafe and Sadeghian (2017) reported that the content of vitamins B_2, B_3, B_5, and folic acid remained unchanged as a result of extrusion [73].

Other important micronutrients in RB are phospholipids or lecithin (one of the major classes of lipid in RB oil), phytosterols, squalene, and phenolic acids. The main phytosterols are β–sitosterol (c. 50%), campesterol (c. 20%), stigmasterol (c. 15%), and isofucosterol (c. 5%) [115]. Phytosterols are known for their cholesterol-lowering properties due to the inhibition of cholesterol absorption in the small intestines. Squalene, which is a triterpene and intermediate metabolite of cholesterol synthesis, is found in shark fish liver oil and vegetable sources such as olive oil, palm oil, wheat-germ oil, and amaranth oil. RB oil has higher squalene content compared with commercial vegetable squalene derived from olive oil [115]. Pokkanta et al. (2022) found 1252 mg/kg of total phytosterols (stigmasterol, campesterol, β-stigmasterol), 99.55 mg/kg of squalene, 3 mg/kg of cholecalciferol, and 2.45 mg/kg of phylloquinone in RB [46].

RB is also a notable source of phenolics, especially in bound form. Ferulic (1863 μg/g) and *p*-coumaric (647 μg/g) acids are the most abundant phenolic acids in RB [99]. The presence of gallic acid, protocatechuic acid, 4-hydroxybenzoic acid, catechin, vanillic acid, chlorogenic acid, caffeic acid, kaempferol, epigallocatechin, trans–*p*–coumaric acid, syringic acid, and sinapic acid was also reported [46,99,116]. In general, pigmented RB extracts have higher contents of phenolic compounds and antioxidant activity [117].

RB contains about 50% carbohydrates, mainly starch [118]. Total dietary fiber (TDF) content of RB ranges from 22 to 32%, with less than 5–6% as soluble dietary fiber (SDF) [50,73,98,99]. Sharma et al. (2004) reported 10–20% starch, 3–8% reducing sugars, 8–11% hemicelluloses, and 10–12% celluloses in RB [51]. Parboiling may also notably affect the properties of starch in rice grain.

Antinutrients in RB are not widely studied. The phytic acid content of raw RB was 23–93 mg/g, depending on rice cultivar, and significant decreases in phytate content were reported following extrusion [50,73,102,119,120] and IR stabilization [94]. Kaur et al. (2015) reported 53.82 TIU/g of trypsin inhibitor activity in raw RB and found a significant decrease following extrusion treatment [102].

7. Conclusions

In conclusion, RB is a highly nutritious and technologically functional by-product with remarkable health-promoting effects. Due to all these mentioned properties, it deserves much more than simply being used as feed. RB forms about 10% of the paddy milled, which corresponds to a very high quantity, since rice is one of the main crops produced worldwide. Reusing such valuable by-products is a key sustainability strategy. RB should either be stabilized or exposed to a process suitable for the end product to which it will be transformed as soon as possible since it deteriorates very rapidly. Although each stabilization method has unique advantages and disadvantages, most of the stabilization methods reported in the literature failed to irreversibly prevent hydrolytic degradation over a long time. Among the stabilization approaches, extrusion, MW, and IR heating seem more promising with regard to industrial scale processing. In addition, steaming and drying (parboiling) is a common and useful process for the RB oil industry. It is critically important to take measures to keep the moisture content of the stabilized RB low throughout storage to prevent reactivation of lipases.

Funding: This research received no external funding.

Data Availability Statement: Not applicable.

Acknowledgments: I would like to express my gratitude to our core research group, Fatma Korkmaz, and N. Barış TUNCEL for their cooperation and support in the realization of many projects on rice milling by-products over a number of years.

Conflicts of Interest: The author declares no conflict of interest.

References

1. Kahlon, T.S. Rice bran: Production, composition, functionality and food applications, physiological benefits. In *Fiber Ingredients: Food Applications and Health Benefits*; Cho, S., Samuel, P., Eds.; CRC Press: Boca Raton, FL, USA, 2009; pp. 305–321.
2. Basri, M.S.M.; Mustapha, F.; Mazlan, N.; Ishak, M.R. Optimization of rice husk ash based geopolymers coating composite for enhancement in flexural properties and microstructure using response surface methodology. *Coatings* **2020**, *10*, 165. [CrossRef]
3. Rice Knowledge Bank, IRRIa. Available online: http://www.knowledgebank.irri.org/step-by-step-production/postharvest/rice-by-products/rice-husk (accessed on 16 February 2023).
4. Mir, S.A.; Shah, M.A.; Bosco, S.J.D.; Sunooj, K.V.; Farooq, S. A review on nutritional properties, shelf life, health aspects, and consumption of brown rice in comparison with white rice. *Cereal Chem.* **2020**, *97*, 895–903. [CrossRef]
5. Saleh, A.S.M.; Wang, P.; Wang, N.; Yang, L.; Xiao, Z.G. Brown rice versus white rice: Nutritional quality, potential health benefits, development of food products, and preservation technologies. *Compr. Rev. Food Sci. Food Saf.* **2019**, *18*, 1070–1096. [CrossRef] [PubMed]
6. Selvam, S.; Masilamani, P.; Umashankar, P.T.; Albert, V.A. Opportunities and challenges in marketing of brown rice. In *Brown Rice*; Monickavasagan, A., Santhakumar, C., Venkatachalapathy, N., Eds.; Springer: Cham, Switzerland, 2017; pp. 296–305.
7. Rice Knowledge Bank, IRRIb. Available online: http://www.knowledgebank.irri.org/step-by-step-production/postharvest/milling/milling-systems?tmpl=component&print=1 (accessed on 17 February 2023).
8. Saunders, R.M. Rice bran: Composition and potential food uses. *Food Rev. Int.* **1985**, *1*, 465–495. [CrossRef]
9. Malekian, F.; Rao, R.M.; Prinyawiwatkul, W.; Marshall, W.E.; Windhauser, M.; Ahmedna, M. Lipase and lipoxygenase activity, functionality, and nutrient losses in rice bran during storage. *Bull. La. Agric. Exp. Stn.* **2000**, *870*, 1–68.
10. Bhardwaj, K.; Raju, A.; Rajasekharan, R. Identification, purification, and characterization of a thermally stable lipase from rice bran. A new member of the (phospho) lipase family. *Plant Physiol.* **2001**, *127*, 1728–1738. [CrossRef]
11. Food and Agriculture Organization of the United Nations, FAOSTAT. 2021. Available online: https://www.fao.org/faostat/en/#data/QCL (accessed on 11 February 2023).
12. Espinales, C.; Cuesta, A.; Tapia, J.; Palacios-Ponce, S.; Penas, E.; Martinez-Villaluenga, C.; Espinoza, A.; Caceres, P.J. The effect of stabilized rice bran addition on physicochemical, sensory, and techno-functional properties of bread. *Foods* **2022**, *11*, 3328. [CrossRef]
13. Doan, N.T.T.; Lai, Q.D.; Vo, H.V.; Nguyen, H.D. Influence of adding rice bran on physio-chemical and sensory properties of bread. *J. Food Meas. Charact.* **2021**, *15*, 5369–5378. [CrossRef]
14. Gul, K.; Yousuf, B.; Singh, A.K.; Singh, P.; Wani, A.B. Rice bran: Nutritional values and its emerging potential for development of functional food—A review. *Bioact. Carbohydr. Diet. Fibre* **2015**, *6*, 24–30. [CrossRef]
15. Tuncel, N.B.; Yılmaz, N.; Kocabıyık, H.; Uygur, A. The effect of infrared stabilized rice bran substitution on physicochemical and sensory properties of pan breads: Part I. *J. Cereal Sci.* **2014**, *59*, 155–161. [CrossRef]
16. Tuncel, N.B.; Yılmaz, N.; Kocabıyık, H.; Uygur, A. The effect of infrared stabilized rice bran substitution on B vitamins, minerals and phytic acid content of pan breads: Part II. *J. Cereal Sci.* **2014**, *59*, 162–166. [CrossRef]
17. Yılmaz Tuncel, N.; Kaya, E.; Karaman, M. Rice bran substituted Turkish noodles (Erişte): Textural, sensorial, and nutritional properties. *Cereal Chem.* **2017**, *94*, 903–908. [CrossRef]
18. Yılmaz, N.; Tuncel, N.B.; Kocabıyık, H. The effect of infrared stabilized rice bran substitution on nutritional, sensory, and textural properties of cracker. *Eur. Food Res. Technol.* **2014**, *239*, 259–265. [CrossRef]
19. Ayoub, W.S.; Zahoor, I.; Dar, A.H.; Anjum, N.; Pandiselvam, R.; Farooq, S.; Rusu, A.V.; Rocha, J.M.; Trif, M.; Jeevarathinam, G. Effect of incorporation of wheat bran, rice bran and banana peel powder on the mesostructure and physicochemical characteristics of biscuits. *Front. Nutr.* **2022**, *9*, 1016717. [CrossRef] [PubMed]
20. Akter, D.; Begum, R.; Rahman, M.N.; Talukder, N.; Alam, M.J. Optimization of extraction process parameter for rice bran protein concentrate and its utilization in high protein biscuit formulation. *Curr. Res. Nutr. Food Sci.* **2020**, *8*, 596–608. [CrossRef]
21. Albayrak, B.B.; Tuncel, N.B.; Yılmaz Tuncel, N.; Masatcioglu, M.T. Extrusion cooking of immature rice grain: Under-utilized by-product of rice milling process. *J. Food Sci. Technol.* **2020**, *57*, 2905–2915. [CrossRef]
22. Prakash, J. Rice bran proteins: Properties and food uses. *Crit. Rev. Food Sci. Nutr.* **1996**, *36*, 537–552. [CrossRef]
23. Echeverria, L.; da Silva, C.; Danesi, E.D.G.; Porciuncula, B.D.A.; Barros, B.C.B. Characterization of okara and rice bran and their application as fat substitutes in chicken nugget formulations. *LWT—Food Sci. Technol.* **2022**, *161*, 113383. [CrossRef]
24. Jiang, R.S.; Xiao, Z.G.; Huo, J.J.; Wang, H.G.; Li, H.; Su, S.; Duan, Y.M.; Gao, Y.Z. Effects of rice bran content on plant-based simulated meat: From the aspects of apparent properties and structural characteristics. *Food Chem.* **2022**, *380*, 131842. [CrossRef]

25. Chakraborty, M.; Budhwar, S.; Kumar, S. Potential of milling byproducts for the formulation of health drink and detox tea-substitute. *J. Food Meas. Charact.* **2022**, *16*, 3153–3165. [CrossRef]
26. Tran, K.N.; Gidley, M.J.; Fitzgerald, M. Opportunities and challenges in processing of by-product of rice milling protein as a food ingredient. *Cereal Chem.* **2017**, *94*, 369–376. [CrossRef]
27. Wang, N.; Cui, X.; Duan, Y.; Yang, S.; Wang, P.; Saleh, A.S.M.; Xiao, Z. Potential health benefits and food applications of rice bran protein: Research advances and challenges. *Food Rev. Int.* **2021**, in press. [CrossRef]
28. Tu, Y.; Zhang, X.X.; Wang, L. Effect of salt treatment on the stabilization of Pickering emulsions prepared with rice bran protein. *Food Res. Int.* **2023**, *166*, 112537. [CrossRef] [PubMed]
29. Zheng, L.; Regenstein, J.M.; Wang, Z.J.; Zhang, H.J.; Zhou, L.Y. Reconstituted rice protein: The raw materials, techniques and challenges. *Trends Food Sci. Technol.* **2023**, *133*, 267–276. [CrossRef]
30. Magnaye, M.J.F.A.; Mopera, L.E.; Flores, F.P. Effect of rice bran protein concentrate as wall material adjunct on selected physicochemical and release properties of microencapsulated beta-carotene. *Food Sci. Technol. Int.* **2022**, *28*, 653–662. [CrossRef]
31. Bonifacino, C.; Palazolo, G.G.; Panizzolo, L.A.; Abirached, C. Study of emulsifying properties of soluble proteins obtained from defatted rice bran concentrate. *J. Am. Oil Chem. Soc.* **2021**, *98*, 851–860. [CrossRef]
32. Pang, M.; Kang, S.M.; Liu, L.; Ma, T.F.; Zheng, Z.; Cao, L.L. Physicochemical properties and cookie-making performance as fat replacer of wax-based rice bran oil oleogels. *Gels* **2023**, *9*, 13. [CrossRef]
33. Wang, Z.H.; Li, S.Z.; Zhang, M.; Yang, H.Y.; Li, G.; Ren, X.; Liang, S. Optimization of oil extraction from rice bran with mixed solvent using response surface methodology. *Foods* **2022**, *11*, 3849. [CrossRef]
34. Chen, H.S.; He, S.D.; Sun, H.J.; Li, Q.Y.; Gao, K.; Miao, X.Y.; Xiang, J.; Wu, X.J.; Gao, L.M.; Zhang, Y. A comparative study on extraction and physicochemical properties of soluble dietary fiber from glutinous rice bran using different methods. *Separations* **2023**, *10*, 90. [CrossRef]
35. Ismail, N.A.; Zhao, J. Effects of ultrasound and steam explosion treatments on the physicochemical properties of rice bran fibre. *Pertenika J. Trop. Agric. Sci.* **2022**, *45*, 893–918. [CrossRef]
36. Wang, L.; Duan, W.; Zhou, S.; Qian, H.F.; Zhang, H.; Qi, X.G. Effect of rice bran fibre on the quality of rice pasta. *Int. J. Food Sci. Technol.* **2018**, *53*, 81–87. [CrossRef]
37. Gu, X.Y.; Du, L.Y.; Meng, Z. Comparative study of natural wax-based W/O emulsion gels: Microstructure and macroscopic properties. *Food Res. Int.* **2023**, *165*, 112509. [CrossRef] [PubMed]
38. Dent, T.; Hallinan, R.; Chitchumroonchokchai, C.; Maleky, F. Rice bran wax structured oleogels and in vitro bioaccessibility of curcumin. *J. Am. Oil Chem. Soc.* **2022**, *99*, 299–311. [CrossRef]
39. Ghosh, S.; Bollinedi, H.; Krishnan, S.G.; Kundu, A.; Singh, A.; Bhowmick, P.K.; Singh, A.; Nagarajan, M.; Vinod, K.K.; Ellur, R.K.; et al. From farm to plate: Spatio-temporal characterization revealed compositional changes and reduced retention of gamma-oryzanol upon processing in rice. *Front. Nutr.* **2022**, *9*, 1040362. [CrossRef]
40. Zaky, A.A.; Chen, Z.; Qin, M.; Wang, M.Z.; Jia, Y.M. Assessment of antioxidant activity, amino acids, phenolic acids and functional attributes in defatted rice bran and rice bran protein concentrate. *Prog. Nutr.* **2020**, *22*, e2020069.
41. Cosmetic Ingredient Review Expert Panel. Amended final report on the safety assessment of Oryza Sativa (rice) bran oil, Oryza Sativa (rice) germ oil, rice bran acid, Oryza Sativa (rice) bran wax, hydrogenated rice bran wax, Oryza Sativa (rice) bran extract, Oryza Sativa (rice) extract, Oryza Sativa (rice) germ powder, Oryza Sativa (rice) starch, Oryza Sativa (rice) bran, hydrolyzed rice bran extract, hydrolyzed rice bran protein, hydrolyzed rice extract, and hydrolyzed rice protein. *Int. J. Toxicol.* **2006**, *2*, 91–120.
42. Tao, J.; Rao, R.; Liuzzo, J. Microwave heating for rice bran stabilization. *J. Microw. Power Electromagn. Energy* **1993**, *28*, 156–164. [CrossRef]
43. Yılmaz Tuncel, N.; Yılmaz Korkmaz, F. Comparison of lipid degradation in raw and infrared stabilized bran and rice bran oil: Matrix effect. *J. Food Meas. Charact.* **2021**, *15*, 1057–1067. [CrossRef]
44. Ju, Y.H.; Vali, S.R. Rice bran oil as a potential resource for biodiesel: A review. *J. Sci. Ind. Res.* **2005**, *64*, 866–882.
45. Saji, N.; Schwarz, L.J.; Santhakumar, A.B.; Blanchard, C.L. Stabilization treatment of rice bran alters phenolic content and antioxidant activity. *Cereal Chem.* **2020**, *97*, 281–292. [CrossRef]
46. Pokkanta, P.; Yuenyong, J.; Mahatheeranont, S.; Jiamyangyuen, S.; Sookwong, P. Microwave treatment of rice bran and its effect on phytochemical content and antioxidant activity. *Sci. Rep.* **2022**, *12*, 7708. [CrossRef] [PubMed]
47. Rashid, M.T.; Liu, K.; Hen, S.; Jatoi, M.A.; Sarpong, F. Nutritional composition and volatile compounds stability in dry-heat and extruded stabilised rice bran during storage. *Int. J. Food Sci. Technol.* **2023**, *in press*. [CrossRef]
48. Chen, Y.X.; Ma, Y.X.; Dong, L.H.; Jia, X.C.; Liu, L.; Huang, F.; Chi, J.W.; Xiao, J.; Zhang, M.W.; Zhang, R.F. Extrusion and fungal fermentation change the profile and antioxidant activity of free and bound phenolics in rice bran together with the phenolic bioaccessibility. *LWT—Food Sci. Technol.* **2019**, *115*, 108461. [CrossRef]
49. Amarasinghe, B.M.W.P.K.; Kumarasiri, N.C.; Gongodavilage, N.G. Effect of method of stabilization on aqueous extraction of rice bran oil. *Food Bioprod. Process.* **2009**, *87*, 108–114. [CrossRef]
50. Fuh, W.S.; Chiang, B.H. Dephytinisation of rice bran and manufacturing a new food ingredient. *J. Sci. Food Agric.* **2001**, *81*, 1419–1425. [CrossRef]
51. Sharma, H.R.; Chauhan, G.S.; Agrawal, K. Physico-chemical characteristics of rice bran processed by dry heating and extrusion cooking. *Int. J. Food Prop.* **2004**, *7*, 603–614. [CrossRef]

52. Mujahid, A.; Haq, I.; Asif, M.; Gilani, A.H. Effect of various processing techniques and different levels of antioxidant on stability of rice bran during storage. *J. Sci. Food Agric.* **2005**, *85*, 847–852. [CrossRef]
53. Escamillo-Castillo, B.; Varela-Montellano, R.; Sanchez-Tovar, S.A.; Solis-Fuentes, J.A.; Duran-de-Bazua, C. Extrusion deactivation of rice bran enzymes by pH modification. *Eur. J. Lipid Sci. Technol.* **2005**, *107*, 871–876. [CrossRef]
54. Li, Y.; Gao, Y.; Wang, Y.; Fan, M.; Wang, L.; Qian, H. Analysis of the aroma volatile compounds in different stabilized rice bran during storage. *Food Chem.* **2023**, *405*, 134753. [CrossRef]
55. Ramezanzadeh, F.M.; Rao, R.M.; Windhause, M.; Prinyawiwatkul, W.; Tulley, R.; Marshall, W.E. Prevention of hydrolytic rancidity in rice bran during storage. *J. Agric. Food Chem.* **1999**, *47*, 3050–3052. [CrossRef]
56. Ramezanzadeh, F.M.; Rao, R.M.; Windhause, M.; Prinyawiwatkul, W.; Marshall, W.E.; Windhauser, M. Effects of microwave heat, packaging, and storage temperature on fatty acid and proximate compositions in rice bran. *J. Agric. Food Chem.* **2000**, *48*, 464–467. [CrossRef] [PubMed]
57. Patil, S.S.; Kar, A.; Mohapatra, D. Stabilization of rice bran using microwave: Process optimization and storage studies. *Food Bioprod. Process.* **2016**, *99*, 204–211. [CrossRef]
58. Li, B.; Zhao, L.; Xu, B.; Deng, B.; Liu, Y.; Dong, Y. Rice bran real-time stabilization technology with flowing microwave radiation: Its impact on rancidity and some bioactive compounds. *Qual. Assur. Saf. Crops Foods* **2018**, *10*, 25–34. [CrossRef]
59. Lavanya, M.N.; Saikiran, K.C.H.; Venkatachalapathy, N. Stabilization of rice bran milling fractions using microwave heating and its effect on storage. *J. Food Sci. Technol.* **2019**, *56*, 889–895. [CrossRef]
60. Yılmaz, N.; Tuncel, N.B.; Kocabıyık, H. Infrared stabilization of rice bran and its effects on γ-oryzanol content, tocopherols and fatty acid composition. *J. Sci. Food Agric.* **2014**, *94*, 1568–1576. [CrossRef] [PubMed]
61. Yılmaz, N. Middle infrared stabilization of individual rice bran milling fractions. *Food Chem.* **2016**, *190*, 179–185. [CrossRef] [PubMed]
62. Wang, T.; Khir, R.; Pan, Z.; Yuan, Q. Simultaneous rough rice drying and rice bran stabilization using infrared radiation heating. *LWT—Food Sci. Technol.* **2017**, *78*, 281–288. [CrossRef]
63. Irakli, M.; Kleisiaris, F.; Mygdalia, A.; Katsantonis, D. Stabilization of rice bran and its effect on bioactive compounds content, antioxidant activity and storage stability during infrared radiation heating. *J. Cereal Sci.* **2018**, *80*, 135–142. [CrossRef]
64. Yan, W.; Liu, Q.; Wang, Y.; Tao, T.; Liu, B.; Liu, J.; Ding, C. Inhibition of lipid and aroma deterioration in rice bran by infrared heating. *Food Bioprocess. Technol.* **2020**, *13*, 1677–1687. [CrossRef]
65. Lakkakula, N.R.; Lima, M.; Walker, T. Rice bran stabilization and rice bran oil extraction using ohmic heating. *Biosour. Technol.* **2004**, *92*, 157–161. [CrossRef]
66. Loypimai, P.; Moonggarm, A.; Chottanom, P. Effects of ohmic heating on lipase activity, bioactive compounds and antioxidant activity of rice bran. *Aust. J. Basic Appl. Sci.* **2009**, *3*, 3642–3652.
67. Dhingra, D.; Chopra, S.; Rai, D.R. Stabilization of raw rice bran using ohmic heating. *Agric. Res.* **2012**, *1*, 392–398. [CrossRef]
68. Ling, B.; Lyng, J.G.; Wang, S. Effects of hot air-assisted radio frequency heating on enzyme inactivation, lipid stability and product quality of rice bran. *LWT—Food Sci. Technol.* **2018**, *91*, 453–459. [CrossRef]
69. Chen, Y.H.; Yen, Y.F.; Chen, S.D. Effects of radio frequency heating on the stability and antioxidant properties of rice bran. *Foods* **2021**, *10*, 810. [CrossRef] [PubMed]
70. Randall, J.M.; Sayre, R.N.; Schultz, W.G.; Fong, R.Y.; Mossman, A.P.; Tribelhorn, R.E.; Saunders, R.M. Rice bran stabilization by extrusion cooking for extraction of edible oil. *J. Food Sci.* **1985**, *50*, 361–364. [CrossRef]
71. Kim, C.J.; Byun, S.M.; Cheigh, H.S.; Kwon, T.W. Optimization of extrusion rice bran stabilization process. *J. Food Sci.* **1987**, *52*, 1355. [CrossRef]
72. Shin, T.S.; Godber, J.S.; Martin, D.E.; Wells, J.H. Hydrolytic stability and changes in E vitamers and oryzanol of extruded rice bran during storage. *J. Food Sci.* **1997**, *62*, 704–728. [CrossRef]
73. Rafe, A.; Sadeghian, A. Stabilization of Tarom and Domesiah cultivars rice bran: Physicochemical, functional and nutritional properties. *J. Cereal Sci.* **2017**, *74*, 64–71. [CrossRef]
74. Abdul-Hamid, A.; Sulaiman, R.R.R.; Osman, A.; Saari, N. Preliminary study of the chemical composition of rice milling fractions stabilized by microwave heating. *J. Food Compos. Anal.* **2007**, *20*, 627–637. [CrossRef]
75. Ertürk, B.; Meral, R. The impact of stabilization on functional, molecular and thermal properties of rice bran. *J. Cereal Sci.* **2019**, *88*, 71–78. [CrossRef]
76. Thanonkaew, A.; Wongyai, S.; McClements, D.J.; Decker, E.A. Effect of stabilization of rice bran by domestic heating on mechanical extraction yield, quality, and antioxidant properties of cold-pressed rice bran oil (*Oryza saltiva* L.). *LWT—Food Sci. Tech.* **2012**, *48*, 231–236. [CrossRef]
77. Orthoefer, F.T.; Eastman, J. Rice bran and oil. In *Rice: Chemistry and Technology*; Champagne, E.T., Ed.; Agricultural Research Service, Southern Regional Research Center, U.S. Department of Agriculture: New Orleans, LA, USA, 2004; Chapter 19; pp. 569–593.
78. Brunschwiler, C.; Heine, D.; Kappeler, S.; Conde-Petit, B.; Nyström, L. Direct measurement of rice bran lipase activity for inactivation kinetics and storage stability prediction. *J. Cereal Sci.* **2013**, *58*, 272–277. [CrossRef]
79. Atungulu, G.G.; Pan, Z. Infrared radiative properties of food materials. In *Infrared Heating for Food and Agricultural Processing*; Pan, Z., Atungulu, G.G., Eds.; CRC Press: Boca Raton, FL, USA, 2011; pp. 29–35.

80. Kubo, M.T.K.; Siguemoto, E.S.; Funcia, E.S.; Augusto, P.E.D.; Curet, S.; Boillereaux, L.; Sastry, S.K.; Gut, J.A.W. Non-thermal effects of microwave and ohmic processing on microbial and enzyme inactivation: A critical review. *Curr. Opin. Food Sci.* **2020**, *35*, 36–48. [CrossRef]
81. Liao, M.; Damayanti, W.; Xu, Y.; Zhao, Y.; Xu, X.; Zheng, Y.; Jiao, S. Hot air-assisted radio frequency heating for stabilization of rice bran: Enzyme activity, phenolic content, antioxidant activity and microstructure. *LWT-Food Sci. Technol.* **2020**, *131*, 109754. [CrossRef]
82. Shin, T.S.; Godber, J.S. Changes of endogenous antioxidants and fatty acid composition in irradiated rice bran during storage. *J. Agric. Food Chem.* **1996**, *44*, 567–573. [CrossRef]
83. Masamran, S.; Chookaew, S.; Tepsongkroh, B.; Supawong, S. Impact of gamma irradiation pre-treatment before subcritical water extraction on recovery yields and antioxidant properties of rice bran extract. *Radiat. Phys. Chem.* **2023**, *207*, 110834. [CrossRef]
84. Pourali, O.; Asghari, F.S.; Yoshida, H. Simultaneous rice bran oil stabilization and extraction using sub-critical water medium. *J. Food Eng.* **2009**, *95*, 510–516. [CrossRef]
85. Raghavendra, M.P.; Kumar, P.R.; Prakash, V. Mechanism of inhibition of rice bran lipase by polyphenols: A case study with chlorogenic acid and caffeic acid. *J. Food Sci.* **2007**, *72*, 412–419. [CrossRef]
86. Gopinger, E.; Ziegler, V.; Catalan, A.A.S.; Krabbe, E.L.; Elias, M.C.; Xavier, E.G. Whole rice bran stabilization using a short chain organic acid mixture. *J. Stored Prod. Res.* **2015**, *61*, 108–113. [CrossRef]
87. Yu, C.W.; Hu, Q.R.; Wang, H.W.; Deng, Z.Y. Comparison of 11 rice bran stabilization methods by analyzing lipase activities. *J. Food Process. Preserv.* **2020**, *44*, e14370. [CrossRef]
88. Yu, C.W.; Luo, T.; Xie, T.; Deng, Z.Y. Classified processing of different rice bran fractions according to their component distributions. *Int. J. Food Sci. Technol.* **2022**, *57*, 4052–4064. [CrossRef]
89. Yılmaz, F.; Yılmaz Tuncel, N.; Tuncel, N.B. Stabilization of immature rice grain using infrared radiation. *Food Chem.* **2018**, *253*, 269–276. [CrossRef] [PubMed]
90. Zeng, Z.; Wang, Y.; Xu, G.; Zhou, L.; Liu, C.; Luo, S. Peroxidase inactivation by cold plasma and its effects on the storage, physicochemical and bioactive properties of brown rice. *Food Biosci.* **2023**, *52*, 102383. [CrossRef]
91. Qian, J.Y.; Gu, T.P.; Jiang, W.; Chen, W. Inactivating effect of pulse electric field on lipase in brown rice. *Innov. Food Sci. Emerg. Technol.* **2014**, *22*, 89–94. [CrossRef]
92. Bergonio, K.B.; Lucatin, L.G.G.; Corpuz, G.A.; Ramos, N.C.; Bartolome, J.; Duldulao, J.B.A. Improved shelf life of brown rice by heat and microwave treatment. *J. Microbiol. Biotechnol. Food Sci.* **2016**, *5*, 378–385. [CrossRef]
93. Goffman, F.D.; Bergman, C. Relationship between hydrolytic rancidity, oil concentration, and esterase activity in rice bran. *Cereal Chem.* **2003**, *80*, 689–692. [CrossRef]
94. Goffman, F.D.; Bergman, C. Hydrolytic degradation of triacylglycerols and changes in fatty acid composition in rice bran during storage. *Cereal Chem.* **2003**, *80*, 459–461. [CrossRef]
95. Rattanathanan, Y.; Kanha, N.; Osiriphun, S.; Rakariyatham, K.; Klangpetch, W.; Laokuldilok, T. Changes in content of antioxidants and hydrolytic stability of black rice bran after heat- and enzymatic stabilizations and degradation kinetics during storage. *J. Food Process. Preserv.* **2022**, *46*, e16795. [CrossRef]
96. Andriani, R.; Subroto, T.; Ishmayana, S.; Kurnia, D. Enhancement methods of antioxidant capacity in rice bran: A review. *Foods* **2022**, *11*, 2994. [CrossRef]
97. Liu, Y.Q.; Strappe, P.; Zhou, Z.K.; Blanchard, C. Impact on the nutritional attributes of rice bran following various stabilization procedures. *Crit. Rev. Food Sci. Nutr.* **2019**, *59*, 2458–2466. [CrossRef]
98. Gualberto, D.G.; Bergman, C.J.; Kazemzadeh, M.; Weber, C.W. Effect of extrusion processing on the soluble and insoluble fiber, and phytic acid contents of cereal brans. *Plant Food Hum. Nutr.* **1997**, *51*, 187–198. [CrossRef] [PubMed]
99. Yılmaz, N.; Tuncel, N.B. The effect of infrared stabilisation on B vitamins, phenolics and and antioxidants in rice bran. *Int. J. Food Sci. Technol.* **2015**, *50*, 84–91. [CrossRef]
100. Wang, C.; Xu, F.; Li, D.; Zhang, M. Physico-chemical and structural properties of four rice bran protein fractions based on the multiple solvent extraction method. *Czech J. Food Sci.* **2015**, *33*, 283–291. [CrossRef]
101. Pradeep, P.M.; Jayadeep, A.; Guha, M.; Singh, V. Hydrothermal and biotechnological treatments on nutraceutical content and antioxidant activity of rice bran. *J. Cereal Sci.* **2014**, *60*, 187–192. [CrossRef]
102. Kaur, S.; Sharma, S.; Singh, B.; Dar, B.N. Effect of extrusion variables (temperature, moisture) on the antinutrient components of cereal brans. *J. Food Sci. Technol.* **2015**, *52*, 1670–1676. [CrossRef]
103. Goffman, F.D.; Pinson, S.; Bergman, C. Genetic diversity for lipid content and fatty acid prodifle in rice bran. *J. Am. Oil Chem. Soc.* **2003**, *80*, 485–490. [CrossRef]
104. Punia, S.; Kumar, M.; Siroha, A.K.; Purewal, S.S. Rice bran oil: Emerging trends in extraction, health benefit, and its industrial application. *Rice Sci.* **2021**, *28*, 217–232. [CrossRef]
105. Lerma-Garcia, M.J.; Herrero-Martinez, J.M.; Simo-Alfonso, E.F.; Mendonça, C.R.B.; Ramis-Ramos, G. Composition, industrial processing and applications of rice bran γ-oryzanol. *Food Chem.* **2009**, *115*, 389–404. [CrossRef]
106. Murase, Y.; Lishima, H. Clinical studies of oral administration of gamma-oryzanol on climacteric complaints and its syndrome. *Obstet. Gynecol. Prac.* **1963**, *12*, 147–149.
107. Xu, Z.; Godber, J.S. Antioxidant activities of major components of γ-oryzanol from rice bran using a linoleic acid model. *J. Am. Oil Chem. Soc.* **2001**, *78*, 645–649. [CrossRef]

108. Juliano, C.; Cossu, M.; Alamanni, M.C.; Piu, L. Antioxidant activity of γ-oryzanol: Mechanism of action and its effect on oxidative stability of pharmaceutical oils. *Int. J. Pharm.* **2005**, *299*, 146–154. [CrossRef] [PubMed]
109. Yokoyama, W.H. Plasma LDL cholesterol lowering by plant phytosterols in a hamster model. *Trends Food Sci. Technol.* **2004**, *15*, 528–531. [CrossRef]
110. Kim, J.; Godber, J.S.; Prinyawiwatkul, W. Inhibition of cholesterol autoxidation by the nanosaponifiable fraction in rice bran in an aqueous model system. *J. Am. Oil Chem. Soc.* **2001**, *78*, 685–689. [CrossRef]
111. Seetharamaiah, G.S.; Krishnakantha, T.P.; Chandrasekhara, N. Influence of oryzanol on platelet aggregation in rats. *J. Nutr. Sci. Vitaminol.* **1990**, *36*, 291–297. [CrossRef] [PubMed]
112. Bergman, C.J.; Xu, Z. Genotype and environment effects on tocopherols, tocotrienols and gamma-oryzanol contents of Southern US rice. *Cereal Chem.* **2003**, *80*, 446–449. [CrossRef]
113. Irakli, M.; Lazaridou, A.; Biliaderis, C.G. Comparative evaluation of the nutritional, antinutritional, functional, and bioactivity attributes of rice bran stabilized by different heat treatments. *Foods* **2021**, *10*, 57. [CrossRef]
114. Britz, S.J.; Prasad, P.V.V.; Moreau, R.A.; Allen, L.H.; Kremer, D.F.; Boote, K.J. Influence of growth temperature on the amounts of tocopherols, tocotrienols, and γ-oryzanol in brown rice. *J. Agric. Food Chem.* **2007**, *55*, 7559–7565. [CrossRef]
115. Singanusong, R.; Garba, U. Micronutrients in rice bran oil. In *Rice Bran and Rice Bran Oil. Chemistry, Processing and Utilization*; Cheong, L.-Z., Xu, X., Eds.; AOAC Press: London, UK, 2019; Chapter 5; pp. 125–159.
116. Peanparkdee, M.; Iwamoto, S. Bioactive compounds from by-products of rice cultivation and rice processing: Extraction and application in the food and pharmaceutical industries. *Trends Food Sci. Technol.* **2019**, *86*, 109–117. [CrossRef]
117. Jun, H.I.; Song, G.S.; Yang, E.I.; Youn, Y.; Kim, Y.S. Antioxidant activities and phenolic compounds of pigmented rice bran extracts. *J. Food Sci.* **2012**, *77*, 759–764. [CrossRef]
118. Nagendra Prasad, M.N.; Sanjay, K.R.; Shravya Khatokar, M.; Vismaya, M.N.; Nanjunda Swamy, S. Health benefits of rice bran: A review. *J. Nutr. Food Sci.* **2011**, *1*, 1000108. [CrossRef]
119. Kumar, A.; Sahu, C.; Panda, P.A.; Biswal, M.; Sah, R.P.; Lal, M.K.; Baig, M.J.; Swain, P.; Behera, L.; Chattopadhyay, K.; et al. Phytic acid content may affect starch digestibility and glycemic index value of rice (*Oryza sativa* L.). *J. Sci. Food Agric.* **2020**, *100*, 1598–1607. [CrossRef] [PubMed]
120. Sapwarobol, S.; Saphyakhajorn, W.; Astina, J. Biological functions and activities of rice bran as a functional ingredient: A review. *Nutr. Metab. Insights* **2021**, *14*, 11786388211058559. [CrossRef] [PubMed]

MDPI AG
Grosspeteranlage 5
4052 Basel
Switzerland
Tel.: +41 61 683 77 34
www.mdpi.com

Foods Editorial Office
E-mail: foods@mdpi.com
www.mdpi.com/journal/foods

www.ingramcontent.com/pod-product-compliance
Lightning Source LLC
Chambersburg PA
CBHW041354210926
51850CB00013B/3078